The Riddle of Organismal Agency

The Riddle of Organismal Agency brings together historians, philosophers, and scientists for an interdisciplinary re-assessment of one of the long-standing problems in the scientific understanding of life.

Marshalling insights from diverse sciences including physiology, comparative psychology, developmental biology, and evolutionary biology, the book provides an up-to-date survey of approaches to non-human organisms as agents, capable of performing activities serving their own goals such as surviving or reproducing, and whose doings in the world are thus to be explained teleologically. From an Integrated History and Philosophy of Science perspective, the book contributes to a better conceptual and theoretical understanding of organismal agency, advancing some suggestions on how to study it empirically and how to frame it in relation to wider scientific and philosophical traditions. It also provides new historical entry points for examining the deployment, trajectories, and challenges of agential views of organisms in the history of biology and philosophy.

This book will be of interest to philosophers of biology; historians of science; biologists interested in analysing the active roles of organisms in development, ecological interactions, and evolution; philosophers and practitioners of the cognitive sciences; and philosophers and historians of philosophy working on purposiveness and teleology.

Alejandro Fábregas-Tejeda is Postdoctoral Research Fellow at the Centre for Logic and Philosophy of Science, Institute of Philosophy, KU Leuven. He obtained a Ph.D. in Philosophy at Ruhr University Bochum and was a Writing-Up Fellow at the Konrad Lorenz Institute for Evolution and Cognition Research. His research concentrates on the history and philosophy of biology, with a specific emphasis on how organism–environment interactions are construed in evolutionary biology and ecology.

Jan Baedke is Professor at the Department of Philosophy I, Ruhr University Bochum. His research is characterised by an integrated approach to the history and philosophy of the life sciences, with a focus on evolutionary biology and microbiology. He is the author of *Above the Gene, Beyond Biology: Toward a Philosophy* *of Epigenetics* (2018) and PI of the German Research Foundation-funded research group "ROTO" (The Return of the Organism in the Bio-sciences: Theoretical, Historical, and Social Dimensions).

Guido I. Prieto is Postdoctoral Research Fellow at the Individualisation in Changing Environments research association (InChangE), Department of Philosophy, Bielefeld University. A biologist by training and self-taught scientific illustrator, he obtained a PhD in Philosophy at Ruhr University Bochum. His research focuses on the philosophical elucidation of the concepts of "organism" and "biological individual" and their roles within and outside biology. He is also interested in the philosophy and practice of visual representations in the sciences.

Gregory Radick is Professor of History and Philosophy of Science at the University of Leeds. His books include *Disputed Inheritance: The Battle over Mendel and the Future of Biology* (2023), *The Simian Tongue: The Long Debate about Animal Language* (2007), and, as co-editor, *The Cambridge Companion to Darwin* (2003; 2nd edition 2009). He has served as President of the British Society for the History of Science and the International Society for the History, Philosophy and Social Studies of Biology.

History and Philosophy of Biology

Series Editor: Rasmus Grønfeldt Winther is Professor of Humanities at the University of California, Santa Cruz (UCSC).

This series explores significant developments in the life sciences from historical and philosophical perspectives. Historical episodes include Aristotelian biology, Greek and Islamic biology and medicine, Renaissance biology, natural history, Darwinian evolution, Nineteenth-century physiology and cell theory, Twentieth-century genetics, ecology, and systematics, and the biological theories and practices of non-Western perspectives. Philosophical topics include individuality, reductionism and holism, fitness, levels of selection, mechanism and teleology, and the nature-nurture debates, as well as explanation, confirmation, inference, experiment, scientific practice, and models and theories vis-à-vis the biological sciences.

Authors are also invited to inquire into the "and" of this series. How has, does, and will the history of biology impact philosophical understandings of life? How can philosophy help us analyze the historical contingency of, and structural constraints on, scientific knowledge about biological processes and systems? In probing the interweaving of history and philosophy of biology, scholarly investigation could usefully turn to values, power, and potential future uses and abuses of biological knowledge.

The scientific scope of the series includes evolutionary theory, environmental sciences, genomics, molecular biology, systems biology, biotechnology, biomedicine, race and ethnicity, and sex and gender. These areas of the biological sciences are not silos, and tracking their impact on other sciences such as psychology, economics, and sociology, and the behavioral and human sciences more generally, is also within the purview of this series.

The Riddle of Organismal Agency
New Historical and Philosophical Reflections
Edited by Alejandro Fábregas-Tejeda, Jan Baedke, Guido I. Prieto and Gregory Radick

For more information about this series, please visit: www.routledge.com/History-and-Philosophy-of-Biology/book-series/HAPB

The Riddle of Organismal Agency

New Historical and Philosophical Reflections

Edited by
**Alejandro Fábregas-Tejeda,
Jan Baedke, Guido I. Prieto and
Gregory Radick**

Routledge
Taylor & Francis Group

LONDON AND NEW YORK

First published 2025
by Routledge
4 Park Square, Milton Park, Abingdon, Oxon OX14 4RN

and by Routledge
605 Third Avenue, New York, NY 10158

Routledge is an imprint of the Taylor & Francis Group, an informa business

British Library Cataloguing-in-Publication Data
A catalogue record for this book is available from the British Library

Library of Congress Cataloging-in-Publication Data
Names: Fábregas-Tejeda, Alejandro, editor. | Baedke, Jan, editor. |
Prieto, Guido I., editor. | Radick, Gregory, editor.
Title: The riddle of organismal agency : new historical and philosophical reflections / Alejandro Fábregas-Tejeda, Jan Baedke, Guido I. Prieto and Gregory Radick.
Description: Abingdon, Oxon ; New York, NY : Routledge, 2024. |
Series: History and philosophy of biology |
Includes bibliographical references and index.
Identifiers: LCCN 2024015554 (print) | LCCN 2024015555 (ebook) |
ISBN 9781032537269 (hardback) | ISBN 9781032537276 (paperback) |
ISBN 9781003413318 (ebook)
Subjects: LCSH: Biology–Philosophy. | Consciousness in animals. |
Speciesism. | Animal behavior. | Animal intelligence.
Classification: LCC QH331 .R455 2024 (print) |
LCC QH331 (ebook) | DDC 570.1–dc23/eng/20240419
LC record available at https://lccn.loc.gov/2024015554
LC ebook record available at https://lccn.loc.gov/2024015555

ISBN: 9781032537269 (hbk)
ISBN: 9781032537276 (pbk)
ISBN: 9781003413318 (ebk)

DOI: 10.4324/9781003413318

Typeset in Times New Roman
by Newgen Publishing UK

Contents

Figures

Contributors

Bendik Hellem Aaby is Senior Lecturer at the Department of Philosophy, Classics, History of Art and Ideas, University of Oslo (Norway). Previously, he was a Postdoctoral Researcher at the Centre for Logic and Philosophy of Science, Institute of Philosophy, KU Leuven. He is currently working on the concept of agency and its applicability to organisms outside the animal kingdom. He has previously worked and published on questions related to evolutionary theory, niche construction theory, and teleology.

Gunnar Babcock is Postdoctoral Associate in the Department of Biology at Duke University (USA). He works in the philosophy of biology, focusing on issues related to teleology and biological individuality. Currently, he is developing with Dan McShea a new account of goal-directed systems called *field theory*. His research has appeared in journals such as the *Biological Journal of the Linnean Society*, the *British Journal for the Philosophy of Science*, and *Synthese*. He received his Ph.D. from the University at Albany, SUNY.

Jan Baedke is Professor at the Department of Philosophy I, Ruhr University Bochum (Germany). His research interests include the history and philosophy of the life sciences (especially evolutionary biology and microbiology), philosophical anthropology, and the concept of race. He is the author of the book *Above the Gene, Beyond Biology: Toward a Philosophy of Epigenetics* (2018) and PI of the German Research Foundation-funded research group "ROTO" (The Return of the Organism in the Biosciences: Theoretical, Historical and Social Dimensions).

Maurizio Esposito is Senior Research Fellow at the Interuniversity Centre for History of Science and Technology at the University of Lisbon (Portugal). He studied philosophy at the University of Bologna and earned a Ph.D. at the University of Leeds in History and Philosophy of Science in 2012. He has taught at UNAM, Mexico, at the University of Santiago of Chile, and at the Federal University of ABC in Brazil. He has published widely in the area of History and Philosophy of the Life Sciences. He is the author of *Romantic Biology, 1890–1945* (2014).

Alejandro Fábregas-Tejeda is Postdoctoral Research Fellow at the Centre for Logic and Philosophy of Science, Institute of Philosophy, KU Leuven (Belgium). He obtained a Ph.D. in Philosophy at Ruhr University Bochum and was a Writing-Up Fellow at the Konrad Lorenz Institute for Evolution and Cognition Research. His research has centred on diverse topics such as the holobiont concept, the organism–environment relationship, the history and historiography of theoretical biology and Anthropocene studies, and the Extended Evolutionary Synthesis debate.

Gregory M. Kohn is Assistant Professor of Comparative Psychology at the University of North Florida (USA). He received his Ph.D. in Animal Behavior from Indiana University under the supervision of Meredith West. His research looks at the reciprocal relationships between social organisation and behavioural development in birds, with a theoretical focus on how organismal agency can inform research on the development, structure, and function of animal behaviour.

Daniel W. McShea is Professor of Biology at Duke University (USA). His main research interest for years was long-term trends in evolution, especially trends in the complexity and hierarchical structure of organisms. Recently, however, he has been working with Gunnar Babcock developing a new theory of teleology—called "field theory"—which encompasses all goal-directed systems, from simple organismal tropisms to wanting and intending in humans and other animals. His books include *Biology's First Law* (2010) and *The Missing Two-Thirds of Evolutionary Theory* (2020), both co-authored with Robert N. Brandon, and he edited with Alex Rosenberg *Philosophy of Biology: A Contemporary Introduction* (2007).

Francesca Michelini is Senior Research Fellow at the University of Kassel (Germany), where she teaches philosophy. Her main fields of research are anti-reductionist theories of life and bridging continental philosophy and science. She is the author of many publications on philosophical anthropology, philosophy of the life sciences, teleological explanations in nature, and autonomy in biology, for example, *Il vivente e la mancanza. Scritti sulla teleologia* (The Living and the Deficiency. Essays on Teleology, 2011). She recently co-edited the volume *Jakob von Uexküll and Philosophy: Life, Environments, Anthropology* (2020).

Kevin J. Mitchell is Associate Professor of Genetics and Neuroscience at Trinity College Dublin (Republic of Ireland). He has studied the genetic basis of brain wiring and its relevance to variation in human faculties. His current research focuses on the biology of agency and the nature of genetic and neural information. He is the author of *INNATE—How the Wiring of Our Brains Shapes Who We Are* (2018) and *Free Agents: How Evolution Gave Us Free Will* (2023).

Auguste Nahas is a Ph.D. candidate at the Institute for the History and Philosophy of Science and Technology, University of Toronto (Canada). He is interested in diverse issues in the history and philosophy of biology of the twentieth and twenty-first centuries. His current research traces the shifting meaning of "teleology" in biology over the twentieth century and brings this historical account to bear on contemporary debates about the nature of organisms as agents. He has published, among other topics, on the complicated legacies of the Kantian view of teleology in the life sciences.

Henry D. Potter is a Ph.D. candidate in Genetics at Trinity College Dublin (Republic of Ireland). His research explores whether philosophical theories of agency can be reconciled with empirical findings from neuroscience, genetics, biology, and psychology in order to develop a naturalistic framework for understanding the role that agency may play in living systems. His recent work has been published in *Entropy*.

Guido I. Prieto is Postdoctoral Research Fellow at the Individualisation in Changing Environments research association (InChangE), Department of Philosophy, Bielefeld University. A biologist by training and self- taught scientific illustrator, he obtained a PhD in Philosophy at Ruhr University Bochum. His research focuses on the philosophical elucidation of the concepts of "organism" and "biological individual" and their roles within and outside biology. He is also interested in the philosophy and practice of visual representations in the sciences.

Gregory Radick is Professor of History and Philosophy of Science at the University of Leeds. His books include *Disputed Inheritance: The Battle over Mendel and the Future of Biology* (2023), *The Simian Tongue: The Long Debate about Animal Language* (2007), and, as co-editor, *The Cambridge Companion to Darwin* (2003; 2nd edition 2009). He has served as President of the British Society for the History of Science and the International Society for the History, Philosophy and Social Studies of Biology.

Sonia E. Sultan is Professor of Biology and Professor of Environmental Studies at Wesleyan University (USA). She holds a B. A. *summa cum laude* from Princeton University in History and Philosophy of Science and a Ph.D. in Organismic and Evolutionary Biology from Harvard University. Her research group studies plant developmental plasticity; she also works on the theoretical integration of plasticity into evolutionary biology.

Marion Thomas is Lecturer at the Department of History of Life and Health Sciences of the University of Strasbourg (France). She specialises in the history of the modern life sciences, particularly animal behaviour and primatology. Between 2013 and 2016, she co-directed the project "Politics of the Living. A Study on the Emergence and Reception of the Cell Theory in France and Germany, *ca.* 1800–1900." She is currently working on a book manuscript titled *Chimpanzees in Modern Science: An Animal-Centered History of the Pasteur Institutes in Paris and in Kindia (Guinea), 1903–1965.*

Louis Virenque is a Ph.D. student in Philosophy of Biology at the University Paris 1 Panthéon-Sorbonne, affiliated to the Institute for the History and Philosophy of Science and Technology (IHPST) and the French National Centre for Scientific Research (CNRS). He is working under the supervision of Matteo Mossio (IHPST/CRNS) and in close collaboration with Denis Walsh (University of Toronto) on the topic of agency in biology. His goal is to study the legitimacy of this concept in modern biology. He has published on the concept of agency within Autonomy Theory.

Denis M. Walsh is Professor in the Department of Philosophy, the Institute for the History and Philosophy of Science and Technology, and the Department of Ecology and Evolutionary Biology at the University of Toronto (Canada). He holds a Ph.D. in Biology from McGill University and a Ph.D. in Philosophy from King's College, University of London. His current work investigates the nature of natural agency. He is the author of *Organisms, Agency, and Evolution* (2015).

Acknowledgements

The organisation and execution of the workshop "The Riddle of Organismal Agency: New Historical and Philosophical Reflections" (9th RUB Workshop in the History and Philosophy of the Life Sciences) held in March 2022, alongside the preparation of this volume, received generous support from many individuals and institutions. Our foremost gratitude extends to all the speakers, including those who, due to various reasons, were unable to contribute to the final edited collection, for their insightful ideas on the history and philosophy of organismal agency in the life sciences, and for kindly collaborating in this project. Likewise, we thank all the attendees for enriching the discussion atmosphere of the workshop. We acknowledge the contributions of additional scholars who joined the line-up of the volume further down the road of the editorial process, and the aid of numerous experts on agency and teleology who, although not mentioned by name, provided detailed reviews that helped improve the submitted chapters; their valuable work should not go unnoticed. We also express our gratitude to Routledge for their interest in our book proposal, including the vote of confidence from an anonymous reviewer, and, particularly, to Rasmus Grønfeldt Winther, editor of the History and Philosophy of Biology book series, and Lakshita Joshi, Senior Editorial Assistant, for their thoughtful support and patience throughout the publication process. We also thank Alexander Böhm, Friederike Andrae, and Yağmur Metin for helping to create the index of this volume. A special note of appreciation is reserved for the German Research Foundation (DFG) for generously providing financial support for A.F.T., J.B., and G.I.P. as part of the Emmy Noether Research Group "ROTO" ("The Return of the Organism in the Biosciences: Theoretical, Historical and Social Dimensions"; project no. BA 5808/2-1). Additionally, we are thankful to the Research Foundation Flanders (FWO) for supporting A.F.T. in the last stretch of the editorial work (grant G070122N).

Alejandro Fábregas-Tejeda, Jan Baedke,
Guido I. Prieto, and Gregory Radick,
Leuven, BE, Bochum, GER, and Leeds, UK,
January 2024

1 Organismal Agency

A Persistent Riddle in the History and Philosophy of the Life Sciences

*Alejandro Fábregas-Tejeda, Jan Baedke,
Guido I. Prieto, and Gregory Radick*

1.1 The Riddle, Old and New

With the aim of advancing a diagnosis on the big questions that the scientific study of life was still confronting in his time, the Norwegian marine biologist Johan Hjort asserted in his 1921 book *The Unity of Science: A Sketch*, "In biology, the greatest interest has centred about the attitude which science should adopt towards the purposiveness of living things. We find in the literature a host of attempts to solve this ancient question" (Hjort, 1921, p. 2). Perhaps hyperbolically, Hjort referred to it as "[…] *the fundamental problem of biology*" (Hjort, 1921,p. 3; emphasis added). Even if it was expressed through a rhetorical magnification, Hjort, as many of his contemporaries did, kept pondering on this issue for decades to come. "Every biologist must find the question of purposiveness one of enormous difficulty," he continued, "the deeper we go into the problem, the more do we feel that it is really itself a part of the whole great structure which science and thought have erected" (Hjort, 1921, p. 4).[1]

Indeed, the question surrounding the alleged purposiveness and agency of organisms is "one of enormous difficulty." It is probably not misguided to compare it with something akin to a *riddle* that re-emerges in every generation of biologists and philosophers, without an apparent solution and, more often than not, with many attempts at its dissolution and elimination. Although it might seem intuitive at first glance, a riddle is a statement or question designed to challenge comprehension through ambiguity, inviting individuals to decipher its hidden meaning and to come up with possible ways out of it. Are (non-human) organisms *bona fide* agents capable of performing activities that serve their own intrinsic goals (e.g., surviving) or does this teleological way of thinking about them constitute a detrimental *faux pas* for biological science? What would it mean to resolve the question, and what exactly is being posited? This is one of the difficulties that emerge with riddles: it is only when one has the solution at hand that one knows how to take the question, and what it is for it to have a satisfactory answer.

Riddles abound in philosophical reflection: famous examples include, among many others, the diachronic identity of the Ship of Theseus, the sorites paradox, Nelson Goodman's "New Riddle of Induction" concerning the projectibility of strong, law-like generalisations *vs.* weak, non-law-like generalisations (Goodman,

DOI: 10.4324/9781003413318-1

1955, chapter III), and the broad comparison between many philosophical problems and unsolvable riddles that Wittgenstein (1921/1922) advanced in the *Tractatus Logico-Philosophicus*, paired with the idea that we would be better off by disavowing directionless conundrums in our thinking. Riddle-talk is also common in the natural sciences. For example, Ernst Haeckel (1899/1901) spoke of a set of "world-riddles" that he addressed in his book, *Die Welträthsel* (translated into English as "The Riddle of the Universe at the Close of the Nineteenth Century"), including the evolution of our species, gradations of mindedness in the history of life, and the quarrelsome relationship between science and religion. But few riddles cross-cut both the domains of science and philosophy in such a deep way as the riddle of organismal agency does. The very concept of *organismal agency* carries a curious profile: it presents itself as capturing complex and enigmatic phenomena at the heart of life's nature and biology's overarching explanatory agenda. However, while seemingly graspable, a clear and unified understanding of the agency of organisms (or its lack thereof) has proven elusive, demanding nuanced explorations of its intricacies and diverse theoretical viewpoints to comprehend it.

From a historical perspective, organismal agency can be said to be a *nomadic concept*, not bound to any particular scientific discipline, but transversally dispersed among many of them.[2] In alignment with recent perspectives offered by historians of biology (see Surman et al., 2014), the notion of nomadic concepts departs from the traditional historiographic approach of tracking the development of concepts within a single discipline or localised socio-material context. Instead, we need to recognise that concepts of this kind are inherently dynamic, undergoing continual reification and transformation as they traverse and intersect various fields of enquiry. This is one of the justifications for espousing a transversal vantage point for investigating the conceptual history of organismal agency in the life sciences, such as the one presented in this collective book.

Agency has played fundamental roles in the history of philosophy and the sciences. For instance, it has been interpreted through the notions of "action" and "intention," with ramifications in long-standing disputes on determinism and free will, personhood, moral responsibility, or the nature of causation in human affairs, among other topics—famous examples include Aristotle's analysis of what is a voluntary or intentional action, the criticism of David Hume regarding the causal powers of human agents, and pioneering developments in the analytic field of philosophy of action by scholars such as Elizabeth Anscombe and Donald Davidson. In the larger study of life and its evolution, the concept of agency has ignited debates on the ontological status of organisms and the activities they undertake in the world. In particular, the observations that organisms have the ability to actively react to environmental changes, autonomously construct and maintain their organisation and identity despite changes in material composition and form, self-regenerate, reproduce, consistently find shelter and food, etc. have long puzzled philosophers and scientists. How do we account for the apparent purposiveness of organismal activities? Do all organisms have agency and pursue goals of their own? Is the scope of agency restricted to goal-directed behaviours or can it be also ascribed to certain developmental processes? What evolutionary consequences obtain, if any, from the agential activities of organisms? How do we make sense of organismal agency and productively integrate it into biological theories and practices?

Throughout the proximate and remote history of the life sciences, manifold answers to these thorny questions have been adduced. In the last two decades, we have seen a surge of scholarship addressing the relevance of the posit of "intrinsic purposiveness" as a constitutive feature of organisms in the nineteenth century and in preceding periods, which was hailed as a cornerstone for the articulation of biology *qua* integrated science of life (see, e.g., Cheung, 2000, 2014; Richards, 2002; Toepfer, 2004; Riskin, 2016; Nassar, 2016; Zammito, 2018; Steigerwald, 2019; Jones, 2023). At our current juncture, after an extended caesura in the mid-twentieth century (Baedke & Fábregas-Tejeda, 2023), we see yet another resurgence of this persisting riddle. The concept of "agency," together with related notions such as "goal-directedness," "purposiveness," and "teleology," is arguably one of the most intensively debated topics in contemporary philosophy of the life and cognitive sciences. Besides its intrinsic philosophical relevance, it is also inextricably linked to several hotly debated topics in present-day sciences, such as contentious developments within evolutionary theory (e.g., niche construction theory and the study of plasticity-led evolution) and 4E approaches to cognition (e.g., embodied and enactive approaches), and it stands out as a cross-disciplinary nexus that connects biological themes with the behavioural and psychological sciences.

For instance, many special issues, book collections, and monographs have tackled organismal agency in recent years (e.g., Moczek & Sultan, 2023; Corning et al., 2023; Walsh, 2015; Moreno & Mossio, 2015; Jones, 2016; Riskin, 2016; Okasha, 2018; Tomasello, 2022; Mitchell, 2023; Rupik, 2024). Its prominence in academic contexts, evidenced not only by the increasing number of publications devoted to the topic but also by the establishment of several international research projects (e.g., Love & Dresow, 2022) and meetings on the issue, strongly suggests that agency will remain a matter of debate in the years to come.

In particular, present-day controversies on the agential status of organisms in ecology and developmental and evolutionary biology (e.g., Walsh, 2020; Aaby & Desmond, 2021; Sultan et al., 2022; Fábregas-Tejeda & Baedke, 2023; Patten et al., 2023; Nuño de la Rosa, 2023; Jaeger, 2024), as well as in the cognitive sciences (e.g., Barandiaran et al., 2009; Fulda, 2017; Segundo-Ortin, 2020; Sims, 2021; Stankozi, 2023), re-open a bundle of old questions regarding the ontological and epistemological dimensions of organismal agency (for discussion, see Gambarotto & Nahas, 2022). Is agency a capacity that belongs to organisms or is it a heuristic tool for scientists to momentarily deal with the intricacies of organismal development and its evolution while mechanistic research takes off? Can we dispense with it for explaining biological phenomena, or is it an inescapable outcome of our rational makeup without which we cannot fully grasp or make intelligible the properties of living beings? Through which criteria can we distinguish agential activities from non-agential ones? Furthermore, what is the structure of teleological explanations and what kind of relations should they trace when whole organisms are considered explanatory *relata*? In addressing these questions, it is necessary to contextualise organismal agency with respect to related concepts such as teleology,

purposiveness, goal-directedness, control, holism, behaviour, normativity, organisation, autonomy, and autopoiesis.

1.2 The Volume

This collective volume reacts to the renewed interest in organismal agency and
"internal teleology" in the biosciences,[3] tying in with these lively discussions by
broadening them both in terms of topics and disciplinary scope. In contrast to previous initiatives, which have mostly focused on scientific and broader approaches
to agency and teleology (i.e., these have not always been restricted to organismal
agency and teleology but attempt to encompass agency and teleology *sensu lato*),
our edition has an explicit Integrated History and Philosophy of Science (&HPS)
orientation to enquire into the problem of organismal agency, which, we believe,
contributes to a better conceptual and theoretical understanding of this notion in the
life sciences (e.g., of its epistemic roles in experimental and explanatory practices).
Moreover, it provides *new historical entry points* to understand the deployment
and challenges of agential views of organisms in the history of the life sciences
and in the history of philosophy.[4] This includes addressing how these historical
episodes and approaches could contribute to better understand and contextualise
what scientists from multiple disciplinary standpoints are aiming for when calling
for a comeback of agency today.

As highlighted by Nahas and Sachs (2023), the discourse surrounding the integration of organismal agency into biological science delves into the epistemic
issue of whether adopting an organismal approach that untangles internal teleology would enhance our understanding and explanation of biological phenomena
(e.g., of certain developmental and evolutionary processes). However, as they also
point out, this enquiry represents a distinctive metatheoretical motivation of particular communities of biologists and philosophers of biology, separate from the
philosophical quandaries concerning the consequences of naturalising teleology
for our conceptions of nature, mind, and the place of *Homo sapiens* in the natural
world. The debate over the naturalisation of teleology is a multi-faceted one, as
it encompasses diverse scientific and philosophical objectives, each with its own
unique significance, implications, and success criteria. Here, we restrict our focus
to exploring organismal agency and its relationship to the life sciences writ large.

Specifically, the main objective of this edited volume is to jointly address the
riddle of organismal agency through the lenses of philosophy, history, and the biological sciences. By adopting an &HPS perspective, this book aims to (i) clarify
the epistemological and ontological underpinnings of organismal agency and teleology; (ii) contextualise this problem in the history of the life sciences and the
history of philosophy in fruitful new directions by focusing on underexplored
periods and authors, and by exploring the continuities and discontinuities of
diverse approaches that have construed and studied agency in dissimilar ways; and
(iii) investigate the consequences of embracing organismal agency for the study of
development, behaviour, and evolution, and its formal integration into biological
theories and practices.

(i) The relevant philosophical questions concern past and present issues about *agentic ontologies* and *agential epistemologies*, i.e., different ways to understand organismal agency ontically (e.g., from vital forces, to capacities, and special organisational regimes) and how to generate and justify knowledge about it or explain it within biological frameworks. The latter deal with issues about the explanatory, methodological, or heuristic usefulness of teleological considerations in the study of organisms, their (non-)reducibility to, for example, mechanistic explanations, and the structure of teleological explanations. Connecting ontological and epistemological dimensions would require clarifying how notions such as organisation, autonomy, and control matter or not for advancing our theoretical grasp of organismal agency (for discussion, see Virenque, this volume; Babcock & McShea, this volume; Potter & Mitchell, this volume; Aaby, this volume). In addition, any theory of organismal agency needs to answer what we call the "activity scope problem," namely whether agency mainly refers to goal-directed behaviours (Nahas, this volume; Aaby, this volume), to development in general (Virenque, this volume; for a critical counterpoint, see Potter & Mitchell, this volume), or to a subset of behavioural and developmental processes (e.g., behavioural innovations and adaptive phenotypic plasticity; see Kohn, this volume; Walsh & Sultan, this volume).[5] For example, does organismal agency only refer to purposive behavioural responses to novel or challenging environmental stimuli or also to self-organising and self-maintaining developmental processes? This, in turn, is related to a fundamental question regarding the referents of the agential standpoint: Can agency be predicated of organisms that (presumably) lack intentionality or do not possess a nervous system (e.g., plants) (Baedke, this volume; Aaby, this volume)? What is the phylogenetic scope of agential capacities?

In addition to the pressing question of the scope of organismal agency, we also collected a range of different theoretical and metaphysical frameworks to parse out the agential status of organisms and the activities they undertake in the world: from perspectives on the ontology of agency grounded in the self-maintenance of closed organisational regimes (Virenque, this volume) and the modulation of organism–environment interactions (Walsh & Sultan, this volume) to approaches that construe agency as a second-order ability (i.e., as a capacity) and frame it with respect to debates in philosophy of action (Aaby, this volume), to different views on the self-individuation and positionality of living beings (Michelini, this volume), and theoretical viewpoints that take goal-directed, agential dynamics as *explananda* and internal control and field theory as *explanantia* (Babcock & McShea, this volume). Which of these existing frameworks is metaphysically more suitable to capture organismal agency, whilst still being apt for scientific practice? These are some of the exciting questions that are unbottled through this volume.

(ii) We highlight that the above philosophical questions cannot be discussed in a vacuum but instead must be embedded in a strong awareness of their long historical roots for both the history of philosophy and the history of biology. We argue that creating this historical consciousness does not muddy the waters but actually increases clarity of conceptual frameworks and theoretical positions. It helps avoid losing one's way or reaching an impasse that was already identified by other scholars in the past. Organismal agency and teleology are complex issues—so we should learn from those that ventured before us to figure them out, gather their insights on these quandaries, and avoid their mistakes. Thus, we urge that recent trends in the life and cognitive sciences and the philosophy of science to recognise and conceptualise organismal agency (*vide supra*) must be understood against this rich history. For instance, identifying the problem space where organismal agency originated from and discerning its intersections with other crucial debates in the history of biology holds immense significance (Esposito, this volume).

Undeniably, one central answer to the riddle of organismal purposiveness was provided by Immanuel Kant (1790/2000) in his *Critique of the Power of Judgement* (for discussion, see Esposito, this volume; Fábregas-Tejeda, this volume; Baedke, this volume). Some historians have contended that many morphologists and physiologists, before and after Kant, sought to explain purposiveness as something constitutive of living organisms and not just as a regulative maxim of the reflective power of judgement with heuristic value for scientific research. Hence, beyond Kantian and neo-Kantian positions, what answers were given to the riddle of organismal agency in biological research (e.g., in debates about vitalism, animal psychology, or the transmutation of species)? Were these ideas linked to particular scientific developments? These are some of the questions explored in the following pages (Baedke, this volume; Radick, this volume; Thomas, this volume). Moreover, the historiographical work of this book reveals that several practising scientists took seriously the idea of organismal agency in early twentieth-century biology, for example, in animal behaviour studies, botany, embryology, and physiology. In this context, we ask the following: How was organismal agency construed within the international organicist and holistic movement of the early twentieth century, and what positive and sceptical positions concerning organismal agency were advanced by theoretical and experimental biologists during that time (Fábregas-Tejeda, this volume; Baedke, this volume; Nahas, this volume)? What were the continuities and discontinuities of frameworks that emphasised the agency of organisms with approaches of the second half of the twentieth century, for example, those that relied on the cybernetic account of teleology (Nahas, this volume), or advanced other notions to ground (and explain away) purposiveness through an appeal to genetic programmes and natural selection? How was organismal agency treated in disciplines such as evolutionary biology at the end of the nineteenth century and in the long twentieth century (Radick, this volume)? And, finally, how do we account for the renewed interest on organismal agency in contemporary biology against this capacious history? How do these past debates shed light on reoccurring challenges in current scientific projects to investigate organisms *qua* agents? Will the study of organismal agency inevitably end up being framed as a "protest science" (*sensu* Radick, 2017) or is there a way to circumvent this designation?

(iii) In this volume, philosophical and historical topics are discussed in close contact with different scientific disciplines and the insight of practising scientists when dealing with teleological phenomena. In fact, teleology is not as alien to scientists as some philosophers might assume. Agency-talk is pervasive in the life and cognitive sciences, ranging from descriptions and explanations of behaviour, the dynamics of homeostatic physiological processes, and the robust trajectories of development to the outcomes of adaptive scenarios in evolution. But in which instances is the attribution of agency warranted? Some biologists regard many of these appellations as careless wordings that should be replaced by non-agential variants or mechanistic explanations (e.g., turning to natural selection has been a way to "screen off" organismal agency; for discussion, see Desmond, 2023, p. 471). Others consider them as harmless and even useful shortcuts paving the way for rigorous scientific research (e.g., Birch, 2012; Ågren & Patten, 2022). Still others argue that they are legitimate and irreplaceable (e.g., Toepfer, 2012). In any case, agency is interlinked with biological practice and the methodologies adopted by scientists in multifarious disciplines. Thus, we aim to explore how agency can be productively studied, and in which way it can be used to explain biological phenomena. Can agential explanations be reduced to non-agential (e.g., selectionist) explanations? How is the environment related to agential dynamics and which role does it play in agency-based explanations (e.g., in the framework of norms of reaction and phenotypic plasticity responses) (Walsh & Sultan, this volume; see also Michelini, this volume)? Is phenotypic plasticity research a good empirical *locus* for strengthening or undermining the agential stance (Potter & Mitchell, this volume)? Does the study of behavioural innovations offer an opportunity to build an empirical science of agency (Kohn, this volume)? What is the relationship between agency and the individuation of organisms, and how does this presumed relationship affect how biological systems should be empirically studied and what knowledge can be extracted from this (Michelini, this volume)? What roles have research organisms, foremost as agents with intrinsic normativity and particular goals, played in the development and success of scientific practices? Are organisms "collaborators" or "resisters" of the epistemic aims and agendas of the scientists studying them (Thomas, this volume)? How can organismal agency be integrated into the theoretical edifice of biology in methodologically and heuristically fruitful directions?[6]

1.3 The Workshop

To facilitate connections among these three dimensions, we initially organised the 9th RUB-Workshop for History and Philosophy of the Life Sciences ("The Riddle of Organismal Agency: New Historical and Philosophical Reflections"). This gathering provided a platform of interaction for historians of science, philosophers, and biologists working on these topics, fostering new reflections on long-standing problems, testing systematic contributions against contemporary scientific practices, and nurturing cross-disciplinary dialogues. Held over two days

Figure 1.1 Speakers and participants of the workshop "The Riddle of Organismal Agency: New Historical and Philosophical Reflections" at the Beckmanns Hof, Ruhr University Bochum. (1) Jan Baedke, (2) Guido I. Prieto, (3) Alejandro Fábregas-Tejeda, (4) Samar Nasrullah Khan, (5) Hugh Desmond, (6) Gregory Radick, (7) Azita Chellappoo, (8) Daniel S. Brooks, (9) Caroline Stankozi, (10) Saana Jukola, (11) Daniel Liu, (12) Thomas Reydon, (13) Anne Sophie Meincke, (14) Francesca Michelini, (15) Denis Walsh, (16) Bendik Aaby, (17) Vera Straetmanns, (18) Kevin J. Mitchell, (19) Gregory Kohn, (20) Abigail Nieves Delgado, (21) Andrea Gambarotto, and (22) Alexander Böhm. Photograph taken on March 25, 2022.

from March 24 to 25, 2022, this workshop convened ten international speakers alongside local scholars from Ruhr University Bochum (Figure 1.1).

The contributions compiled in this volume have been curated to align with its thematic &HPS focus. Additionally, a few extra chapters from invited authors, although not part of the workshop, complement this thematic concentration. After the workshop, a collaborative effort between the authors and editors was undertaken to revise individual contributions, ensuring coherence and eliminating potential overlaps. Each chapter underwent anonymous review by expert colleagues, and authors made further revisions to enhance the overall quality of the volume.

1.4 The Contributions

The volume is organised into four distinct parts. Part I, "Trajectories of Organismal Agency in the History of the Life Sciences and Philosophy," deals with past accounts of and discussions on this nomadic concept in the late eighteenth and nineteenth centuries, and especially in the early twentieth century. From Kant's critical philosophy and romantic science to vitalist, mechanicist, and organicist positions in interwar biology, as well as debates on plant development and cognition and on purposive behaviourism in the pre-cybernetics era, the chapters included in this part investigate important facets of the intricate history of how philosophers and scientists have reasoned about organismal agency.

In the opening chapter, Maurizio Esposito begins by recalling the Aristotelian thoughtworld in which, from antiquity through the early modern period, agency posed no explanatory challenges at all. It was literally of the essence of things, from elements to organisms, to change in ways that fulfilled their natural purposes. Only with the rejection of Aristotelianism, and so the banishing of teleology from the approved explanatory toolkit of natural philosophy, did apparent purposiveness become a problem. Esposito conducts the reader through the scientific and philosophical changes that made the new tension ever more conspicuous—Kant figures importantly here—before concentrating on the Romantic response to it, especially as glossed by Georges Gusdorf and other scholars of Romantic science and technology. Although somewhat neglected now, Gusdorf helpfully delineated three emphases as characteristic of Romantic science: on the problematic relations between "wholes" and "parts"; on the dynamic connection between "external" and "internal" factors; and on "creativity" as distinguished from "determinism." Esposito goes on to show how well this threefold taxonomy highlights otherwise easily missed features in the collaborative work of two Scottish biologists active in the decades around 1900, the polymath Patrick Geddes and the populariser John Arthur Thomson. After World War II, Esposito suggests, Geddes and Thomson's romantic synthesis, and the centrality of "agency" (their term) in their vision of biology, was overwhelmed by highly successful alternatives that conceived organisms as epiphenomena of intracellular activities.

Continuing with the historiographic space opened by Esposito, Alejandro Fábregas-Tejeda contends that there is a "historiographic gap" concerning how organismal purposiveness and agency were discussed and framed by experimental and theoretical biologists during the first decades of the twentieth century. As a first stride towards addressing this issue, Fábregas-Tejeda offers a systematic reconstruction of contrasting positions at play in diverse disciplines. In his chapter, he argues that it is possible to discern seven distinct stances embraced by scientists with respect to the riddle of organismal purposiveness and agency: (i) the neo-Aristotelian position, (ii) the Drieschian answer, (iii) the heuristic approach, (iv) the holistic alternative to purposiveness, (v) purposiveness eliminativism, (vi) organismal agency as noetic principle, and (vii) organismal purposiveness and agency as *explananda* for equilibria-related research. Some views solely focused on ontological dimensions, underlining or disallowing organismal purposiveness;

others prioritised integrating teleological ideas into the epistemology of biological science; and a subset navigated both ontological and epistemic domains. This taxonomy reveals that the conventional tripartite classification of mechanism–organicism–vitalism does not adequately encompass the breadth of viewpoints on organismal purposiveness and agency in the early twentieth century.

Jan Baedke's chapter explores the theoretical progression of discussions on teleological phenomena in plants since the late eighteenth century. This history is marked by a persistent tension between denying plants certain attributes linked to animals, such as vitality and active perception, and recognising their diverse agential capacities, including plasticity, responsiveness, regenerative abilities, and reproductive versatility. Baedke traces this historical trajectory from Kant's exploration of trees through the nineteenth century's mechanism–vitalism debate, addressing challenges in understanding plants as integrated units of purpose. He examines the holistic approach *(Ganzheitsbiologie)* to plants in the early twentieth century, by focusing on plant physiologist Emil Ungerer. Ungerer rejected psycho-vitalist anthropomorphic frameworks and proposed instead an empirically access-ible plant teleology, centred on the investigation of plants' dynamic equilibrium and physiological organisation. The chapter concludes by analysing the post-World War II shift towards mechanistic accounts and reflects on the dialectical and cyc-lical nature of the historical progression of plant agency debates—from holism to mechanism, to vitalism, and back again.

In his chapter, Auguste Nahas adeptly advances the historical contextualisation of the cybernetic view of teleology, situating it within prior debates on behaviour and holism. Central to Nahas' enquiry is Lawrence K. Frank's enigmatic assertion that the cybernetic rendition of teleology is simultaneously holistic (non-reductive) and mechanistic and that this entails no contradiction. Nahas traces the lineage of the cybernetic teleological approach, including the distinction made between a behaviouristic and a functional mode of study, in the thought of holistic thinkers in behaviouristic psychology and philosophy, focusing on the links that Norbert Wiener and some of his contemporaries established with "purposive behaviourists" such as Edwin Holt, Ralph B. Perry, and Edward Tolman. Nahas proceeds to argue that within their non-reductive depictions of purpose, problematic ambiguities pertaining to the differentiation between intrinsic and extrinsic purposiveness held sway. Moreover, he maintains that these same tensions persisted and surfaced within cybernetics, posing analogous challenges in the understanding of purpose within this framework.

Part II, "Evolutionary Perspectives on Agency," unearths some historical undercurrents of the contentious discourse surrounding organismal agency within evolutionary biology while putting forward novel theoretical ideas on how the agential plastic responses of organisms could be understood. It places the focus on debates surrounding plasticity-led evolution in the form of the so-called "Baldwin effect" and the concept of "norms of reaction." This part also includes a critically balanced discussion of an emerging preference among advocates of agential biology for construing development itself through an agen-tial lens, as shaping individual-level adaptedness and adaptive population change in a manner that might even supersede explanations based on natural selection's action on genetic variation.

In many discussions of agent-centred biology, it is customary for Darwinism to serve as a foil—an example of a scientifically and even culturally prestigious biological science in which organisms and so lineages are but the playthings of law-governed, agency-denying causes. But as Gregory Radick reminds us in his chapter, the later nineteenth century saw a great deal of creative theorising about agency *within Darwinism*. A major stimulus was Darwin's theory of sexual selection, by which he set great store, and which was controversial among his contemporaries precisely because it granted a central evolutionary role to the agency of choosy animals, indeed female choosy animals. These controversies did much to put field studies of animal behaviour in motion. But there was also new interest in clarifying the psychological nature of the choices being made, with an especially brilliant contribution by the Bristol-based Conwy Lloyd Morgan in his landmark *Introduction to Comparative Psychology*, published in 1894. No less far-reaching was Morgan's intervention in another biological debate undergoing similarly careful, creative, and agency-friendly rethinkings at the same moment, on the relations between natural selection and the inheritance of acquired characters—so-called Lamarckism. Within a few years, Morgan became, along with the Americans James Mark Baldwin (a psychologist) and Henry Fairfield Osborn (a palaeontologist), a co-discoverer of what posterity remembers as the "Baldwin Effect": a process by which adaptive changes by or within individual organisms can eventually acquire a hereditary basis. Radick's survey of these developments goes a long way to suggesting how badly distorted our image of biological science *circa* 1900 has become due to the fame of the triple convergence in that year on Mendel's deterministic work. Keeping in mind the triple convergence on the Baldwin Effect can help us recall potentialities that even today remain instructive.

Moving from past to present, Denis Walsh and Sonia Sultan's chapter emphasises the significance of agency in understanding "norms of reaction," which capture the phenomena whereby genetically identical organisms exhibit diverse phenotypes in different environments. Contrary to the traditional view of the norm of reaction as a dispositional property of a genotype, the authors argue for interpreting it as a manifestation of organismal agency. They propose that it reflects the developing organism's capacity to influence its phenotypic expression, challenging the notion that genes alone determine the phenotype. The authors distinguish between gene-centred and agential views on evolution, surveying diverse empirical evidence from recent ecological developmental biology that supports the latter. They contend that individuals capable of transmitting a functionally adaptive capacity to respond to their environments to their offspring exhibit a higher-order norm of reaction, comprising a range of possible norms rather than a genetically predetermined one. The chapter concludes by underscoring the impact on evolutionary dynamics of transmitting an adaptive higher-order norm of reaction, and by emphasising the implicit invocation of agency when explaining evolution through the adaptiveness of an organism.

In dissent from the preceding chapter, Henry D. Potter and Kevin J. Mitchell sceptically appraise the "agential stance" on development, which posits organisms as active agents shaping their own developmental trajectories towards adaptive outcomes. This perspective challenges established theories by combining the ideas of holistic causation and goal-directedness and extracting three key implications: the prioritising of organism-centred frameworks, the accentuating of individual-level adaptedness based on the organism's agency, and the proposing of an adaptive population change model centred on organismal agency rather than solely on natural selection. In contrast, Potter and Mitchell's examination reveals a lack of support for a literal interpretation of the agential stance. Specifically, their critique reins in the ancillary arguments in support of this stance and tones down its radical implications. Instead, they argue that traditional perspectives can account for observed developmental phenomena like robustness and plasticity as evolved traits without attributing real-time executive control of development to organisms. Consequently, Potter and Mitchell contend that the agential stance lacks empirical backing and epistemic utility, and thus that a more cautious view, where agency is only attributed to goal-directed behaviour, should be preferred.

Part III, "Behaviour, Scientific Practice, and Self-Individuation," broaches organismal agency in three important domains: the study of behaviour in developmental and evolutionary timescales, the implications of agency for and from scientific practice, and the links between past and present outlooks to construe organismal agency with respect to self-individuation and entanglements with the environment. The first domain scrutinises how organismal agency could become integrated into the models and methods of practising biologists, while the second one accentuates how the very organisms that are studied by scientists actively react to the experimental settings in which they are embedded. The third domain explores the resonances and theoretical divergences between Helmuth Plessner's early twentieth-century musings and autopoietic enactivism, including their links with empirical science and views regarding living organisms.

In his chapter, Gregory M. Kohn argues that organisms are radically creative (e.g., animals demonstrate remarkable and immediately effective behaviours, spanning from unique individual motoric patterns to the collective co-construction of new social systems); however, a crucial question arises: Does behavioural creativity fundamentally alter the evolutionary function of behaviour, or is it more akin to variation around optimally adaptive behaviours? He proposes that behavioural creativity reflects neither random actions nor programmed responses, but *bona fide* agency, wherein organisms continually produce novelties in service of their goal of sustaining and creating far-from-equilibrium states within and across generations. Throughout his discussion, Kohn presents various examples illustrating how existing biological systems undergo functional reorganisation, resulting in behaviours that depart from their phylogenetic and developmental histories. These behaviours serve specific purposes within immediate contexts, contributing to the attainment of the goal of self-maintenance. His argument pivots on the idea that investigating behavioural creativity offers an invaluable opportunity to construct an empirical science of agency in animal behaviour. Along these lines,

Kohn emphasises that organismal agency serves as a foundational element for both the generation of adaptive variations and their subsequent selection.

Over recent years, scholars in the interdisciplinary field known as "animal studies" have shown how revealing it can be for historians to take seriously the individuality—including capacity for agency—that the scientists who study animal behaviour take seriously. Marion Thomas' chapter uses animal agency as an analytical tool to investigate the ways in which colonial chimpanzees enrolled in biomedical and psychological experiments in French science in the heyday of empire should be understood as "resisters" or "collaborators" or sometimes both. She examines a number of scientific studies, including Élie Metchnikoff's work on syphilis and Paul Guillaume and Ignace Meyerson's studies on ape intelligence at the Pasteur Institute in Paris in the early twentieth century, to show that the Pastorian chimpanzees not only featured as passive objects acted upon by their human researchers but interacted with their (scientifically-built) environments, including the equipment they were exposed to and the staff who administered them. As Thomas introduces the reader one by one to a series of individual chimpanzees—Edwige, Edouard, Nicole, Farce, Moos, Kambi, and Bimba—and their various fates, she also examines the ways in which their agency contributed to the development and presentation of the scientists' work and public image.

In her chapter, Francesca Michelini recovers Helmuth Plessner's philosophical biology for current debates about organismal agency. Michelini asserts that Plessner's anti-reductionist stance and his insights into living beings align closely with the principles of autopoietic enactivism. Noteworthy parallels emerge between "autopoiesis" and Plessner's conceptualization of self-individuation and organismal boundaries, as well as between enactivism's tenet of "sense making" and J. J. Gibson's concept of "affordances" and Plessner's notion of "environmental intentionality (*Umweltintentionalität*)," which he applied, together with J. J. Buytendijk, to animal behaviour. While recognising these convergences, Michelini also points to enduring divergences that resist complete assimilation between the two perspectives. Nevertheless, she contends that these differences not only reveal unresolved questions within enactivism but also contribute to the ongoing relevance and applicability of Plessner's philosophical biology in contemporary philosophical discourse. Michelini's comparison invites a nuanced understanding of organismal agency, bridging historical perspectives with present-day debates with the aim of enriching the evolving dialogue on the agency of living systems.

Finally, Part IV, "Theoretical and Metaphysical Frameworks for Organismal Agency," outlines some of the extant approaches that have been articulated, alongside the ecological account advanced by Denis Walsh and Fermín Fulda (see, for instance, Walsh, 2015, 2018; Fulda, 2017, 2023; Walsh & Sultan, this volume), to spell out a rigorous foundation for understanding the agency of organisms: (i) field theory, (ii) the autonomy school, and (iii) de-intellectualised philosophy of action. These approaches offer a range of different epistemic advantages and virtues (e.g., generalisability and projectibility, cogency, or simplicity) and tackle organismal agency from distinct ontological vantage points (e.g., introducing "fields" into the ontology of goal-directedness, relying on focal ideas such as "organisational closure," or subscribing to a metaphysics of powers and "capacities" for action), so it is important to recognise their distinctive strengths and potential shortcomings.

In their chapter, Gunnar Babcock and Daniel McShea argue that teleological systems (in their view, goal-directed entities guided by a nested series of upper-level fields) can be, but need not be, agential, and they enquire about what the extra step is that makes a teleological system agential. For instance, they discuss how a falling rock, albeit exhibiting marginal goal-directedness, lacks a substantial internal hierarchical structure, thus rendering it minimally agential. In their field theoretic, positive account of agency, they put forward the notion that agency relies on "internal control." According to them, agency manifests as a teleological system's capacity to exert control over itself, irrespective of the purportedly deterministic nature of causal chains affecting it or some of the fields impinging on it. In line with their overarching perspective advocating continuous variation across diverse domains, biological and non-biological alike, they assert that agency is intrinsically graded. It exists along a spectrum, with degrees of agency varying based on the presence and complexity of hierarchical structures and external fields, such as the ones that make up an organism.

Aiming to ground the study of phenomena such as niche construction and phenotypic plasticity, Louis Virenque's chapter maintains that agency should be naturalised through autonomy theory as an ability of organisms. Virenque recounts that "agency" simpliciter commonly refers to the intentionality of human behaviour: human actions are performed for a reason, and representing the motivating reasons appears to be the sole method to naturalise agency without resorting to final causes and will. He contends that the use of this concept in the philosophy of action and the philosophy of mind has led philosophers of science to overlook its application beyond representationalism. For instance, new mechanism's causal explanations of biological phenomena leave room for entities and processes without invoking any form of agency. Virenque argues that in contrast to these two perspectives, the theory of autonomy provides a robust theoretical foundation for the naturalisation of organismal agency. This stance asserts that living beings are self-organised systems instantiating a circular organisation distinct from artefacts and representational systems. This particular form of organisation enables them to functionally and purposively interact with their environment, maintaining their intrinsic *telos* as their construed organisation.

In his chapter, Bendik Hellem Aaby argues that agency can be construed as a second-order ability or, in the terminology Aaby borrows from Anthony Kenny, as a "capacity." In particular, Aaby considers agency as the capacity of a biological entity to engage in goal-directed behaviour (to some degree) under its control, distinct from random occurrences or mere "happenings." Under this view, a putative agent possesses a range of abilities whose combined exercise constitutes an instance of goal-directed behaviour. Aaby stresses that examining organismal capacities and abilities offers valuable insights for understanding agency in non-human organisms: his approach commences with human agency as a foundational model and phylogenetically extends "downwards," employing concepts and distinctions from the philosophy of action to establish generalisable conditions for agency applicable across a broader range of critters. In this sense, he counters assertions that organisms lacking nervous systems are incapable of agency, attributing such misconceptions to a failure to recognise agency as a capacity. To differentiate agential activities from non-agentive processes like flinching, digestion, shivering, and eye blinking, he proposes four distinct conditions: (i) the organism must possess goals and demonstrate goal-oriented behaviours; (ii) these goals should stem from the intrinsic needs or wants of the organism itself; (iii) agential behaviours must be credited to the organism acting as a unified entity, not just to its individual parts; and (iv) the organism, as a whole, should exercise some level of control over the behaviour in question. Finally, Aaby claims that some of these conditions are partly psychological in nature, and it is precisely the fact that organismal agency is a partly psychological capacity which allows us to properly differentiate machines and artefacts from living, behaving organisms.

Apart from the philosophical, historical, and scientific perspectives on organismal agency outlined in Section 1.2, some guiding questions stood at the beginning of the workshop and this publication project—not all of which are completely answered yet, leaving open many paths of enquiry for the future. We invite the reader to engage with the following contributions while keeping these questions and topics in mind. First, we are interested in asking how the large and ever-growing body of literature on teleology and agency can (and should) be systematised best: How can we bring together quite disparate topics and phenomena (like phototropism in bacteria and intentional human actions)? More important, in the *ca.* 2000 years of debates on teleology, we see an increasing diversification of theoretical resources to understand organismal agency, from (neo-)Aristotelian approaches, to (neo-) Kantian ones, to (neo-)vitalism, holism, and organicism, to (self-)organisational and thermodynamic views, to more recent schools, from autopoietic and autonomy accounts to field theory and ecological, affordance-based, and enactivist movements. Does this increasing theoretical plurality towards organismal agency result from, or lead to, a better understanding of the different facets that make up teleological processes (or how we conceptualise and study them)? Or does it display an increasing fragmentation of knowledge about organismal agency that, in fact, hinders grasping the core of this difficult riddle? Should we instead rather try to treat all teleological phenomena under the same theoretical and conceptual umbrella?

Second, we identify across the contributions to this volume (and the larger literature on the topic) recurring tensions and alleged dichotomies: these include, among others, human agency *vs.* non-human agency, external *vs.* internal teleology, purposes *vs.* causes, persistence and maintenance *vs.* creativity and novelty, developmental and physiological agency *vs.* behavioural agency, and agency exerted towards oneself *vs.* agency exerted towards one's environment. How should we recognise and pay tribute to these tensions when constructing a comprehensive theory of organismal agency? Do we have to implicitly accept dualistic and often dialectical forms of reasoning when approaching agency? Or should we actively try to reduce these tensions and integrate opposing viewpoints?

Third, we urge the reader to take history seriously. Teleology and organismal agency are not topics in which paying lip service to past scholarship is enough. This is the case because teleological debates have deeply shaped biology over time to turn it into the field it is today. Most importantly, when discussing agency with biologists, one quickly realises that the ghost of substantive vitalism is still around, and so is the Kantian fear of anthropomorphising organisms through teleological reasoning. As a consequence, (re-)introducing agency into today's biology is not only a conceptually challenging enterprise, but one that wrestles with significant parts of the field's own history. For some, agency still seems to advocate spurious vital forces and rejects a naturalist worldview. For others, internal purposiveness can be fully translated into external teleology (e.g., through an appeal to natural selection-based explanations). And yet others were educated on the idea that any purpose-talk can be nicely replaced by the concept of "functions." We suggest taking the contributions to this volume as a stimulus to rediscovering past positions to enrich the contemporary debate. They will hopefully make the reader realise that the riddle of organismal agency is not just an empirical problem, but a theoretical and conceptual one that grew and evolved over a long period of time.

Finally, the reader might be interested in reasoning with the authors about good practices on how to study agency scientifically. By which criteria do we individuate agents in a warranted way? What kind of empirical settings allow us to trace and test agential properties and activities in and of organisms? Can we, for example, transfer studies on animal behaviour to plants to investigate their agential capacities? What are the limitations to present model systems and experimental conditions to study agency, say, in developmental or evolutionary biology?

Readers are encouraged to formulate their own opinions on the extent to which the aforementioned guiding topics effectively shape the perspectives presented in each chapter. Although the authors did not provide simplistic or easily generalisable answers to these challenges, keeping them in mind may still offer valuable guidance for readers.

We hope that this &HPS excursus into the riddle of organismal agency in biology significantly contributes to the ongoing conversation on this fuzzy, nomadic concept. We also hope that this volume offers enough incentives to try to build more bridges in the future between how organismal agency is framed and discussed within the biological and the cognitive sciences, and to expand these enquiries into other domains where a good conceptual grasp of organismal agency is fundamental and bears important practical consequences, such as conservation biology and environmental ethics (for recent discussions, see, e.g., Heyd, 2005; Bekoff, 2013).

If Hjort was right, the problem of organismal agency, "one of enormous difficulty," will certainly re-emerge in the future, prompting renewed reflections among scientists and philosophers. This book adds to our generation's response to the riddle, now part of the history of attempts to confront it and crack it open.

Notes

1 By a path similar to that which led the German embryologist Hans Driesch to reach the rudiments of his vitalist standpoint, Hjort got interested in the topic of organismal purposiveness while doing experimental research on the development of ascidians and discovering two radically different ontogenetic series leading to the same robust post-gastrulation phenotypes (see, for instance, Hjort, 1896).
2 The same is true, we think, for agency simpliciter.
3 For discussion on the disambiguation of "internal teleology" (i.e., teleology referring to the intrinsic or immanent purposiveness of evolved entities like organisms and not to the general purposiveness of evolution or nature *in toto*) with respect to other concepts of teleology (e.g., external, intentional, and cosmic teleology), see Toepfer (2011) and Fábregas-Tejeda and Baedke (2023).
4 For instance, we do not devote a full chapter to exploring the topic of "teleonomy," which has been covered by historical work several times before (e.g., Krieger, 1998; Dresow & Love, 2023), nor we foreground the widely debated contrast between organisms and machines in terms of their purposiveness (for recent discussions, see, e.g., Nicholson, 2013; Esposito, 2019). We opted for fresh entryways to historicise how organismal agency has been approached.
5 Barham (2012) has distinguished between two different problems within the larger project to naturalise agency and normativity: firstly, what he calls the "Scope Problem," concerned with delineating the appropriate extent within nature for applying the concept of normative agency; and secondly, the "Ground Problem," focused on rationalising normative agency in relation to our broader understanding of the natural world. Barham's scope problem is different from ours: he is calling for the clarification of the *phylogenetic scope* of agency and normativity beyond humans (i.e., which kinds of organisms are agents that act according to their own endogenous norms and whether it is even possible to make a cut somewhere along the tree of life), and we are highlighting the importance of recognising the scope of organismal activities that can be deemed agential and evaluating the reasons and arguments mobilised to sustain these judgements. We believe that solving the activity scope problem could be a promising initial step in tackling the phylogeny scope problem.
6 All of these questions are especially important against the backdrop of current debates on organism–environment interactions and the organism's active roles in evolutionary processes (Nicholson, 2014; Baedke, 2019).

References

Aaby, B. H., & Desmond, H. (2021). Niche construction and teleology: Organisms as agents and contributors in ecology, development, and evolution. *Biology & Philosophy*, *36*(5), 47. https://doi.org/10.1007/s10539-021-09821-2

Ågren, J. A., & Patten, M. M. (2022). Genetic conflicts and the case for licensed anthropomorphizing. *Behavioral Ecology and Sociobiology*, *76*(12), 166. https://doi.org/10.1007/s00265-022-03267-6

Baedke, J. (2019). O organism, where art thou? Old and new challenges for organism-centered biology. *Journal of the History of Biology*, *52*(2), 293–324.

Baedke, J., & Fábregas-Tejeda, A. (2023). The organism in evolutionary explanation: From early twentieth century to the extended evolutionary synthesis. In T. E. Dickins & B. J. A.

Dickins (Eds.), *Evolutionary biology: Contemporary and historical reflections upon core theory* (pp. 121–150). Springer International Publishing. https://doi.org/10.1007/978-3-031-22028-9_8

Barandiaran, X. E., Di Paolo, E., & Rohde, M. (2009). Defining agency: Individuality, normativity, asymmetry, and spatio-temporality in action. *Adaptive Behavior, 17*(5), 367–386.

Barham, J. (2012). Normativity, agency, and life. *Studies in History and Philosophy of Science Part C: Studies in History and Philosophy of Biological and Biomedical Sciences, 43*(1), 92–103.

Bekoff, M. (Ed.). (2013). *Ignoring nature no more: The case for compassionate conservation.* University of Chicago Press.

Birch, J. (2012). Robust processes and teleological language. *European Journal for Philosophy of Science, 2*(3), 299–312.

Cheung, T. (2000). *Die Organisation des Lebendingen. Die Entstehung des biologischen Organismusbegriffs bei Cuvier, Leibniz und Kant.* Campus.

Cheung, T. (2014). *Organismen: Agenten zwischen Innen- und Außenwelten 1780-1860.* Transcript.

Corning, P. A., Kauffman, S. A., Noble, D., Shapiro, J. A., & Vane-Wright, R. I. (Eds.). (2023). *Evolution "on purpose": Teleonomy in living systems.* MIT Press.

Desmond, H. (2023). The generalized selective environment. In A. du Crest, M. Valković, A. Ariew, H. Desmond, P. Huneman, & T. A. C. Reydon (Eds.), *Evolutionary thinking across disciplines: Problems and perspectives in generalized Darwinism* (pp. 453–476). Springer International Publishing.

Dresow, M., & Love, A. C. (2023). Teleonomy: Revisiting a proposed conceptual replacement for teleology. Biological Theory, 18(2), 101–113.

Esposito, M. (2019). En el principio era la mano: Ernst Kapp y la relación entre máquina y organismo. *Revista de Humanidades de Valparaíso,* 14, 117–138.

Fábregas-Tejeda, A., & Baedke, J. (2023). Teleology, organisms, and genes: A commentary on Haig. In T. E. Dickins & B. J. A. Dickins (Eds.), Evolutionary biology: Contemporary and historical reflections upon core theory (pp. 249–264). Springer International Publishing. https://doi.org/10.1007/978-3-031-22028-9_15

Fulda, F. C. (2017). Natural agency: The case of bacterial cognition. *Journal of the American Philosophical Association, 3*(1), 69–90.

Fulda, F. C. (2023). Agential autonomy and biological individuality. *Evolution & Development, 25*(6), 353–370.

Gambarotto, A., & Nahas, A. (2022). Teleology and the organism: Kant's controversial legacy for contemporary biology. *Studies in History and Philosophy of Science, 93,* 47–56.

Goodman, N. (1955). *Fact, fiction, and forecast.* Harvard University Press.

Haeckel, E. (1901). *The riddle of the universe at the close of the nineteenth century.* Harper & Brothers Publishers (Original work published 1899).

Heyd, T. (Ed.). (2005). *Recognizing the autonomy of nature: Theory and practice.* Columbia University Press.

Hjort, J. (1896). Germ-layer studies based upon the development of Ascidians. In H. Mohn & G. O. Sars (Eds.), *The Norwegian North—Atlantic Expedition 1876–1878* (pp. 1–72). Grøndahl and Søn.

Hjort, J. (1921). *The unity of science: A sketch.* Gyldendal.

Jaeger, J. (2024). The fourth perspective: Evolution and organismal agency. In M. Mossio (Ed.), *Organization in biology* (pp. 159–186). Springer International Publishing.

Jones, A. (2023). *How Kant matters for biology: A philosophical history*. University of Wales Press.

Jones, D. M. (2016). *The biological foundations of action*. Routledge.

Kant, I. (2000). *Critique of the power of judgment*. Cambridge University Press (Original work published 1790).

Krieger, G. J. (1998). Transmogrifying teleological talk? *History and Philosophy of the Life Sciences*, 20(1), 3–34.

Love, A. C., & Dresow, M. (2022). Organizing interdisciplinary research on purpose. *BioScience*, 72(4), 321–323.

Mitchell, K. J. (2023). *Free agents: How evolution gave us free will*. Princeton University Press.

Moczek, A. P., & Sultan, S. E. (2023). Agency in living systems. *Evolution & Development*, 25(6), 331–334.

Moreno, A., & Mossio, M. (2015). *Biological autonomy: A philosophical and theoretical enquiry* (Vol. 12). Springer Netherlands.

Nahas, A., & Sachs, C. (2023). What's at stake in the debate over naturalizing teleology? An overlooked metatheoretical debate. *Synthese*, *201*(4), 142. https://doi.org/10.1007/s11 229-023-04147-w

Nassar, D. (2016). Analogical reflection as a source for the science of life: Kant and the possibility of the biological sciences. *Studies in History and Philosophy of Science Part A*, *58*, 57–66.

Nicholson, D. J. (2013). Organisms≠machines. *Studies in History and Philosophy of Science Part C: Studies in History and Philosophy of Biological and Biomedical Sciences*, *44*(4, Part B), 669–678.

Nicholson, D. J. (2014). The return of the organism as a fundamental explanatory concept in biology. *Philosophy Compass*, *9*(5), 347–359.

Nuño de la Rosa, L. (2023). Agency in reproduction. *Evolution & Development*, *25*(6), 418–429.

Okasha, S. (2018). *Agents and goals in evolution*. Oxford University Press.

Patten, M. M., Schenkel, M. A., & Ågren, J. A. (2023). Adaptation in the face of internal conflict: The paradox of the organism revisited. *Biological Reviews*, *98*(5), 1796–1811.

Radick, G. (2017). Animal agency in the age of the modern synthesis: W.H. Thorpe's example. *BJHS Themes, 2*, 35–56.

Richards, R. J. (2002). *The romantic conception of life: Science and philosophy in the age of Goethe*. University of Chicago Press.

Riskin, J. (2016). *The restless clock: A history of the centuries-long argument over what makes living things tick*. University of Chicago Press.

Rupik, G. (2024). *Remapping biology with Goethe, Schelling, and Herder: Romanticizing evolution*. Routledge.

Segundo-Ortin, M. (2020). Agency from a radical embodied standpoint: An ecological-enactive proposal. *Frontiers in Psychology*, *11*. https://doi.org/10.3389/fpsyg.2020.01319

Sims, M. (2021). A continuum of intentionality: Linking the biogenic and anthropogenic approaches to cognition. *Biology & Philosophy*, *36*(6), 51. https://doi.org/10.1007/s10 539-021-09827-w

Stankozi, C. (2023). A hermeneutical back-and-forth between different approaches to agency. *Spontaneous Generations*, *11*(1), 1–13.

Steigerwald, J. (2019). *Experimenting at the boundaries of life: Organic vitality in Germany around 1800*. University of Pittsburgh Press.

Sultan, S. E., Moczek, A. P., & Walsh, D. (2022). Bridging the explanatory gaps: What can we learn from a biological agency perspective? *BioEssays*, *44*(1), 2100185. https://doi.org/10.1002/bies.202100185

Surman, J., Stráner, K., & Haslinger, P. (2014). Nomadic concepts in the history of biology. *Studies in History and Philosophy of Science Part C: Studies in History and Philosophy of Biological and Biomedical Sciences, 48*, 127–129.

Toepfer, G. (2004). Zweckbegriff und Organismus: Über die teleologische Beurteilung biologischer Systeme. Königshausen & Neumann.

Toepfer, G. (2011). Zweckmäßigkeit. In G. Toepfer (Ed.), Historisches Wörterbuch der Biologie: Geschichte und Theorie der biologischen Grundbegriffe. Band 3: Parasitismus—Zweckmäßigkeit (pp. 786–834). J.B. Metzler.

Toepfer, G. (2012). Teleology and its constitutive role for biology as the science of organized systems in nature. *Studies in History and Philosophy of Science Part C: Studies in History and Philosophy of Biological and Biomedical Sciences, 43*(1), 113–119.

Tomasello, M. (2022). *The evolution of agency: Behavioral organization from lizards to humans*. MIT Press.

Walsh, D. M. (2015). *Organisms, agency, and evolution*. Cambridge University Press.

Walsh, D. M. (2018). Objectcy and agency: Towards a methodological vitalism. In D. J. Nicholson & J. Dupré (Eds.), *Everything flows: Towards a processual philosophy of biology* (pp. 167–185). Oxford University Press.

Walsh, D. M. (2020). Teleology in evo-devo. In L. Nuño de la Rosa & G. Müller (Eds.), *Evolutionary developmental biology: A reference guide* (pp. 1–14). Springer International Publishing.

Wittgenstein, L. (1922). *Tractatus Logico-Philosophicus*. Harcourt, Brace & Company, Inc. (Original work published 1921).

Zammito, J. H. (2018). *The gestation of German biology: Philosophy and physiology from Stahl to Schelling*. University of Chicago Press.

Part I

Trajectories of Organismal Agency in the History of the Life Sciences and Philosophy

2 The Problem of "Organismal Agency" in the History of the Life Sciences

Maurizio Esposito

2.1 Introduction

What is the real cause of animal movement? Aristotle sought to answer the question in his treatise *De Motu Animalium* (Movement of Animals) more than 2000 years ago. He assumed that the motion we observe in non-living things was qualitatively different from what we see in living beings. His basic insight was that animals, unlike rocks, can start new lines of motion that do not entirely depend on external, and previous, motions. The animal's movement presupposes a limited sort of "freedom" and, therefore, some kind of inner "deliberation," exceeding external stimuli. As Aristotle himself put it,

> For all lifeless things are moved by something else, and the origin for all the things moved in this way is something that moves itself. Among things of this sort, we have spoken already about animals [...]. Now we see that the movers of the animal are reasoning and *phantasia* and choice and wish and appetite. And all of these can be reduced to thought and desire. For both *phantasia* and sense-perception hold the same place as thought, since all are concerned with making distinctions—though they differ from each other in ways we have discussed elsewhere.
>
> (Aristotle, as translated in Nussbaum, 1978, pp. 34–38)

The passage is thick, and I will only comment on it sparingly.[1] Aristotle noted that any inert thing moves in virtue of something else that, ultimately, cannot be a lifeless thing. This something else is either the *Prime Mover*—the first necessary source of any movement—or is an animal. If it is an animal, then it must be an entity with some special something capable of starting new causal trajectories.[2] Aristotle called this internal something *pneuma* and conceived it as a metaphorical arm of the *psyché* (soul). While the *pneuma* was the proximate cause of movement, the soul was its ultimate one. Thus, the difference maker of the animal movement was the *psyché*, which nowadays we might (although anachronistically) associate with the source of "organismal agency." Among the many meanings of *psyché* we find in Aristotle's treatises—whether concerning animal generation (i.e., development), functional organization (i.e., physiology), or human mind (reason)—there

DOI: 10.4324/9781003413318-3

is in fact a sense of "soul" that refers to the origin of "agential behaviour" (i.e., choice, wish, reasoning).

Yet, Aristotle's "motion" was not the quantifiable "motion" of modern science. Motion was conceived as a thoroughly qualitative process that involved any kind of alterations, from warm to cold; from wet to dry; from marble to the statue; from embryo to adult; from ignorance to wisdom; from Athens to Assos, etc. Aristotle's "movement" described something that changes according to its internal *telos*. Aristotle's encompassing view of movement was coherent with his cosmological perspective. His "ensouled entities" were at home in a world where even physical things had their natural places and where everything could be explained by identifying formal and final causes. Aristotle's notion of *Physis* (nature), as something that has its internal source of movement, made "agency" as an illustrious citizen of his hylomorphic realm. In a thoroughly teleological world, where every object has its internal cause or principle of organization, "agency" was not a mystery but something to be expected. However, in the early modern period, when the Aristotelian hylomorphic view started to crumble under the powerful intellectual gunfire of the "moderns" and gradually replaced by a clockwork cosmos, living beings' supposed "agency" became suspicious. How could something relatively "free" exist in a material, mechanical, and totally determinist universe where every intelligible cause is assumed to be "external"? What was the relationship between genuine "agents" and an inherently predictable physical world? And, crucially, how could a mechanical world generate unmechanical and creative selves? To those unsatisfied with self-congratulatory theological explanations, the question of how real "agents" reproduced and persisted in an ocean of material inertia was reckoned as distinctly urgent.

Immanuel Kant intensely felt the issue in the second half of the eighteenth century. He had inherited a Newtonian universe where matter was conceived as intrinsically passive and stirred by forces ruled by universal laws. The foundational question haunting Kant in his second *Critique* was, in fact: how could genuine moral "agents"—as humans supposedly are—exist in a Newtonian universe populated by material and passive objects moved by external mechanical forces? Could moral agency exist in a non-hylomorphic universe devoid of formal and final causes? His answer was based on his typical transcendental argumentation: to have genuine "agency," we need genuine freedom, and to have freedom in a deterministic world, we need to suppose that such a universe does not entirely determine human action. In other words, if moral deliberation is possible at all, humans must necessarily be sorts of "citadels" of freedom surfing on a sea of predictable determinism. Kant's solution was as widely influential as it was misinterpreted by most romantic readers, who, by the end of the eighteenth and early nineteenth centuries, invoked and invented a new, more powerful, conception of subjectivity and, therefore, a new conception of agency that was much more *catholic*, in the sense that it encompassed human and non-human animals.

In one of his monographs dedicated to the romantic world, titled *L'Homme Romantique*, the French philosopher and cultural historian Georges Gusdorf (1984) argued that the romantic age is the age of the creative activity against any form of automatism, mechanism, and physical determinism. Today Gusdorf has been partially forgotten, yet he provided a seminal reconstruction of romantic science and culture that, I argue, is still valuable to understand the philosophical significance and implications of the concept of "agency." He observed that, in the eighteenth century, the British empiricists and enlightened French philosophers had reduced the subject to a bundle of largely coherent sensations. The subject itself was conceived as an epiphenomenon of an impalpable set of sequential perceptions unified by the weak force of memory. The ego was seen as an abstract, rational, and largely passive entity who interacts with the world through its (supposedly) unprejudiced mind. Much more than Kant, the romantics reacted against this conception of the "passive" self. They replaced such *alienated* universal subjects with individual and creative "agents." The romantic self is not a *camera obscura* or a piece of wax upon which impressions impinge; neither is it a passive object slave of the external environment. Rather, it is an unpredictable entity that interprets and experiences the world through the constitutive intervention of passions, desires, interests, and needs.

Yet, as Gusdorf (1984) added, the conceptual shift we observe on the concept of "subjectivity" and "agency" needs to be understood within the framework of a new romantic philosophy of nature, which saw nature as a productive, organic, and teleological entity. The creative "I" could only dwell in a creative universe. Not surprisingly, passivity and lifelessness were the common enemies of romantic sensibility whenever they referred to consciousness, matter, animals, or plants. Gusdorf (1984) showed how the inert clockwise universe (containing clockwise humans) bequeathed by modern science was replaced by an active and organic cosmos characterized by a recurrent dialectic between internal and external forces, parts and wholes, and creativity versus determinism. The view that Aristotle had glimpsed in his speculations over animals' movement was resurrected in the cogitations of romantic naturalists.

The aim of this chapter is not to track the long philosophical career of the concept of "organismal agency," but to identify some of the philosophical conditions making it plausible. The gist of my short historico-philosophical account is that, however we may define the concept of organismal agency, the definition will hinge upon one persistent and general problem troubling many philosophical minds since Aristotle: to reconcile organisms' apparent behavioural unpredictability with a physical universe that seems to be mostly predictable. I argue that we might situate the modern idea of "agency" in the romantic intellectual space that Gusdorf described throughout his monographs. Then, drawing on Gusdorf's framework, I mention one instructive historical episode that has not received the attention it deserves: the partnership between the polymath Patrick Geddes and the zoologist John Arthur Thomson who, between 1889 and 1931, published four major popular books—*The Evolution of Sex* (1889), *Evolution* (1912), *Biology* (1925), and *Life: Outlines of General Biology* (1931). These books reveal two fundamental points: (1) that the seminal insights of Aristotle on animals' "agency" were far from forgotten; and (2) that the romantic sensibility in the life sciences persisted well into the twentieth century. This sensibility consisted of three related premises: the primacy of the *interiority* over *exteriority* in characterizing living processes, the existence of minimal goal-oriented organic activities shaping the parts in view of an organized whole, and finally, the preponderance of creativity over universal determinism in the natural world. I claim that the concept of "organismal agency" was itself the upshot of those premises.

2.2 Georges Gusdorf and the Making of "Agency"

Gusdorf observed that the romantics promoted what he called a "non-Galilean revolution" (Gusdorf, 1976, 1982, 1984). While the Galilean agenda corresponds to the development of modern science—based on the reduction of nature to mathematical, geometrical, and mechanical elements—the non-Galilean revolution consisted in questioning the idea that nature could be mathematically understood and mechanically predicted. Against "Galileans," the romantics conceived the universe as a *Totalorganismus*. To them, nature could not be conceptualized as a gigantic machine waiting for a supreme watchmaker to put it in motion but as a living, developing, and self-organizing system. No doubt, the opposition between mechanists and antimachinists is widely known. But what makes Gusdorf's analysis original and particularly interesting for us is his view that the difference between Galilean and non-Galilean perspectives went beyond the opposition between mechanism (with its different variants of physicalism or materialism) and holism or vitalism (with all their varieties). For Gusdorf, the crucial difference was whether one gives priority to internal or external causes for explaining the nature and behaviour of any complex entity. He observed that the physicalist perspective, emerging from the scientific revolution and based on a mathematical *episteme*, excluded inner causes in favour of an assorted ensemble of external material forces. Any sort of opaque *interiority* was dismissed, replaced, and reduced to more transparent *exteriorities*.

According to Gusdorf, the overcoming of the Aristotelian *Weltanschauung* and, therefore, the exclusion of any kind of formal or final cause (eventually confining them to the theological debate) introduced a new frame of intelligibility. If no inner force of final principle (e.g., entelechies, vital powers, souls, forms, and archetypes) exists in the natural world, then everything needs to be understood in terms of external (mechanical and material) interactions. Gusdorf (1984) dedicated many pages to show how such a colossal transformation affected the very idea of "subject" and "subjectivity." In the minds of many modern philosophers such as Locke or Hume, La Mettrie or Condillac, the subject simply vanished into a mass of sequential impressions. In his *Traité des sensations* (1754), Condillac's imaginary statue, which "learns" and "acquires" knowledge from a long series of accumulated sensations, demonstrated that innate ideas aren't needed to explain human behaviour. As Leonard (2002) has aptly observed:

> The comparison of the system of sensation with an animated statue externalizes the operations of mind and makes them observable [...] the description of the

statue allows us to visualize how something is put together in order to grasp how it functions.

(p. 493)

If we consider a subject as something similar to Condillac's statue, which is nothing but an entity storing sensations, then we must conclude that *external* stimuli are all we need to understand the self's activities.[3]

For Gusdorf, the external/internal dichotomy could even account for the difference between Lamarck's and the romantics' bio-philosophy. While Lamarck's biology started with simplest living forms, the romantics assumed that life was ubiquitous and cannot be reduced to any form of physical "exteriority." In other words, for the former, life was a property that emerged from matter's organization; for romantics, life was always intrinsic in matter. As Gusdorf (1985) explains,

> Romantic biology does not start with a minimum but with a maximum of life; life itself, from the beginning, is endowed with a specific developmental program; its odyssey in the species responds to an internal finality linked to the project of an external finality [...]. The opposition is particularly evident between the views of Lamarck and Treviranus. For the latter, life corresponds to an activity triggered from the inside and not the outside.

(p. 152)

And, quoting Treviranus, Gusdorf (1985) concluded, "The movements of life are different from mechanical movement insofar the former are not caused by extrinsic causes, but are themselves the outcome of intrinsic causes" (p. 152). In short, the priority of the inner over the external demarcates the space of the romantic intelligibility of life. Organic forms are not simply defined in opposition to a machine; they are defined in opposition to something that is the exclusive result of extrinsic forces. This does not mean that external causes were assumed as irrelevant for organic entities; rather, the internal principle or form of organization had an epistemic priority over the physical environment (whatever it was conceived to be).

Now, the internal/external dichotomy, which replaces or complements the more traditional opposition between vitalism and mechanism, might help to explain why romantics were not necessarily averse to mechanistic explanations. After all, as Jessica Riskin (2016) has recently shown, not all kinds of mechanistic traditions denied agency to matter. Ever since Leibniz introduced the notion of *vis viva*, mechanism could well coexist with vitalist positions. Metaphorically speaking, a machine could be considered either a dead clock requiring external force for working or a clock possessing its "inner power" to move. One could say that the real issue was not whether an organism was a machine but whether an organism could be considered a passive device moved by external causes or an active engine "animated" from within. John Tresch expressed very clearly the distinction when he discriminated between traditional and romantic conceptions of machine in the nineteenth century. While the traditional machine was conceived as an inert structure waiting for an external source of energy to move, the romantic machine was defined as "[...] flexible, active, and inextricably woven into the circuit of both living and inanimate element" (Tresch, 2012, p. XI). Likewise, the romantic organism is not a dead corpse animated by vital fluids but an active, dynamic "entity" that does not exclude the operation of particular mechanisms. By emphasizing the internal dynamic organization over the external world, the romantic biologist avoided the most speculative versions of vitalism while stressing the fundamental intertwinement of matter, organization, and life. A very similar strategy was adopted by many other neo-romantic biologists in the twentieth century, including Geddes and Thomson whom we will consider in greater detail later.

The idea that matter possessed intrinsic capacities for organization in many different forms (from stones to higher organisms) was entangled with another fundamental node of romantic sensibility: the dynamic relation between wholes and parts. If we grant that organic beings, unlike inorganic ones, are characterized by internal principles of organization, then the parts can only be shaped and articulated according to something transcending them from within. Contrary to a rubble of stones where the whole is equal to the sum of individual stones and the whole form depends entirely on external forces (gravity, energy, etc.), the organism shapes itself and its parts as a whole and purposive entity. Such an inner tendency to organize the parts in conformity to a whole makes teleology natural. A *whole* is organized insofar as it includes a *telos* which shapes all its parts. The syllogism was simple: if living beings are organized wholes, and organized wholes are the result of a *telos*, then living beings suppose a *telos*. No *telos*, no organized wholes; and no organized wholes, no living beings.

To elucidate the romantic notion of *telos* would require another essay, so I will not address it further.[4] For the present discussion, however, it suffices to say that the defence of teleology against traditional forms of mechanism needs to be assessed within the framework of the romantic view of matter as active and dynamic where emergent properties (wholes) could not be equated with their components. Indeed, the *telos* was rarely conceived as an external force that "informed" matter; it was rather a sort of internal drive making an organic mechanism a living one. The romantic "organism" was a dynamic entity composed of co-determined parts harmonized by a purposive whole. There are no organized parts without the whole, but there is no whole without the organizing parts. It was the unstable dialectic between parts and whole what constitutes an organic being. The German romantic physiologist Karl Friedrich Burdach exemplified well the view when he defined the organism and life as follows:

A composed totality assembling in itself the parts in reciprocal action and that subsists itself through continuous activities fixed by itself and directed toward a specific end [...]. With "life" we refer to the set of activities proper to an organism [...] everything in life is harmonious: every organ, every activity fits with the others and the whole cooperates according to a common end.

(Burdach, 1842, p. 27)

Once again, as we will see in the following section, Geddes and Thomson's view entirely agreed with Burdach's definition.

Beyond the nodes that constituted the aforementioned dichotomies, i.e., internal/external and parts/wholes, there was a further node that we need to address and that Gusdorf himself explored broadly: the dichotomy between creativity versus determinism. The dichotomy is intertwined with the previous ones and is based on the following issue: is the natural world (including organisms) strictly determined by universal laws? Or is it the result of a creative and partially unpredictable process? According to the Galilean perspective, the answer to the first question would be positive and, to the second, largely negative: the world is the outcome of eternal mathematical laws, and no entity can escape those laws, even when we think some entities apparently do. Like a perfect and reliable clock, the Galilean universe "ticks" according to the movement of the eternal cogs that God himself had manufactured and put in motion. Laplace's famous demon, who was imagined to be able to predict any future state of the universe, could only exist in the Galilean domain of intelligibility: i.e., a "clockwork" form of intelligibility. But in a non-Galilean world, the demon would be hopeless.

For most romantic biologists, the austere and predictive world feeding Laplace's ambitions did not account for the intrinsic creativity we observe in the living realm. To make sense of living things, they replaced the Galilean clock-world with another, non-mechanical, metaphor. They conceived the universe itself as a growing organism, an epigenesis of proliferating forms, a theatre of creative processes. The sweeping metaphor put forward by the German romantic physician Gotthilf Heinrich Schubert in his *Geschichte der Steele* (1830; History of the Soul) was a telling example: the cosmos was an epic of life's creation, a historical metamorphosis that from the simplest living forms conduced to human consciousness. Of course, like many other *Naturphilosophen*, Schubert had been influenced by Johann Gottfried Herder, who had emphasized the primacy of dynamism over inert mechanisms and the overarching creativity of nature throughout cosmological history. As he famously stated in his widely known *Outlines of a Philosophy of the History of Man*,

> From stones to crystals, from crystals to metals, from plants to brutes, and from brutes to man, we here see the form of organization ascend; and with this, the powers and propensities of the creatures have become more various, till at length they have all united in the human frame.
>
> (Herder, 1800, p. 107)

As we will see, Geddes and Thomson also entirely subscribed to a creative view of the cosmos and organic evolution.

To Gusdorf, these three nodes, namely, inner activity, dynamic relation between parts and whole, and intrinsic creativity of natural processes, defined the intellectual space in which Biology as a discipline emerged.[5] Biology was not a new scientific field investigating particular kinds of phenomena escaping the too-coarse lenses of unspecialized naturalists; it was the upshot of a new eclectic philosophy of nature that had introduced a new form of non-Galilean intelligibility. As Gusdorf (1985) observed,

> Romantic organicism made possible the constitution of general biology, the science of general characters of life in all its different forms. The emergence of a discipline presupposes the definition of an epistemic space [*espace épistémologique*] serving as the theatre of operations of the new research.
>
> (p. 258)[6]

From Herder to Goethe, from Schelling to Kielmeyer, the new epistemic space arose from a new basic awareness: life is something primordial that challenges and assimilates *externalities*. Life is a conjunction of processes generating new processes over time and a creative force that could never be reduced to physico-chemical elements. Gusdorf's basic argument is that Biology, as a specific and unique field, and the notion of "organism" itself were the consequence of the romantic intelligibility of nature as built on the categorical rejection of the Galilean worldview.[7] The primacy of the internal over the external, the dynamic relation between parts and whole, and the intrinsic creativity of matter in an evolving universe defined the space of intelligibility in which the notion of "life," the notion of "organism," and the science of Biology emerged.

The romantic tradition did not end in the nineteenth century. Gusdorf showed how it survived World War I and named figures such as Kurt Goldstein, Adolf Portmann, Frederik Buytendijk, the Jesuit palaeontologist Teilhard de Chardin, and philosophers such as Édouard Le Roy and Henri Bergson, whom he deemed as the real heirs of the romantic tradition in the twentieth century.[8] I argue that the form of romantic intelligibility, as described by Gusdorf, not only defined the eighteenth- and nineteenth-century romantic tradition giving rise to Biology as a science but persisted well into the twentieth century and inspired many reactions against more recent forms of mechanism and reductionism (from genetic determinism to molecular biology, from behaviourism to neo-Darwinism). The very notion of "agency" needs to be put in the context of these debates and reactions. After all, however we define organismal agency, the concept itself supposes that *interiority* has some primacy over *exteriority*, that some goal-oriented dynamics are in place, and that some basic sort of creativity in the universe exists. An "agent" that is entirely shaped and determined by external mechanical forces could hardly be defined as a genuine "agent." We can certainly ascribe some kind of "agency" to a human-made machine. But we know that this mechanical "agency" ultimately refers to human "agency" itself. As Aristotle himself would argue, the ultimate cause of a machine is not a machine.

In what follows, I explore more in detail these issues by recovering the productive partnership of Geddes and his pupil Arthur Thomson, who wrote some remarkable neo-romantic works that have been largely overlooked by historians and philosophers alike.[9] It will be shown that there is a fascinating link between neo-romantic views and the very first twentieth-century notions of "organismal agency."

2.3 "Agency" in Twentieth-Century Romantic Biology: Geddes and Thomson

I consider the partnership between Geddes and Thomson as an instructive instance of the revival of romantic biology in the twentieth century.[10] Both were born in Scotland in the second half of the nineteenth century, and both contributed to a biological perspective based on anti-mechanist, holistic, Spencerian, and Bergsonian views.[11] Since 1889, they published four monographs: *The Evolution of Sex* (1889), *Evolution* (1911), *Biology* (1925), and *Life: Outlines of General Biology* (1931). The books provided a highly original and accessible compendium of the life sciences between the end of the nineteenth and the first decades of the twentieth century. The authors felt themselves the heirs of a venerable tradition stretching back to Goethe and Treviranus and updated by Spencer and Bergson's evolutionism, Haldane's organicism, and Kropotkin's optimistic views of evolution through cooperation.

Both Geddes and Thomson came from the deep countryside. The former was born in Ballater, a small village in the County of Aberdeen, in 1854, while the latter was born near East Saltoun, 20 miles east of Edinburgh. In his 20s, Geddes studied at the Royal School of Mines in London under Thomas Huxley's supervision (Kitchen, 1975). He taught zoology at the University of Edinburgh and botany at University College Dundee before travelling to India, where he held a chair in Sociology at the University of Bombay.[12] Thomson got his formal education in Edinburgh, Jena (under Haeckel's supervision), and Berlin and spent his professional life as a Professor of Natural History at the University of Aberdeen. A convinced supporter of the Free Church of Scotland, he had pondered becoming a minister before falling under the spell of Geddes himself. With Geddes, Thomson embarked on an ambitious programme of science popularization. As the science editor of the Home University Library, Thomson was "[…] one of the most prolific science writers of the age" (Bowler, 2009, p. 45). The volume they published in 1911, *Evolution*, sold more than 20,000 copies in Britain and the US in the first two years (Bowler, 2009, p. 129). While Thomson and Geddes did not make any fundamental discovery, they both were broadly known for their efforts to popularize Biology and were widely read on both sides of the Atlantic.

Since their first book, *The Evolution of Sex*, Geddes and Thomson emphasized that life itself was pure activity, an unstable flux that could not be reduced to some specific physicochemical mechanism. The book was an interesting mixture of Spencer's views on sexual reproduction—in particular it used Spencer's distinction between anabolic and katabolic processes—with German *Naturphilosophie*. From the latter, Geddes and Thomson seized the idea that the phenomena of life could not be reduced to physicochemical processes. While anatomists could dissect and compare the anatomy of different organisms highlighting the relations between inert structures, the real biologist had to be a physiologist, namely, a scholar studying the activities of living beings. In fact, life was neither a thing nor a material structure, but a patterned process spelled out by rhythmical changes that could not be explained purely by extrinsic causes. To use their Spencerian terminology, organic entities were ultimately based upon dialectic oscillations between anabolic (constructive) and katabolic (disruptive) processes. The different equilibria between anabolism and katabolism drove organic reproduction, growth, evolution, and behaviour. Thus, at the bottom of the grand biological synthesis that Geddes and Thomson framed lay the unstable polarity between activity (internal) and passivity (external).[13] Every living process—from development to evolution, from reproduction to behaviour—was essentially characterized by a polarity between construction and destruction, action and reaction. In reaffirming the idea that life itself resulted from the irreducible polarity of opposite forces, Geddes and Thomson supplied their conception of life science with one of the great themes of romantic biology. Unsurprisingly, they identified in one of Friedrich Schiller's 1795 poems, *Die Weltweisen* (The Wordly Wise), the poetic essence of their philosophical conception of life: "While philosophers are disputing about the government of the world, hunger and love are performing the task" (Schiller, as cited in Geddes & Thomson, 1889, p. 332).

We understand the importance of Schiller's poem for Geddes and Thomson if we grasp the meaning and place of "hunger" and "love" in their neo-romantic view. The dichotomy mentioned above, anabolism/katabolism, was part of an even larger opposition that the Scottish naturalists identified with nutrition (hunger) and reproduction (love) (see Figure 2.1).

As we can see in Figure 2.1, while the nutritional function included the general metabolic opposition between constructive (anabolic) and disruptive (katabolic) processes, the reproductive function referred to the differentiation of the two sexes in which the preponderance of anabolic processes gave rise to female, while the contrary produced males. The figure does not represent a hierarchical distribution of functions and processes structured in a static order. It is a scheme where

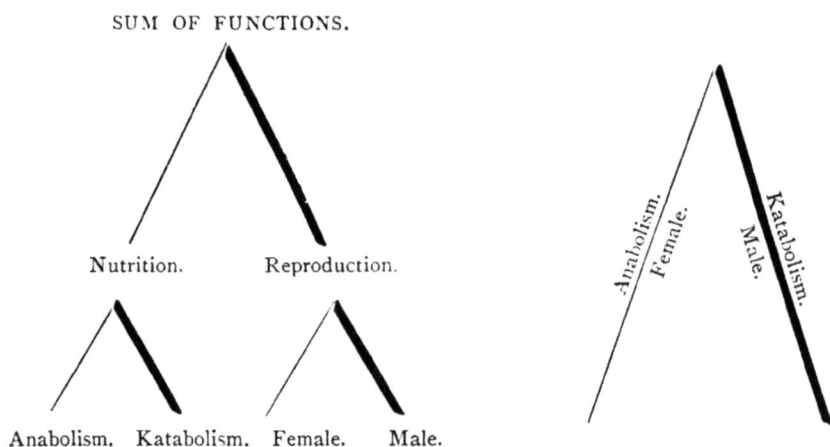

SUM OF FUNCTIONS.

Nutrition.   Reproduction.

Anabolism. Katabolism. Female.   Male.

Anabolism. Female. Male. Katabolism.

Figure 2.1 The opposition between nutrition and reproduction as represented by Geddes and Thomson (1889, p. 237).

everything is connected with everything else. The most general dichotomy, nutrition/reproduction, was itself explicable as an opposition between anabolic and katabolic processes because, for them, organic growth (development) was intimately associated with nutrition (therefore construction), and reproduction was closely related to the increase of katabolic processes:

> The early growth of the cell, the increasing bulk of contained protoplasm, the accumulation of nutritive material, correspond to a predominance of protoplasmic processes, which are constructive or *anabolic* [...]. Yet the life, or general metabolism, continues, and this entails a gradually increasing preponderance of destructive processes, or *katabolism* [...]. The limits of growth, when waste has overtaken and is beginning to exceed the income or repair, corresponds in the same way to the maximum of katabolic preponderance consistent with life.
>
> (Geddes & Thomson, 1889, p. 223; emphasis in original)

Nutrition and reproduction, the Schillerian "hunger and love," were systemically integrated into a general metabolic process of construction and destruction. Additionally, reproduction itself was triggered by the progressive decrease of anabolic processes during development.

The dialectic between reproduction and nutrition, anabolism and katabolism, and activity and passivity also had significant consequences for evolution. Geddes and Thomson (1889) referred to Treviranus, "[...] a biologist too much neglected both in his lifetime and since" (p. 301), as the scholar who first emphasized the importance of sexual reproduction for spurring creativity in evolution. To them, Treviranus had glimpsed, in his own way, the relevant link between reproduction, as the engine of novelty, and the historical metamorphoses we observe in phylogeny. They also mentioned Kant (who is oddly put in the evolutionist camp), Oken, and Goethe, who, despite the excesses of their speculations, had understood that evolution had to be conceived as a dynamic intercourse between organisms and their environments:

> Oken (1809) saw the light of the evolution idea dancing like a will-o' the wisp in the midst of his "Urschleim" speculations and seemed chiefly to interpret the organic process in terms of action and reaction between the organism and its surrounding; while in the noble epic of evolution which we owe to his contemporary Goethe, the adaptive influence of the environment is clearly recognized.
>
> (p. 302)

In their following book dedicated to evolutionary biology, *Evolution*, Spencer's framework was partially replaced by Bergson's *Creative Evolution*, a book published in France in 1907 and translated to English in 1911.[14] Following Bergson, they developed and expanded the idea that there was a fundamental hiatus between observing nature organically or mechanistically, processually or substantially. They argued that both Lamarck and Darwin's theories of evolution were associated with two important historical moments: the Age of Enlightenment and the Industrial Revolution. While the former, spawned by beliefs in social and moral progress, produced the general "doctrine of evolution," the latter, marked by industries' competition and population explosion, shaped "the doctrine of natural selection," which ultimately led to a "commercial" view of evolution. Both doctrines had a grain of truth. The first emphasized the force of the environment and the internal striving of organisms, while the second stressed the struggle and selection of the fittest individuals. However, the neo-Lamarckian and neo-Darwinian doctrines leaned towards a mechanistic and materialist understanding of life and could only be partial accounts of how evolution works (Geddes & Thomson, 1911).

Interestingly, Geddes and Thomson linked the contrast between mechanist and organicist views of evolution to what they called "urban" and "rustic" overviews. The first was essentially determinist and quantitative, whereas the latter was organicist and qualitative: "[…] while in our town herbaria, we distinguish grasses and orchids essentially by their post-mortem structure, the true physiologists, knowing their difference as lives, the grass so vegetative that cattle and farm and city all live upon its surplus" (Geddes & Thomson, 1911, p. 242). Although the urban perspective was more precise and detailed, the rustic approach unified what the urban overview had separated. Geddes and Thomson's metaphorical dichotomy hinged on the distinction between enlightenment and romanticism, mechanism versus organic synthesis. It reminded of Bergson's fundamental distinction between intuitive and rational (intelligent) knowledge.[15] Geddes and Thomson believed that Biology had to be mainly "rustic" and thus organicist. Where the urban citizen saw stable structures and inert entities—cogs or whole machines that could be, in principle, dismantled and reassembled like an industrial product—the "rustic farmer" perceived irreducible, dynamic, and creative processes. The urban citizen sees life as derived from matter; the rustic farmer sees life as a primordial force tangled with matter.

As practising farmers, Geddes and Thomson (1911, p. 232) maintained that living entities were creative and purposive systems, and life, accordingly, could not be reduced to the physical order. They observed that the inner and purposeful creativity of life could not be grasped by purely mechanistic explanation. Life itself had to be conceived as a protean force defying a determinist universe and could only be comprehended under the category of creative *agency*: "Living creatures are agents; they thrust as well as parry; they act on their surroundings, modifying them; they are ever seeking out a new environment, and conquering them" (p. 198). The dynamic relationship of organisms with their environment was so constitutive and intertwined that any separation conducted to considering Biology as "necrology" insofar as a living entity without its *milieu* is a dead entity. Drawing again on Bergson, Geddes and Thomson observed that such creativity or *unpredictability* of life was the outcome of a dynamic and internal "impetus" pushing organic evolution towards higher forms of organization and adaptation to the environment.[16] Like Bergson, they did not confuse the existence of an irreducible living "impetus" informing matter with vitalism. They believed that the crucial philosophical alternative was not between mechanism and vitalism but between inert mechanism and organicism. The organicist view, in fact, while keeping with the ideas that life was driven by internal forces and the organism was an integral agential unit, avoided the speculative excesses of most vitalist proposals. Furthermore, Geddes and Thomson (1911) stressed the centrality of the individual organism in evolution, which they called "[...] the organismal factor in evolution" (p. 206). By anticipating one of the great controversies of evolutionary biology in the second half of the twentieth century well resumed by Ernst Mayr's distinction between typological and population thinking, Geddes and Thomson lamented that evolution could neither be reduced to the reshuffling of *determinants* (against Weismann) nor the mere selective force of the environment (because, after all, they understood natural selection to be only a filter). Instead, evolutionary thinking had to move along Goethean coordinates, given that Goethe himself was the first who "[...] discerned that [...] the moving spirit in the drama of evolution is the Organism" (Geddes & Thomson, 1911, p. 191). If the organism was the leading actor on the stage of the evolutionary drama, then the origins of novelties had to be sought within the organism itself. Geddes and Thomson speculated that species could emerge, once again, from the opposition of two forces, anabolic and katabolic processes:

> Growth and arrest, giant and dwarf, rest and movement, sleep and waking, even female and male, are contrasts all physiologically akin; and this single and simple rhythm of metabolisms, of passivities and activities, goes on into compound and re-compounding rhythms, like the figures of the pendulograph. The forms of life are thus distinct and definite, because harmoniously unified. They have a certain stability, great or small, yet they are anew transformable, like musical variations, like singing flames.
>
> (Geddes & Thomson, 1911, p. 245)

Like Treviranus, who had only presumed what Schiller had poetically hinted, Geddes and Thomson claimed that evolution needed to be understood as a dialectical process between reproduction and nutrition (the Schillerian love and hunger).

2.4   From Organismal "Agency" to Social "Agency"

Geddes and Thomson held an idiosyncratic conception of Biology, conceived as a discipline connecting physics, chemistry, and human sciences. In the last two monographs they wrote together, *Biology* (1925) and *Life: Outlines General Biology* (1931), they set the stage for assessing the dominion and limits of twentieth-century life science. To them, the basic question any serious biologist should start with was how living things differed from non-living ones. The question was far from obvious because, as they claimed, we observe a continuity in nature that does not reveal, immediately at least, any fundamental break. The cosmos itself is a dynamic and ever-transforming process that has nothing to do with a monotonous clock managed by an eternal watchmaker. But even though everything seemed to be, in some sense, alive, living organisms exhibited properties that could not be reduced to inanimate objects. Geddes and Thomson did not believe that a general and simple definition of life could be provided. Yet, they argued that a series of elements could outline the vast domain of the organic realm. First, "life" is enduring activity: "Life's image is thus the burning bush, flaming away and yet not consumed. Its very activity maintains it, to abide the same" (Geddes & Thomson, 1925, p. 12). "Life" is also about growth as it develops and reproduces. Furthermore, "life" is variability. It is memory, experience, and behaviour. But "life" is also what Geddes and Thomson called "insurgence," namely, the ability to spread everywhere and cover any corner of the earth. "Life" is beauty because "organisms are like works of art, of all various schools and levels [...]" (Geddes & Thomson, 1925, p. 29). All those different elements demarcated the realm of living from the non-living one. "Life," more generally, is "protean" tree because of "[...] its potency, its achievement, and its unending promise of Evolution" (Geddes & Thomson, 1925, p. 34).

Geddes and Thomson adventured in a general and idiosyncratic formalization of the life processes that could somewhat summarize all the above: "life" as L is equal to an unknown x so that $L = x$. They assumed that the variable x included two forms of environments and two forms of organic behaviour: active or passive. We have environment E (active) and e (passive), and organism O (active) and o (passive), and all could be included in the function f. Hence, we have the function Efo (E → f → o) when the environment acts on the organism, and conversely, we have the function Ofe (O → f → e) when the organism transforms the environment. L could therefore be defined as the result of the dialectical processes between these two functions so that $x = Efo:Ofe$. Thus, x was both agent and effect of the environment so that L could be defined in terms of the whole dynamic interaction between organism and environment. The conclusion was that the variable x is not a property of the organism itself, but a process emerging from the diachronic interface between organic activity and environment.

Geddes and Thomson believed such dialectical processes could include the human sciences. In fact, the same scheme connecting organisms to the environment could be extended to human social life because organic and social life share the basic and constant interaction with the environment. However, while in the organic realm, the interaction between organisms and environment could be sufficient to explain organic evolution, the social domain required new concepts. They singled out three essential notions which outlined human existence: Geography (as a place), Economy (as work), and Anthropology (as folk). All together constituted a sort of Comtean Sociology, conceived as the supreme human science. Here the formal scheme consisted of connecting the three elements, place, work, and folk, in the formula Pwf: Fwp. Namely, the dynamic combination of place, work, and folk produced particular kinds of societies, as well as the combination of organism and environment produced particular kinds of species. With this idiosyncratic formal synthesis, Geddes and Thomson observed that

> [...] we do justice to environmental impressions and experiences on one side of our notation, yet to organismal and social impulse and expression on the other. We see, then, in the process and progress of life, the alternation of stimulus and

response, of passivities and activities, in unending yet varying rhythm; with the
latter on the whole as increasingly potent and thus directive, even *telic.*

(Geddes & Thomson, 1925, p. 243; emphasis in original)

"Life" was organic and social; it struggled according to an inner creative force that
conduced to higher forms of organization from protozoa onwards. To Geddes and
Thomson, the Schopenhauerian's "will to live," updated by Bergson's *Élan vital*,
constituted the red line linking the evolution of simple organisms to the evolution of
higher civilizations. The internal "impetus" struggled and eventually domesticated
the external forces. New whole and irreducible entities were created in the course of
evolution, and the entire process was essentially and unequivocally unpredictable
and, therefore, creative all along. Most *Naturphilosophen* of the nineteenth century
would have felt Geddes and Thomson's broad synthesis entirely intelligible and
convincing. The Gusdorfian scheme—based on the idea that the romantic view
followed the emphasis on inner causes thwarted by external forces, the dynamic
relation parts/whole, and the essential creativity of life—is entirely applicable to
Geddes and Thomson's assertive—and for us highly idiosyncratic—speculations.
According to their sensibilities and knowledge, the two Scottish naturalists updated
a romantic tradition that, despite its continuous metamorphoses, still maintained a
well-recognizable conceptual shape in the early twentieth century.

2.5 Conclusion

Aristotle had glimpsed the riddle of animal "agency" more than 2000 years ago.
While animals and stones might share the physical principle of change, their motions
were consistently different. Animals' motions exhibited a kind of unpredictability
that hinted at some kind of internal freedom. In an Aristotelian world dominated by
natural events (as based on inner principles of motions and rest), an *agent* could
set forth "unnatural" changes, but no inorganic thing could produce "unnatural"
motions. In the hylomorphic universe of Aristotle, the existence of agency was
something to be expected and not a surprising aberration. However, in the mechanical, anti-Aristotelian world picture emerging in the early modern period, agency
becomes nothing short of a superstition. We had to wait for Kant's philosophical
imagination to redeem agency from a purely mechanical and deterministic world.
Yet, Kant's notion of *agency* was tailored for human moral action. It aimed to
preserve human freedom against physical determinism. The romantic followers
of Kant, unconvinced by the diplomatic solution of the Prussian philosopher, who
divided the world between a determinist (physical) and an indeterminist (moral)
realm, reunited the two worlds into a unique teleological realm where agency could
find a comfortable place again. The crucial Aristotelian insight that animals were
kinds of *agents* was resurrected in the late eighteenth century and developed amid
a new indeterminist metaphysics of the universe and a new epistemology of life
science. Gusdorf described the moment when many post-Kantian philosophers and
naturalists moved against the mechanical universe that had left no space for real
agency. To Gusdorf, the non-Galilean revolution replaced passivity, determinism,

and mechanism with creativity and purposefulness. A new notion of organismal agency was a natural corollary of this neo-Aristotelian and neo-romantic world. The non-Galilean revolution reached its philosophical zenith in Henri Bergson's overarching philosophy in the twentieth century.

We have seen that Geddes and Thomson put forward a neo-romantic view of life as filtered through Spencer and then Bergson's philosophical frameworks. While the contents of their books might appear to us as eccentric, wildly speculative, and largely outdated, even for the years in which they were published, their books were popular, widely read, and hid some philosophical gems (see Bowler, 2001, 2009). Their biological synthesis put together many different sources to outline a new form of organicist biology fit for the twentieth century. They followed many nineteenth-century naturalists in emphasizing the relevance of inner potencies over external forces, the power of organic whole over its parts, and the primacy of creativity over determinism in the physical and living world. Gusdorf's three unstable dichotomies that characterized romantic thinking not only informed Geddes and Thomson's bio-philosophy, but they also informed most neo-romantic agendas in the twentieth century, from the works of Yves Delage to John Scott Haldane, from William Emerson Ritter to Kurt Goldstein (see Esposito, 2017).

There are at least three interesting conclusions we can extract from the history outlined here. First, the modern concept of "agency" cannot be easily disconnected from the romantic "space of intelligibility." Second, notions of "agency" are intertwined with particular philosophies of nature (whether Aristotelian, romantic, or neo-romantic). Third, there is an illuminating connection between the three Gusdorfian elements characterizing the non-Galilean revolution: i.e., *interiority*, *teleology*, and *creativity*, and the notion of "organismal agency." After all, we might wonder whether any genuine "agency" could exist without accepting some version of them.

Acknowledgements

I thank Greg Radick and the anonymous reviewer for their very insightful comments on the first chapter's version. This research was possible thanks to the support of the Fundação para a Ciência e a Tecnologia (FCT), Project N. CEECIND/02290/2018 and through the funding (PeX) 2022.05256.PTDC. DOI: 10.54499/2022.05256. PTDC.

Notes

1 For a thorough commentary, see Nussbaum (1978) and Rapp and Primavesi (2020).
2

> For if we exclude the motion of the universe, living creatures are responsible for the motion of everything else, except such things as are moved by each other through striking against each other. Hence all their movements have a limit; for so do the motions of living creatures.

(Aristotle, as translated in Nussbaum, 1978, pp. 36–38)

3 Gusdorf also exemplified the dichotomy between internal and external forms of intelligibility with a comparison between Newton's *Optics* and Goethe's *Farbenlehre.* While the former excluded the perceiving subject by focusing on the physical features of light, the latter prioritized the subjective perception of light (and thus colour) over its physical nature. In short, for Newton, the explanation went from the external phenomena to their internal representation; for Goethe, internal sensibility was the condition for grasping the external, physical phenomena. The physical "exteriority" was in fact only a special case of a pervasive and encompassing subjectivity.

4 See, in particular, Gusdorf (1985), and, in English, Lenoir (1989) and Zammito (2018).

5 On the origins of Biology as scientific discipline in the German context, see Zammito (2018).

6 By "epistemic space," Gusdorf meant what is intelligible, comprehensible, and representable in a particular system of knowledge.

7 Gusdorf had followed the lead of Ernst Hirschfeld, who in his 1930's study on romantic medicine stressed that

> The notion of organism is the most elevated romantic concept in all domains of knowledge, not only in the natural sciences. It is the concept inaugurated by Herder, defined by Kant, applied by Kielmeyer, sealed by Schelling, and superbly developed by Goethe. Without such a concept it is impossible to understand romanticism itself.
>
> (Hirschefeld, as cited in Gusdorf, 1984, p. 258)

8 Elsewhere I tracked down how such a tradition inspired British and American biologists, physiologists, morphologists, and embryologists and was behind the emergence of a new form of organicism in twentieth-century biological sciences (Esposito, 2016).

9 With the important exception of Bowler (2001, 2009).

10 I am not claiming here that Geddes and Thomson were romantic biologists in the sense of nineteenth-century scholars. My only claim is that they took very seriously some of the basic tenets of eighteenth- and nineteenth-century romantic biology as filtered through the pervasive influence of Spencer. There are, of course, many scholars who held romantic views even when they were generally "unromantic." Spencer is undoubtedly one of them and, in the twentieth century, there were a few others of that sort (see Radick, 2017). But when we go through Geddes and Thomson's books, we realize that the continuity between romantic scholars and their views is evident, not only because there is an affinity of ideas but, more crucially, because Geddes and Thomson used romantic scholars as a meaningful intellectual source.

11 Chris Renwick (2009, pp. 36–57) saw Geddes as the epitome of Spencerian science, although I would add that after the 1910s, Geddes and Thomson were increasingly influenced by Bergson. Peter Bowler (2001, 2009) considered Thomson as one of the most important science popularisers in the early twentieth century.

12 Today Geddes is mainly remembered as a sociologist and social and town planner both in India and Palestine, although, as we will see further on, his sociology was an extension of his conception of Biology. On Geddes' biography, see Radick and Gooday (2004).

13 Insofar as the organism only responds to external stimuli.

14 For a very recent translation, see Bergson (2023).

15 Back in 1903, in his *Introduction to Metaphysics*, Bergson argued that reality itself is a process, which can be either partially grasped through analysis, and thus, conceptual fragmentation, or grasped absolutely through *intuition*, which was the only faculty that could get into the continuous flux of reality.

16 Bergson's philosophy of life was seen by Geddes and Thomson as an effective synthesis of Goethe, Treviranus, and Lamarck.

References

Bergson, H. (2023). *Creative evolution* (D. Landes, Trans.; 1st ed.). Routledge.
Bowler, P. (2001). *Reconciling science and religion: The debate in early-twentieth-century Britain*. Chicago University Press.
Bowler, P. (2009). *Science for all: The popularization of science in early twentieth century*. University of Chicago Press.
Burdach, K. F. (1842). *Blicke ins Leben: Bd. Comparative Psychologie*. Voss.
Esposito, M. (2016). *Romantic biology –1890-1945*. Routledge.
Esposito, M. (2017). The organismal synthesis: Holistic science and developmental evolution in the English-speaking world, 1915–1954. In R. Delisle (Ed.), *The Darwinian tradition in context* (pp. 219–241). Springer International.
Geddes, P., & Thomson, A. (1889). *The evolution of sex*. Walter Scott.
Geddes, P., & Thomson, A. (1911). *Evolution*. Henry Holt.
Geddes, P., & Thomson, A. (1925). *Biology*. Williams & Norgate, LTD.
Geddes, P., & Thomson, A. (1931). *Life: Outlines of general biology*. Herper & Brothers.
Gusdorf, G. (1976). *Naissance de la Conscience Romantique au Siécle des Lumiéres*. Payot.
Gusdorf, G. (1982). *Fondements du Savoir Romantique*. Payot.
Gusdorf, G. (1984). *L'Homme Romantique*. Payot.
Gusdorf, G. (1985). *Le Savoir Romantique de la Nature*. Payot.
Herder, J. G. (1800). *Outlines of a philosophy of the history of man*. Bergman Publishers.
Kitchen, P. (1975). *A most unsettling person: An introduction to the ideas and life of Patrick Geddes*. Gollancz.
Lenoir, T. (1989). *The strategy of life: Teleology and mechanism in nineteenth century biology*. University of Chicago Press.
Leonard, D. (2002). Condillac's animated statue and the art of philosophizing: Aesthetic experience in the Traité des sensations. *Dalhousie Review, 82*, 491–513.
Nussbaum, C. M. (1978). *Aristotle's De Motu Animalium*. Princeton University Press.
Radick, G. (2017). Animal agency in the age of modern synthesis: W.H. Thorpe's example. *BJHS Themes, 2*, 35–56.
Radick, G., & Gooday, G. (2004). Geddes P. In B. Lightman (Ed.), *The dictionary of nineteenth century British scientists* (Vol. 2) (pp. 764–768). Thoemmes Continuum.
Rapp, C., & Primavesi, O. (2020). *Aristotle's De Motu Animalium: Symposium Aristotelicum*. Oxford University Press.
Renwick, C. (2009). The practice of Spencerian science: Patrick Geddes's biosocial program, 1876-1889. *Isis, 100*(1), 36–57.
Riskin, J. (2016). *The restless clock: A history of the centuries-long argument over what makes living things tick*. University of Chicago Press.
Tresch, J. (2012). *The romantic machine: Utopian science and technology after Napoleon*. University of Chicago Press.
Zammito, J. (2018). *The gestation of German biology: Philosophy and physiology from Stahl to Schelling*. University of Chicago Press.

3 Charting Contrasting Stances on Organismal Purposiveness and Agency in Early Twentieth-Century Biology

Alejandro Fábregas-Tejeda

3.1 Probing a Historiographic Gap

In 1923, the Aristotelian Society hosted a debate featuring biologist and philosopher Edward Stuart Russell, physiologist John Scott Haldane, and psychologist Leslie McKenzie. During this event, the issue of "organismal purposiveness" emerged as a central point of contention (Haldane et al., 1923). In a separate address on this topic, Russell underscored the centrality of organismal purposiveness for biological science and argued:

> [The] individualized activity [of organisms] is not stereotyped or unchanging, but shows definite tendency or striving towards an end. Think, for instance, of a salmon ascending a stream, or the growth and differentiation of a seedling plant. The activities of a living thing are coordinated to achieve some end related to its development, persistence or reproduction.
>
> (Russell, 1923, p. 143)

Russell and his colleagues are but an example of numerous early twentieth-century scholars interested in this problem, some belonging to organicist and holist trenches, while others positioned themselves along the mechanism-vitalism divide. The period saw numerous debates and varying standpoints on organismal purposiveness, spanning dissimilar fields of inquiry. For this reason, reconstructing how organismal purposiveness and its cognate concept "agency" were discussed by experimental and theoretical biologists during the first decades of the twentieth century poses a considerable challenge for historians. How can we effectively address this diversity?

While an increasing number of scholars have examined mechanistic, vitalistic, organicist, and holistic perspectives during the interwar period, drawing attention to commonalities, frictions, and unique viewpoints across various local contexts (see, e.g., Nicholson & Gawne, 2015; Esposito, 2016; Peterson, 2016; Baedke, 2019; Donohue & Wolfe, 2023), a systematic exploration of how practicing scientists framed organismal purposiveness and agency is hitherto lacking. In this sense, it could be contended that there is a *historiographic gap* in our current reconstructions of early twentieth-century biology.

DOI: 10.4324/9781003413318-4

Many approaches could be countenanced to broach this issue, but as a first steppingstone for future historical reconstructions, I focus on the general contrasting stances at play in different biological disciplines that scientists customarily assumed. Along these lines, I argue that it is possible to identify—at least—seven distinct stances apropos organismal purposiveness and agency: (i) the neo-Aristotelian position; (ii) the Drieschian answer; (iii) purposiveness eliminativism; (iv) the heuristic approach; (v) the holistic alternative to purposiveness; (vi) organismal purposiveness and agency as *explananda* for dynamic equilibria-based research; and (vii) organismal agency as noetic principle. I cover each of them individually in Sections 3.2–3.8, and then I offer some concluding remarks on this taxonomic endeavor (Section 3.9).

3.2 The Neo-Aristotelian Position

In 1929, zoologist Edwin Grant Conklin attempted to summarize the purported central problems that the science of embryology was confronting in his time: the progressive differentiation of parts in development, the coordination and regulation of ontogenetic phenomena, and the problem of the teleology of development. On this last point, he said: "[...] how can one explain the apparent teleology of development where the end seems to be in view from the beginning? [...] From beginning to end it appears that development is moving toward a goal" (Conklin, 1929, p. 31). Conklin was convinced, as many of his embryologist peers were, that accounting for, or not to gloss over, the intrinsic purposiveness of development was a real problem for biologists: "To refuse to recognize the teleological aspect of development does not eliminate it; we can not explain it by merely explaining it away, for the problem remains and future generations will deal with it if we can not" (Conklin, 1929p. 32). In another publication, Conklin resorted to affiliate his views rhetorically and intellectually with those of Aristotle:

> Aristotle, "the master of those who know," was great in every field of knowledge, but greatest of all in zoology. In his "De Partibus Animalium" he maintained that the essence of a living animal is found not in *what* it is, or *how* it acts, but *why* it is as it is and acts as it does. This question *why* is a hard one to answer [...], but it is one that cannot be wholly disregarded, and, by facing it frankly, we may be saved from much shallow thinking, cocksureness and narrowmindedness.
>
> (Conklin, 1944, p. 127; emphasis in original)

If one peruses early twentieth-century embryological literature, one finds that Conklin was not the only scientist granting the constitutive purposiveness of developing organisms and associating this idea to Aristotle (see also Esposito, this volume). For example, in an article on the morphogenesis of muscoid flies, the Canadian entomologist William Robin Thompson, later author of the book *Science and Common Sense: An Aristotelian Excursion* (see Thompson, 1937), asserted:

> The vital movement, taken as a whole, is *essentially adaptive*, or, in other words, directed to the attainment of ends advantageous to the organism itself and to

its maintenance in existence. This perpetual adjustment to changing conditions both within and without the organism has always been acknowledged to be the characteristic feature of vital action, admirably defined by Aristotle [...] in the statement that life is an "immanent movement."

(Thompson, 1929, p. 237; emphasis in original)

Yet another example is afforded by the Scottish theoretical biologist D'Arcy Wentworth Thompson, who acted as translator of the biological writings of the Stagyrite and enshrined his legacy in the scientific study of life (see Thompson, 1913). Thompson, like Aristotle, accentuated the centrality of "form," and he contended that a recourse to final causes should be part of the toolkit of the biologist (for discussion, see Plochmann, 1953).

Conklin and the two Thompsons are representative of a larger trend in early twentieth-century biology: we can identify a definite neo-Aristotelian stance toward the problem of purposiveness, wherein this is postulated as an intrinsic, salient property of what makes organisms what they *essentially* are.[1] For biologists who assumed the neo-Aristotelian vantage point, the idea of resorting to purposiveness and final causes only as heuristic devices was not convincing and this had nothing to do with the postulation of non-physical forces acting on developmental trajectories (see Haldane et al., 1917, p. 459). In contrast to this, we reach the second stance with respect to organismal purposiveness in the landscape of early twentieth-century biology.

3.3 The Drieschian Answer

A common misassumption among scholars is that an ontological posit of purposiveness entails, *ipso facto*, a descent into a metaphysical form of vitalism. This conflation is conceptually mistaken, and many early twentieth-century biologists were acutely aware of that. For instance, Conklin (1937, p. 232), a staunch mechanist, averred that purposiveness is a mark of organisms while maintaining that mechanistic explanations should be sought for all purposive phenomena. The German embryologist and philosopher Hans Driesch insisted on this point:

The main question of Vitalism is not whether the processes of life can properly be called purposive: it is rather the question if the purposiveness in those processes is the result of a special *constellation of factors* known already to the sciences of the inorganic, or if it is the result of an *autonomy* peculiar to the processes themselves.

(Driesch, 1914, p. A; emphasis in original)

We are all familiar with the experiment of Driesch (1892) separating sea urchin blastomeres, finding out that each one of them developed into a complete, albeit smaller larva, something that contradicted earlier studies of Wilhelm Roux that suggested that inactivating a frog blastomere with a heated needle thwarted its development. These results led Roux to a strong conclusion: each individual blastomere possesses the remarkable ability of self-differentiation. In contrast, Driesch considered organisms to be harmonious-equipotential systems with the capacity to *re*-produce the whole if disturbed. This plastic, self-regulatory capacity was designated by Driesch as "equifinality." He also experimented with adult sea squirts of the genus *Clavellina*, and he showed that some parts could form a complete individual if detached (Driesch, 1902). In contrast, a part that is not separated from the whole organism does not realize this potential. What is responsible for this? Driesch considered the analogical correspondence between organisms and machines and thus the possible physico-chemical options to explain this phenomenon, even the idea that some parts of a machine could contain the potentiality to become a new whole. However, he sequentially crossed out the possibilities, all except one: in the form of a disjunctive proposition, he reached the idea that the factor E that explains purposiveness was non-physical and non-chemical; in other words, he presented an *argument per exclusionem*:

> E is either this, or that, or the other, and it was shown that it could not be any of all these except one [...]; something new and elemental must always be introduced whenever what is known of other elemental facts is proved to be unable to explain the facts in a new field of investigation.
>
> (Driesch, 1908, p. 142)

Driesch later renamed this factor as *entelechy*, borrowing the term from Aristotle but recognizing it meant something different under his guise. Allen (2008, p. 52) maintained that entelechy was not "the blueprint of an organism's organization, nor the creative agent that brings it about, but a kind of a mediator [...] that protects the tendency of the system from being disrupted by extraneous factors," such as an ablation. For Driesch, the existence of an entelechy did not violate the physical laws of nature as it was not a means of creating energy outside the laws of thermodynamics, it was simply a non-material, vital force that kept development on its route (for discussion, see Churchill, 1969).[2]

In the first decades of the twentieth century, Driesch's vitalism amassed some followers, but it ignited a strong opposition as well: biologists from many disciplines rejected his standpoint but, in doing so, vetoed any reference to organismal purposiveness.[3] Even though purposiveness and entelechy are conceptually distinct (e.g., through the *explanandum-explanans* distinction), some authors dismissed both altogether. And with this move we reach the third viewpoint in this survey.

3.4 Purposiveness Eliminativism

In several fields of early twentieth-century biology, the stance I call here "purposiveness eliminativism" held sway. An illustrative example is *Entwicklungsmechanik*. Roux and colleagues proclaimed that all purposiveness in development is just superficial and apparent (*Scheinbar*), and thus misleading for embryology (Roux et al., 1912, p. 460). They maintained that biology can perfectly do without agentic notions. The only concession they made was that the use of the idiom of "purposiveness" is exclusively warranted in cases involving human conscious intention.

Purposiveness eliminativism was not exclusive of *Entwicklungsmechanik*. An example of this rationale in a different field is embodied in the work of British plant morphologist William Henry Lang, who renounced viewing organisms as exhibiting intrinsic purposiveness:

> The course of development to the adult condition can be looked upon as the manifestation of the properties of the specific substance under certain conditions. This decides our attitude as morphologists to the functions of the plant and to teleology. […] [U]ntil purpose can be shown to be effective as a causal factor it is merely an unfortunate expression for the result attained.
>
> (Lang, 1915, p. 784)

Other botanists also disallowed intrinsic purposiveness and insisted on its epistemic irrelevance: according to them, conceptualizing self-regulation processes of developing plants as "purposive" only "obscured" sequences of highly coordinated physiological events (e.g., Farmer, 1903, p. 219).

Another prominent example of purposiveness eliminativism comes from physiologist Jacques Loeb and his tropism theory in animal behavior research. Loeb investigated so-called reflex-actions and construed them as reactions of isolated parts inside an organic body. In an analogous fashion, he defended that the reaction of whole organisms can be reduced to physicochemical triggers. For Loeb, a stumbling block for the scientific study of animal behavior was to grant that organisms pursue intrinsic goals; in particular, he alleged that a difficulty for the widespread acceptance of his tropism theory was

> created by the fact that the *Aristotelian viewpoint* still prevails to some extent in biology, namely, that an animal moves only for a purpose […]. *The analysis of animal conduct only becomes scientific in so far as it drops the question of purpose* and reduces the reactions of animals to quantitative laws.
>
> (Loeb, 1919, pp. 17–18; emphasis added)

Needham (1930, p. 192) concurred with Loeb's outlook and declared that "[the neo-mechanistic position] knows teleology to be an unquantitative category, and banishes it from the laboratory to the domain of the philosophers, who are quite capable of dealing with it." Many scientists of the time also harmonized with this standpoint by claiming that organismal purposiveness was only a pseudo-problem (*Scheinproblem*) that scientists should not bother to solve (e.g., Heikertinger, 1917).

If not as a constitutive element of what makes organisms what they are, did some early twentieth-century biologists grant to purposiveness at least a heuristic role that could impact their research? This question leads us to our next stance.

3.5 The Heuristic Approach

In his discussions on the segmentation patterns of animal embryos, the US-American zoologist and geneticist Edmund Beecher Wilson recovered a heuristic role for purposiveness:

> we cannot comprehend the specific forms of cleavage without reference to the end-result of the formative process [...]. The teleological aspect of cleavage thus suggested has been recognized [...] most by Lillie, who has urged that with this principle in mind "one can thus go over every detail of the cleavage, and knowing the fate of the cells, can explain all the irregularities and peculiarities displayed". [...] Such a conclusion need involve no mystical doctrine of teleology or of final causes.
>
> (Wilson, 1925, p. 1005)

Wilson's musings allude to a prevalent approach employed by embryologists such as C. O. Whitman and Frank Lillie, who frequently utilized teleological language in their investigations of cell lineages (Churchill, 1969). Theoretical biologist Joseph Henry Woodger, a defender of intrinsic purposiveness as a constitutive feature of organisms, cited the case of Wilson as a representative example of a larger trend in interwar biology to treat purposiveness in a heuristic manner. Moreover, Woodger (1929) attempted to clarify what exactly could be this heuristic role: "[...] in regard to explanation a knowledge of the *outcome* of a biological process is often more illuminating than knowledge of what went before" (p. 432; emphasis added).

The heuristic approach has a long philosophical pedigree and was most famously articulated by Immanuel Kant in his *Critique of the Power of Judgment*.[4] Besides noticing the extrinsic fit of "organized beings" within their surroundings in a way that could foreshadow design in a larger system of nature, Kant also emphasized the intrinsic fit between the parts of organisms that allows them to be functional wholes, *organisierte Wesen* with generative and nutritive processes. In that sense, organisms, for Kant, are perceived as natural purposes (*Naturzwecke*) "in which everything is a purpose and reciprocally also a means" (KdU, 5: 376).[5]

According to Kant, the acknowledgment of the teleological organization of organisms seems to transcend the boundaries that define acceptable scientific discourse. For him, scientific explanations should adhere to a mechanistic framework. However, he also acknowledged that biological phenomena cannot be fully encompassed by mechanistic explanations. As a result, scrutinizing the concept of "purposiveness" and the proper and improper ways it is employed in science are crucial (Körner, 1955, pp. 197–198).

Consequently, this led to the emergence of the widely known "antinomy of the teleological power of judgment": the thesis "All generation of material things and their forms must be judged as possible in accordance with merely mechanical laws" is juxtaposed with the anti-thesis "[s]ome products of material nature cannot be judged as possible according to merely mechanical laws" (KdU, 5: 387). Kant's chosen stance to resolve this dilemma is that teleological judgment is valid only for cognitively limited beings like ourselves as a regulative ideal for inquiry, operating as a reflective judgment that does not delve into the ontological nature of the objects under consideration (Illetterati & Michelini, 2008, pp. 4–5). According to Kant, organismal purposiveness is not a mere projection that can be subjectively chosen or dismissed. Once biological entities become the subject of investigation, there is no alternative for us, given our finitely rational make-up, but to see them as purposive (for discussion, see Desmond & Huneman, 2020).

Kant's influential answer rippled throughout the late nineteenth and early twentieth centuries (for discussion, see, e.g., Baedke, this volume). Until recently, the mainstream historiographic view was that, after Kant, appeals to regulative teleology shaped the incipient life sciences through a tradition dubbed by Lenoir (1989) as "teleomechanism." However, scholars have challenged Lenoir's reconstructions and have adduced evidence to contend that many scientists, before and after Kant, sought to explain "purposiveness" as something constitutive of organisms, and not as a regulative maxim of the reflective power of judgment with heuristic value for biological research (e.g., Zammito, 2012). Most often than not, biologists who were heavily inspired by Kant's critical philosophy decided to part ways with him regarding the issue of organismal purposiveness, and this is particularly clear in the next stance I recount.

3.6 The Holistic Alternative to Purposiveness

John Scott Haldane is a prime example of someone embodying a different outlook regarding organismal purposiveness in the first decades of the twentieth century. In an early piece co-authored with his brother, the lawyer, and philosopher Richard Burdon Haldane, they argued that, for the sake of building a better science and philosophy of science, "[…] the mastery of the critical investigations of Kant and Hegel […] will be absolutely essential. But such a discipline can form simply a part […] of *preparatory culture*" (Haldane & Haldane, 1883, p. 66; emphasis added). Indeed, when turning his attention to the study of organisms, J. S. Haldane was dissatisfied with Kant's treatment:

> At the time of Kant and his immediate successors biology had hardly begun to be conscious of her strength. Living organisms seemed, as it were, to be at the best only dotted about here and there in the midst of a totally foreign physical universe. Kant assigned to them a very doubtful place in his philosophy.
>
> (Haldane, 1921, p. 99)

In brief, Haldane believed Kant's conclusions were premature: "As regards life, […] Kant was mistaken" (Haldane, 1929, p. 65).

In the aforementioned piece, the Haldane brothers considered that

> the distinguishing feature of vital activity is self-preservation […]; and this is just as true of the most complicated actions of the human body as of the movement of the amoeba towards a source of nourishment. […] The fact is that

every part of the organism must be conceived as actually or potentially acting on and being acted on by the other parts and by the environment, so as to form with them a self-conserving system.

(Haldane & Haldane, 1883, pp. 54–55)

A leitmotif in Haldane's work would be the emphasis on the importance of the environment in the self-maintenance of life. Embracing an ontological co-constitution perspective, Haldane viewed the organism and its environment as intricately entangled, forming an interconnected whole that resists trouble-free separation (see Baedke et al., 2021).

Within this holistic position, Haldane reasoned that the concept of purposiveness loses its coherence. "What is ideally implied in such words as 'function,' 'purpose,' 'means,' and 'end,'" the Haldane brothers argued, "is that we are looking at the organism, not as acted on by things outside it, but as in teleological connection with that which is different from, but not existent independently of it" (Haldane & Haldane, 1883, p. 58). Later on, Haldane laid down his vantage point clearly:

It seems to me that as mere biologists *we have no need to make use of the concepts of either memory or purpose*. What we observe in all lives is simply their tendency to maintain and reproduce themselves as co-ordinated wholes.

(Haldane, 1931, p. 161; emphasis added)

Other examples of authors who argued for replacing the concept of purposiveness with that of "wholeness" were the German botanist and philosopher Emil Ungerer and the Dutch ecologist Cornelis van der Klaauw. In his studies on plant regulation (see Baedke, this volume), Ungerer (1919) suggested that if an organism preserves itself during development as a whole, the processes that seem to condition the preservation of this wholeness can only be described as "purposive" in a mere descriptive, almost void sense. Following a similar line of thought, van der Klaauw (1935) also disagreed with the Kantian interpretation of *Zweckmäßigkeit* and sought to counteract what he saw as a substantial limitation. He proposed replacing the concept of "purposiveness" with "wholeness," aligning with the perspective previously presented by Ungerer (Trienes, 1992, p. 14).

The holistic standpoint on purposiveness was widely debated in those years. Baedke et al. (2024) have suggested that it replaced neo-Kantian positions after they peaked in significance in early twentieth-century debates. For example, in the Seventh International Congress of Philosophy held at Oxford in 1930, a whole session was devoted to the topic of organismal purposiveness and the holistic stance dominated there. Biologists and philosophers, such as Haldane, Ungerer, Wildon Carr, Alfred Hoernlé, and others, agreed that self-maintenance could not be described as purposive. In the summary of this session, Hoernlé expressed:

THE ANTITHESIS of mechanism and purpose is out of date. [...] Evidences of this shift are: (a) The substitution (of the concept of purpose by that of the whole) [...]. Whole is a concept equally applicable to plant, animal, man [...].

(b) The biological protagonists [...] argued primarily on methodological, not on metaphysical, grounds. The crucial question was: What is the most profitable technique of investigation in biology? [...] [The conclusion was:] *Biological processes must be holistically conceived, in order to be mechanistically studied.*
(Hoernlé, 1931, p. 44; emphasis in original)

The holistic alternative was sometimes paired with purposiveness eliminativism. A clear example of this can be found in the work of the German neurologist Kurt Goldstein. He agreed with Ungerer that the concept of wholeness should be given priority in the explanation of organisms and that all purpose-talk should be sidestepped:

> I agree with Ungerer so far as to reduce the term "teleological" to this mere descriptive use and, furthermore, with this demand that the term "purposive" would be best avoided altogether. [...] The idea of an intended task is super-fluous for an understanding of the organism, but that of a definite end [...] may be very fruitful [...]. Yet the idea of "end" must also be taken only as a guiding notion for the procedure of knowledge rather than in a metaphysical sense [...]. In this sense, one can describe the concept of wholeness, as a category, as the category that substantiates and encompasses the subject matter of biology.
> (Goldstein, 1995, p. 324)

In contrast, proponents of the viewpoint I introduce below, situated within the same problem domain as the holistic approach, introduced novel *explanantia* and reinstated the significance of the concept of organismal intrinsic purposiveness.

3.7 Organismal Purposiveness and Agency as *Explananda* for Dynamic Equilibria-Related Research

In a 1915 article, the Canadian physiologist Ralph Stayner Lillie addressed the problem of organismal agency head on:

> Apparently, *the general "purpose" of most animal actions is to take some advantage of conditions existing in the environment, or to modify the relations between the individual and the environment in some way favorable to the species.* [...]. This is why they impress us as "purposive." The "teleological" characteristic of living beings appears most conspicuously in this aspect of their life. But from the physiological point of view it is necessary to reach some purely objective or physicochemical definition of the term "purposive" as applied to such actions.
> (Lillie, 1915, p. 589; emphasis added)

Lillie wanted to answer if purposive behavior is an essential attribute of material systems characterized by the unique organization of living organisms and their distinctive interactions with their surrounding conditions: "[...] My procedure and methods of reasoning will be those of objective natural science purely; and purposive actions [...] will be considered simply as events in external nature, disregarding their possible conscious or psychic accompaniment" (Lillie, 1915, p. 590). Lillie deliberated that purposive actions obtain in complex physico-chemical systems like organisms that (i) exhibit metabolism and (ii) preserve an equilibrium with a changing environment.

His argument depended on a physiological understanding of "adaptation," which he defined in a very general way as a "species conserving characteristic." He went on: "An organism is said to be adapted to its environment when it exhibits physical properties, structural characters and activities of such a kind that its development, growth, and continued existence in that environment are ensured" (Lillie, 1915, pp. 590–591). Lillie suggested that all organisms convert matter and energy sourced from their environment into their organization, and they engage in activities that aim for self-preservation. For this, Lillie claimed that an organism needed a "metabolic equilibrium," a balance between gain and loss of materials and energy. In this sense, adaptation "as a condition characterizing the relations between organism and environment, is thus defined [as] *the maintenance of the organic equilibrium*; adaptive features of external structure or behavior are those which contribute to this end" (Lillie, 1915, p. 592; emphasis added). From such a vantage point, Lillie reasoned that

> [T]he organism is to be regarded as a physicochemical system of a special kind, exhibiting a dynamic equilibrium with its surroundings, i. e., an equilibrium in which two sets of processes, one constructive or constitutive, the other destructive or dissipative, balance each other. Many other so-called "stationary" systems—characterized by a continual and balanced interchange of material and energy with the surroundings—exist in nature; [...] hence, the comparison of a living organism with a vortex or candle-flame is traditional, and serves to make clear certain fundamental peculiarities of the living condition. One of the most interesting general properties of such systems is a certain power of regulatory adjustment to changes of condition.
>
> (Lillie, 1915, p. 593)

Finally, we can see how Lillie rounded up his argument. He expounded that organismal purposive actions have the consequence of securing an energy supply that can be utilized for future requirements. Any action that facilitates the continued existence of an organism can thus be deemed as purposive. Furthermore, Lillie (1915) ratiocinated that all instances of adaptation, whether reliant on static characteristics of form or structure, or on specific organismal activities, fundamentally represent organic equilibria of varying complexity from a physicochemical perspective. In essence, he postulated that purposive actions arise through the establishment of a dynamic equilibrium between the organism and its ever-changing environment.

Over the following years, Lillie remained steadfast in advocating this perspective, but he ultimately took a different, biopsychist turn (e.g., Lillie, 1945). For his former idea of adaptation as a dynamic equilibrium between organism and environment, he was heavily inspired by the German physiologist Paul Jensen, who was convinced that organismal purposiveness and agential behaviors could be understood within the bounds of physiological science (Jensen, 1907).

Indeed, Jensen and Lillie were not isolated advocates of this view in early twentieth-century biology (see also Nahas, this volume). Sumner (1919, p. 356) also embraced a physiological view of adaptation and argued that organismal purposive activities, from developmental phenomena such as regeneration to behavioral activities, can be explained by *"the reattainment of a condition of equilibrium which has been overthrown"* (emphasis in original). Additionally, Hooker (1919) argued that Le Chatelier's principle of chemical equilibrium was enough to account for organismal purposiveness. This principle states that if a system in chemical equilibrium is subjected to a constraint that shifts the equilibrium, a reaction takes place which partially annuls its effects and restores the equilibrium. Hooker claimed that

> *every system in equilibrium is teleological.* The means that produce the reaction are directed to a definite end, to overcome the constraint, and the reaction might be said to take place in order that the system may be preserved. *This is evidently the source of the "purposefulness," that has occasioned endless biological discussion.*
>
> (Hooker, 1919, p. 509; emphasis added)

In the next years, biologists recognized that living systems, even when resting, can never fully be in a state of "equilibrium," in which the total flow of energy is zero. Hill (1930) pointed out that biological systems are, rather, in a "steady state." The Austrian theoretical biologist Ludwig von Bertalanffy contented that an equilibrium can only occur in closed systems, while in organisms, paradigmatic open systems, the increase in entropy is compensated by an influx of free energy from the environment, yielding a thermodynamic state of stationary non-equilibrium (*Fliessgleichgewitch*; for an overview, see Bertalanffy, 1953). Bertalanffy, and others like Burton (1939), directed the steady-state view to embryonic development and behavior. Bertalanffy argued that this solved the riddle of intrinsic purposiveness that had long puzzled Driesch:

> Organic processes are explicable by means of two assumptions: (1) that the internal forces of living systems are directed towards states of equilibrium, and (2) that this direction holds for the system as a whole. This self-regulation of the organism is wonderful enough in its details, but, in principle, every connected system of the inorganic world in which the groupings of internal forces is directed towards equilibrium, behaves in the same way (Principle of Le Chatelier [...]) In this way Driesch's paradox is elucidated. Because the organism is a unitary system, what happens at a given place is determined according to its relations in the whole. [...] *We can, however, suppose that the "purposefulness" and "striving towards a goal" of organic processes is nothing else than the outcome*

*of communicating systems of causally determined processes, the inner dynam-
ical conditions of which tend towards equilibrium.*

(Bertalanffy, 1933, pp. 103–104; emphasis added)

Let us cap off with the last position in this historical survey. To comprehend the
postulation of organismal agency as a noetic principle for biological practice, we
must delve into the writings of E. S. Russell.

3.8 Organismal Agency as Noetic Principle

Russell, at the time of publication of his famous *Form and Function* and still under
the influence of his mentors, Patrick Geddes and Arthur Thomson (see Esposito,
this volume), was a convinced defender of the neo-Aristotelian standpoint regarding
organismal purposiveness. In fact, the book ends with a vocal call to, once again,
imbue biological research with Aristotle's insights:

> We need to look at living things with new eyes and a truer sympathy. We shall
> then see them as active, living, passionate beings like ourselves, and we shall
> seek in our morphology to interpret as far as may be their form in terms of their
> activity. This is what Aristotle tried to do, and a succession of master-minds
> after him. We shall do well to get all the help from them we can.
>
> (Russell, 1916, p. 364)

This notwithstanding, I want to argue that Russell subsequently revised his neo-
Aristotelian positioning and put forward a different view, one that, although shared
sympathies with it, has nuances of its own that can be extricated.

In the interwar period, Russell argued that it was important to "[…] regard the
activities of a given individual from its own point of view, in relation to the world
as perceived by it, not in relation to the world as perceived by us" (Russell, 1924,
p. 147). Along these lines, he espoused a dynamical view in which

> a living thing must be able to follow and counter by appropriate response the
> changes in its environment which […] have a vital meaning for it. Now this can
> be achieved only through perception. I use the word here […] in a broad way
> [i.e., not related to conscious states], to cover all degrees of the receptive side
> of vital activity.
>
> (Russell, 1924, p. 57; text inside brackets added)

Russell claimed that separation from the environment is needed by the organism
to have agency, or phrased differently, that individuality is a precondition for agency.
"The living thing can have, no more than the machine, an internal or self-existent
purposiveness," he stated, "unless it has at the same time real individuality and persistence" (Russell, 1924, pp. 16–17). We come to understand now that Russell sought
to carve out organisms as subjects, as agents in the world that act on their own behalf:

[I]t is mainly through perception that life becomes individualized and separates itself out from the environing flux. Through perception the organism clears, as it were, a space around it in which to live [...]. *The living individual is then a subject, or better—if we lay emphasis on action rather than on presentation or representation—an agent*; and there is necessarily implied an object, or more generally, something sensed to which the individual responds. *Individuality has therefore as its necessary complement a sensed environment*, an objective world however dimly presented.

(Russell, 1924, p. 59; emphasis added)[6]

Russell's research encompassed various examples that highlighted the agency of organisms. For instance, he conceptualized some types of fish migration as goal-directed behavior and highlighted the importance of individual experiences in the sea (Russell, 1934, p. 41; 1937).

But how exactly can Russell's standpoint be construed as advancing a noetic principle for biological practice? Here, I am using the term "noetic" to refer to *scientific understanding*, and not as something related to the realm of the mental. For Russell, postulating organismal purposiveness and agency as ontological principles did epistemic work as well, especially for the scientific study of animal behavior. He contrasted the prevalence of causal explanations with his subject matter: the activities and behaviors of organisms. In this regard, he asserted:

In the physical sciences and to a large extent in physiology we are concerned primarily with causal explanations [...]. We do not get this sort of knowledge by using the direct or descriptive method. What we get may be better described as *an understanding of behaviour.* When you see a wasp, for example, standing on a gate post and busily chewing at the wood, you do not understand this action until you follow the wasp up and find out that it uses this material to construct its nest; you go on then to discover what the nest is used for—the care and upbringing of the young. In a word, *to understand the action of the wasp on the gate post you have to integrate this action into the whole directive cycle of activity* [...]. This is the sort of knowledge which we get by studying animals from the [agential] point of view, and it is extremely valuable knowledge too. *Without this knowledge we simply cannot make sense of behaviour;* even the most extreme believer in mechanism must use this sort of knowledge to make behaviour *intelligible* at all.

(Russell, 1934, p. 15; emphasis added)

In other words, the recognition of the goal-directed character of behavior is absolutely necessary for *understanding* it, for making it *intelligible* to scientific scrutiny. Russell's noetic standpoint was not only confined to organismal behavior. He extended it to physiological processes, asserting that they only "become intelligible" when scientists link them up "with one of the main biological ends which the organisms blindly pursues" (Russell, 1945, p. 9).

Russell consistently upheld this stance in his career. For example, he stressed:

> If we disregard ends and the directiveness of activities towards ends, we leave out what is distinctive in life, and we simply amass more and more data about [...] the physico-chemical conditions of life, without connecting them up in a rational way with the functional life-cycle of the organism.
>
> (Russell, 1945, p. 8)

Within Russell's perspective, coming to terms with the purposive nature of organismal activities lies at the heart of rendering them comprehensible for scientific investigations:

> If then it is true, and indeed the chief truth, about the living organism that its activities [...] are directive towards living, reproducing and developing, we must, in our study of these activities, consider first and above all their biological significance [...], their relation to one or other of these biological ends. If we do not do so, but consider them separately, without relation to the life of the organism as a living, developing, reproducing whole, we shall never understand them, even though we succeed in working out their physico-chemical "mechanism" or mode of action. We shall acquire [...] a vast mass of unrelated facts of biochemistry and biophysics, but we shall never build up a real biology.
>
> (Russell, 1945, p. 9)

I consider that a way to spell out Russell's stance is to frame it as a form of transcendental argument. In employing this notion, I draw inspiration from Chang (2008) and Brooks (2021), who have recently attempted to rehabilitate transcendental arguments in naturalist frameworks. A transcendental argument is one that accords to the following form: If we want to engage in a particular epistemic activity (e.g., understanding animal behavior and making it intelligible), *then* it is necessary to presume particular ontological principles (accordingly, organismal purposiveness and agency). An ontological principle, under this conception, is one that makes systematically intelligible at least one kind of epistemic activity. In this case, postulating organismal purposiveness as an ontological principle allows scientific understanding *qua* epistemic activity and allows for the design of experimental interventions and the interpretation of empirical results. Moreover, I think that Russell's standpoint can said to be transcendental insofar as it delves into the *conditions of knowability* (see Van de Vijver et al., 2005) of the behavior of other organisms (i.e., we need to posit their agentic status in order to begin to understand their activities and doings), and from there, a biological science that seeks complementary causal-mechanistic explanations can then take off. In sum, for Russell, the presumption of agency and purposiveness is a *noetic principle* for practicing scientists dealing with developing, behaving organisms.

Russell's view has some echoes with the Kantian interpretation of organismal teleology. For Kant, "[...] seeing an organism as an agent is even a precondition (the transcendental ground) to being able to make a projection onto a natural system. [...] Assuming agency makes ascribing empirical methodology and even (behavioral) property to organisms possible" (Desmond & Huneman, 2020, p. 58). A key difference between Russell and Kant is that the former is interpreting agency ontically and this would be anathema under Kant's eyes.

The emphasis on the importance of granting intrinsic purposiveness to organisms in order to understand them can be glimpsed in Russell's early works: "Teleology in Kant's sense is and will always be a necessary postulate of biology. It does not supply an explanation of organic forms and activities, *but without it one cannot even begin to understand living things*" (Russell, 1916p. 35; emphasis added). Later on, his noetic stance became palpable in his empirical studies on fish behavior and in his valence theory on why only certain stimuli in an animal's sensed environment elicit behavioral responses. For example, a hermit crab (*Pagurus arrosor*) devours or keeps a commensal anemone in its shell if it is hungry, but not as a univocal reaction to the presentation of the environmental stimuli. Russell (1935) made sense of this through his valence theory and argued that a deterministic stimulus-response view fails to apprehend the actual behavior of organisms because it disregards their internal states and needs. In the empirical terrain, Russell performed some detour experiments with sticklebacks. In this setting, the shortest way to a visually attractive object is blocked for the test animal, albeit a roundabout way is left open. Russell tested sticklebacks to see if they could learn to find their way into a small jar filled with food, and then he evaluated if they could also learn to come out from there. In his interpretation, the fish, at first, found the solution by chance, but after a few training trials, the entrance-seeking behavior became purposive, with the clear goal of attaining the nourishing items in its sensed environment. With this, there was a measurable sudden decrease in the time it took the organisms to get the food (Russell, 1931). By presuming that organisms are agents, Russell could begin planning his experiments and make sense of the data he collected. This is a concrete case of his noetic principle in action.

3.9 Concluding Remarks

Here, I recounted how, in the early decades of the twentieth century, biologists from diverse disciplinary trenches conceptualized organismal purposiveness and agency. According to my reconstruction, it is possible to identify seven distinct stances: (i) the neo-Aristotelian position; (ii) the Drieschian answer; (iii) purposiveness eliminativism; (iv) the heuristic approach; (v) the holistic alternative to purposiveness; (vi) organismal purposiveness and agency as *explananda* for dynamic equilibria; and (vii) organismal agency as noetic principle.

Certain stances distinctly emphasized the ontological domain, either to stress or deny organismal purposiveness (e.g., i, iii), whereas others directed their attention toward accentuating the inclusion of teleological notions within the epistemology of biological science (e.g., iv). Meanwhile, a subset of positions transacted in an amalgamated area of onto-epistemic concerns (e.g., v, vi, vii). A historiographic upshot of this taxonomy is to show that the traditional tripartite classification of mechanism-organicism-vitalism *qua* metatheoretical commitments of scientists (see Hein, 1972) is insufficient to capture the diversity of past standpoints on organismal purposiveness and agency.

Charting a taxonomy of positions is a useful first step when facing the theoretical and practice-associated heterogeneity of different biological disciplines, but subsequent detailed historical analyses are still needed to fill the historiographic gap on this subject. Several questions and research paths remain open for exploration. It will be important to determine the prevalence of each identified standpoint, uncover the nuances and sub-varieties of each position, and elucidate instances of overlap. Furthermore, a comprehensive analysis of the network of authors, their connections, mutual influences, and social arenas will prove to be essential for an in-depth investigation into the topic of organismal purposiveness in early twentieth-century biology. Another significant question, with far-reaching implications, is how these discussions related to earlier nineteenth-century debates and late twentieth-century developments.

This rich tapestry of positions could enrich the contemporary debate on organismal agency, which, in the philosophy of biology, features just three primary factions (Walsh, 2015; Moreno & Mossio, 2015; Babcock & McShea, this volume), with the (Kantian) heuristic perspective enduring alongside them (Desmond & Huneman, 2020). Valuable insights could be gleaned from the diversity that may have once seemed "lost."

Acknowledgments

I thank attendees of the "Life-Philosophies: Exploring Heterodox Naturalism" Seminar at Ghent University (December 2021) and "The Riddle of Organismal Agency" workshop at Ruhr University Bochum (March 2022) for their helpful comments and observations on this work. Likewise, I express gratitude to Auguste Nahas and Jan Baedke for their constructive feedback on early versions of this manuscript. I acknowledge the *Deutsche Forschungsgemeinschaft* (project no. BA 5808/2-1) for financially supporting my research.

Notes

1   It is worth noting that not many of these biologists fully recovered Aristotle's conceptual apparatus for comprehending organisms (e.g., *Bios*, *eidos*, and primary substances) as beings enacting a particular kind of way of life, which contains a formal and final principle. The neo-Aristotelian position often functioned more as a rhetorical banner and a shared influences-view to stress organismal intrinsic purposiveness, rather than representing a rehabilitation of Aristotle's original works.
2   Scholars have critically re-evaluated the historiography of vitalism in recent years. This reassessment has debunked the thesis that it was a homogeneous, monolithic position (e.g., Wolfe, 2011; Donohue & Wolfe, 2023). Importantly, vitalism existed along a nuanced continuum of interpretations and varying degrees of metaphysical commitments (Bognon-Küss et al., 2018).
3   Many biologists treated Driesch's vitalism not as a metaphysical heresy but as a legitimate response to the limitations of contemporaneous mechanistic explanations. Chen

(2018) has argued that their refutation of vitalism was based on logical grounds, rather than relying on metaphysical remonstrations.

4   It should be mentioned that not all instances of the heuristic approach are Kantian in a committed philosophical sense, i.e., the Kantian position is an instance of the heuristic approach, but not all heuristic attitudes on organismal teleology are Kantian. For discussion, see Gambarotto and Nahas (2022).

5   Paragraphs from Kant (1790/2000) are herein cited. It must be said, however, that Kant's views on organisms have been the subject of protracted exegetical disagreements between specialized scholars. Taking a stand on how best to interpret Kant is outside the purview of this chapter.

6   Russell acknowledged that his approach was also shared by the Austro-Hungarian botanist Raoul Heinrich Francé, and especially by the Austrian botanist Adolf Wagner (for discussion on these authors, see Baedke, this volume).

References

Allen, G. E. (2008). Rebel with two causes: Hans Driesch. In O. Harman & M. Dietrich (Eds.), *Rebels, mavericks, and heretics in biology* (pp. 37–64). Yale University Press.

Baedke, J. (2019). O organism, where art thou? Old and new challenges for organism-centered biology. *Journal of the History of Biology, 52*(2), 293–324.

Baedke, J., Böhm, A., & Reiners-Selbach, S. (2024). From Kant to holism: The decline of neo-Kantianism and the rise of theoretical biology. In H. Pulte, J. Baedke, D. Koenig, & G. Nickel (Eds.), *New perspectives on neo-Kantianism and the sciences* (pp. 253–277). Routledge.

Baedke, J., Fábregas-Tejeda, A., & Prieto, G. I. (2021). Unknotting reciprocal causation between organism and environment. *Biology & Philosophy, 36*(5), 48. https://doi.org/10.1007/s10539-021-09815-0

Bertalanffy, L. v. (1933). *Modern theories of development*. Oxford University Press.

Bertalanffy, L. v. (1953). *Biophysik des Fließgleichgewichts: Einführung in die Physik offener Systeme und ihre Anwendung in der Biologie*. Vieweg+Teubner.

Bognon-Küss, C., Chen, B., & Wolfe, C. T. (2018). Metaphysics, function and the engineering of life: The problem of vitalism. *Kairos. Journal of Philosophy & Science, 20*(1), 113–140.

Brooks, D. S. (2021). Adaptive design, contingency, and ontological principles for limited beings. *Philosophy of Science, 88*(5), 871–881.

Burton, A. C. (1939). The properties of the steady state compared to those of equilibrium as shown in characteristic biological behavior. *Journal of Cellular and Comparative Physiology, 14*(3), 327–349.

Chang, H. (2008). Contingent transcendental arguments for metaphysical principles. *Royal Institute of Philosophy Supplements, 63*, 113–133.

Chen, B. (2018). A non-metaphysical evaluation of vitalism in the early twentieth century. *History and Philosophy of the Life Sciences, 40*(3), 50. https://doi.org/10.1007/s40656-018-0221-2

Churchill, F. B. (1969). From machine-theory to entelechy: Two studies in developmental teleology. *Journal of the History of Biology, 2*(1), 165–185.

Conklin, E. G. (1929). Problems of development. *The American Naturalist, 63*(684), 5–36.

Conklin, E. G. (1937). What shapes our ends? *The American Scholar, 6*(2), 225–235.

58   *Alejandro Fábregas-Tejeda*

Conklin, E. G. (1944). Ends as well as means in life and evolution. *Transactions of the New York Academy of Sciences, 6*, 125–136.

Desmond, H., & Huneman, P. (2020). The ontology of organismic agency: A Kantian approach. In A. Altobrando & P. Biasetti (Eds.), *Natural born monads: On the metaphysics of organisms and human individuals* (pp. 33–64). De Gruyter.

Donohue, C., & Wolfe, C. T. (Eds.). (2023). *Vitalism and its legacy in twentieth century life sciences and philosophy*. Springer.

Driesch, H. (1892). Entwicklungsmechanische Studien. I. Der Werth der beiden ersten Furchungszellen in der Echinodermenentwicklung. Experimentelle Erzeugung von Theft- und Doppelbildungen. *Zeitschrift für wissenschaftliche Zoologie, 53*, 160–184.

Driesch, H. (1902). Studien über das Regulationsvermögen der Organismen: Die Restitutionen der *Clavellina lepadiformis*. *Archiv für Entwicklungsmechanik der Organismen, 14*(1), 247–287.

Driesch, H. (1908). *The science and philosophy of the organism*. Aberdeen University.

Driesch, H. (1914). *The history and theory of vitalism*. Macmillan and Co., Ltd.

Esposito, M. (2016). *Romantic biology, 1890–1945*. Routledge.

Farmer, J. B. (1903). On stimulus and mechanism as factors in organisation. *New Phytologist, 2*(10), 217–225.

Gambarotto, A., & Nahas, A. (2022). Teleology and the organism: Kant's controversial legacy for contemporary biology. *Studies in History and Philosophy of Science, 93*, 47–56.

Goldstein, K. (1995). *The organism*. Zone Books.

Haldane, J. S. (1921). *Mechanism, life and personality: An examination of the mechanistic theory of life and mind*. John Murray.

Haldane, J. S. (1929). *The sciences and philosophy*. Hodder & Stoughton.

Haldane, J. S. (1931). *The philosophical basis of biology*. Doubleday, Doran & Company, Inc.

Haldane, J. S., Russell, E. S., & Mackenzie, L. (1923). Symposium: The relations between biology and psychology. *Aristotelian Society Supplementary, 3*(1), 56–94.

Haldane, J. S., Thompson, D. W., Mitchell, P. C., & Hobhouse, L. T. (1917). Symposium: Are physical, biological and psychological categories irreducible? *Proceedings of the Aristotelian Society, 18*, 419–478.

Haldane, R. B., & Haldane, J. S. (1883). The relation of philosophy to science. In A. Seth & R. B. Haldane (Eds.), *Essays in philosophical criticism* (pp. 41–66). Longmans, Green, and Co.

Heikertinger, F. (1917). Das Scheinproblem von der Zweckmäßigkeit im Organischen. *Biologisches Zentralblatt, 37*, 333–352.

Hein, H. (1972). The endurance of the mechanism-vitalism controversy. *Journal of the History of Biology, 5*(1), 159–188.

Hill, A. V. (1930). Colloid science applied to biology. A general discussion. Part I. Equilibrium in protein systems. Membrane-phenomena in living matter: Equilibrium or steady state. *Transactions of the Faraday Society, 26*(0), 667–673.

Hoernlé, R. F. A. (1931). Must biological processes be either purposive or mechanistic? In G. Ryle (Ed.), *Proceedings of the Seventh International Congress of Philosophy* (pp. 44–46). Oxford University Press.

Hooker, H. D. (1919). Behavior and assimilation. *The American Naturalist, 53*(629), 506–514.

Illetterati, L., & Michelini, F. (2008). Introduction. In L. Illetterati & F. Michelini (Eds.), *Purposiveness: Teleology between nature and mind* (pp. 1–8). Ontos.

Jensen, P. (1907). *Organische Zweckmäßigkeit, Entwicklung und Vererbung vom Standpunkte der Physiologie*. G. Fischer.

Kant, I. (2000). *Critique of the power of judgment*. Cambridge University Press (Original work published 1790).

Körner, S. (1955). *Kant*. Penguin Books.

Lang, W. H. (1915). Plant morphology. *Science, 42*(1092), 780–791.

Lenoir, T. (1989). *The strategy of life: Teleology and mechanics in nineteenth-century German biology*. University of Chicago Press.

Lillie, R. S. (1915). What is purposive and intelligent behavior from the physiological point of view? *The Journal of Philosophy, Psychology and Scientific Methods, 12*(22), 589–610.

Lillie, R. S. (1945). *General biology and philosophy of organism*. University of Chicago Press.

Loeb, J. (1919). *Forced movements, tropisms, and animal conduct*. J.B. Lippincott.

Moreno, A., & Mossio, M. (2015). *Biological autonomy*. Springer.

Needham, J. (1930). *The sceptical biologist*. W.W. Norton.

Nicholson, D. J., & Gawne, R. (2015). Neither logical empiricism nor vitalism, but organicism: What the philosophy of biology was. *History and Philosophy of the Life Sciences, 37*(4), 345–381.

Peterson, E. L. (2016). *The life organic: The theoretical biology club and the roots of epigenetics*. University of Pittsburgh Press.

Plochmann, G. K. (1953). D'Arcy Thompson: His conception of the living body. *Philosophy of Science, 20*(2), 139–148.

Roux, W., Correns, C., Fischel, A., & Küster, E. (1912*). Terminologie der Entwicklungsmechanik der Tiere und Pflanzen*. Wilhelm Engelmann.

Russell, E. S. (1916). *Form and function: A contribution to the history of animal morphology*. John Murray.

Russell, E. S. (1923). Psychobiology. *Proceedings of the Aristotelian Society, 23*(1), 141–156.

Russell, E. S. (1924). *The study of living things: Prolegomena to a functional biology*. Methuen & Company Limited.

Russell, E. S. (1931). Detour experiments with sticklebacks (*Gasterosteus aculeatus* L.). *Journal of Experimental Biology, 8*(4), 393–410.

Russell, E. S. (1934). *The behaviour of animals: An introduction to its study*. E. Arnold & Company.

Russell, E. S. (1935). Valence and attention in animal behaviour. *Acta Biotheoretica, 1*(1), 91–99.

Russell, E. S. (1937). Fish migrations. *Biological Reviews, 12*(3), 320–337.

Russell, E. S. (1945). *The directiveness of organic activities*. Cambridge University Press.

Sumner, F. B. (1919). Adaptation and the problem of "Organic Purposefulness." II. *The American Naturalist, 53*(627), 338–369.

Thompson, D. W. (1913). *On Aristotle as a biologist*. Clarendon Press.

Thompson, W. R. (1929). A contribution to the study of morphogenesis in the Muscoid Diptera. *Transactions of the Royal Entomological Society of London, 77*(2), 195–244.

Thompson, W. R. (1937). *Science and common sense: An Aristotelian excursion*. Longmans Green & Co.

Trienes, R. (1992). Holism and Kantian teleology in C.J. Van De Klaauw's structuralization of oecology. *Acta Biotheoretica, 40*(1), 11–22.

Ungerer, E. (1919). *Die Regulationen der Pflanzen. Ein System der teleologischen Begriffe in der Botanik*. Julius Springer.

Van de Vijver, G., Van Speybroeck, L., De Waele, D., Kolen, F., & De Preester, H. (2005). Philosophy of biology: Outline of a transcendental project. *Acta Biotheoretica, 53*(2), 57–75.

van der Klaauw, C. J. (1935). Die Bedeutung der Teleologie Kants für die Logik der Ökologie (Ökologische Studien und Kritiken I). *Südhoffs Archiv für Geschichte der Medizin und der Wissenschaften, 27*, 516–588.

Walsh, D. M. (2015). *Organisms, agency, and evolution.* Cambridge University Press.

Wilson, E. B. (1925). *The cell in development and heredity*, 3rd edition. Macmillan.

Wolfe, C. T. (2011). From substantival to functional vitalism and beyond: Animas, organisms and attitudes. *Eidos: Revista de Filosofía de la Universidad del Norte, 14*, 212–235.

Woodger, J. H. (1929). *Biological principles.* Kegan Paul & Co.

Zammito, J. H. (2012). The Lenoir thesis revisited: Blumenbach and Kant. *Studies in History and Philosophy of Science Part C: Studies in History and Philosophy of Biological and Biomedical Sciences, 43*(1), 120–132.

4 Plant Agency

A Short History from Kant to Plant Psychology, Then to Holism, and Back Again

Jan Baedke

4.1 Introduction

Are plants agents? And if they are, how are their agential capacities different from those of animals or humans? This chapter traces the history of these questions and the theoretical approaches put forward to answer them in the history of biology and philosophy. While, in recent years, we have witnessed a strong trend toward understanding plant agency through ideas of plant intelligence and cognition (highlighting that plants, when it comes to learning, memory, or even subjectivity, are not so different after all from beings with central nervous systems), we must historically contextualize this new development to fully understand its emergence and relevance. We need to study it against the background of a long-lasting tension between denying plants certain fundamental attributes typically associated with animals, such as being alive, actively perceiving one's surroundings, or forming an integrated individual with a core, and, conversely, acknowledging plants' diverse agential capabilities, including their plasticity, perceptive responsiveness to the environment, regenerative abilities, behavioral repertoires, and reproductive versatility.

Therefore, in what follows, I will discuss central stages in the discussion of plant teleology since the late eighteenth century. I start with Kant's influential discussion of trees in his '3rd Critique' and the challenges he and later scholars faced when trying to grasp plants as integrated individuals unified by a shared purpose (Section 4.2). This especially refers to the problems of totipotency and autonomy of parts in plants, which led some mechanistic plant physiologists in the nineteenth century to consider plants merely as loosely organized aggregates, rather than unified agents with a *telos*. As a counter-position to this, plant vitalists tried to highlight the guiding influence of the whole plant on its components and structures by invoking vital forces. To conceptualize and exemplify these forces, they drew on different analogies between humans' psychological capacities and plants' traits and behaviors. Until the early twentieth century, this mechanism–vitalism debate was in substantial ways driven by botanists and philosophers who studied teleological phenomena in plants (Section 4.3). This chapter then discusses how holism (*Ganzheitsbiologie*) tried to find a middle way between these two extreme positions in the first half of the twentieth century (Section 4.4). I focus on the influential account of botanist

DOI: 10.4324/9781003413318-5

and philosopher of biology Emil Ungerer, who rejected the anthropomorphic frameworks of plant vitalists and psychologists, while recognizing the teleological importance of larger plant wholes, in contrast to mechanists. Ungerer argued that researchers should replace Kant's notion of purposiveness (*Zweckmäßigkeit*), which he considered to be biased toward human intentional actions. Instead, he developed an empirically accessible framework of plant teleology that focused on the main-tenance of the dynamic equilibrium and organization—wholeness (*Ganzheit*)—of organisms. He claimed that when plants develop and regenerate, they maintain or reestablish their internal organization in a goal-directed manner. Finally, I discuss reasons for the replacement of plant holism through more mechanistic accounts after World War II (Section 4.5), and, against the background of the recent reemergence of another version of plant psychology, I conclude with a general assessment of the history of plant agency (Section 4.6): The historical progression appears to follow a 'dialectical triad' from holism to mechanism, then to vitalism, and back to holism, mechanism, vitalism, and so forth, seemingly forevermore. Uncovering this triad might help us to anticipate future theoretical pathways to attain a better understanding of teleological phenomena in plants.

4.2 Kant on Plants

The status of plants has long been debated in the history of philosophy and biology. In fact, since antiquity, many scholars have questioned whether plants even count as living beings. According to Aristotle, plants have the lower capacities of growth, reproduction, and nutrition, but not the capacities of perception and experience, which for him are crucial to being alive. He argued that plants do not need experi-ence, since they obtain their nourishment from the soil. Thus, for Aristotle, espe-cially in comparison to animals that are capable of locomotion and perception, plants are strictly speaking not alive (Aristotle, 2001, 681a12-15; see Toepfer, 2011; Coren, 2019).

This position remained influential until the late eighteenth century and the emer-gence of the discipline of biology. One central theoretical view on plants, which still struggled with their living status, was developed by Immanuel Kant. In his *Opus Postumum*, Kant describes plants at some point as merely 'vegetating bodies' (*vegetirende Körper*) in contrast to living bodies, but at some other point as 'living beings' (see Kant, 1936/1938, AA XXI, pp. 196, 541). Given this indecisive pos-ition on plants, it is surprising that plants take a central role in Kant's discussion of teleology and purposiveness in his *Critique of the Power of Judgment*. In his transcendental analysis and critique of teleology, Kant famously embraces purpos-iveness as a valid heuristic guiding tool for elucidating phenomena that emerge from the distinctive internal structure of organized bodies (*organisierte Wesen*), or construed differently, of living beings. All living beings appear to purposefully produce and maintain themselves as integrated units, given that their constituent parts interact reciprocally, which leads to the unique teleological patterns observed.

Interestingly, to develop this argument about organized beings and teleology, Kant addresses plants—especially trees—as an example in the central §64 '*Of the peculiar character of things as natural purposes.*' After discussing trees' abilities to reproduce and grow, he states:

> I would say provisionally: a thing exists as a natural purpose, if it is […] both cause and effect of itself. […]. *[E]ach part of a tree generates itself* in such a way that *the maintenance of any one part depends reciprocally on the mainten-ance of the rest. A bud of one tree engrafted on the twig of another produces in the alien stock a plant of its own kind, and so also a scion engrafted on a for-eign stem.* Hence we may regard each twig or leaf of the same tree as merely engrafted or inoculated into it, and so as an independent tree attached to another and parasitically nourished by it. At the same time, while the leaves are products of the tree they also in turn give support to it; for the repeated defoliation of a tree kills it, and its growth thus depends on the action of the leaves upon the stem. *The self-help of nature in case of injury in the vegetable creation,* when the want of a part that is necessary for the maintenance of its neighbours is supplied by the remaining parts; and the abortions or mal formations in growth, in which certain parts, on account of casual defects or hindrances, *form them-selves in a new way to maintain what exists,* and so produce an anomalous crea-ture, I shall only mention in passing, *though they are among the most wonderful properties of organised creatures.*
>
> (Kant, 1790/1914, AA V, §64; emphasis added)

What exact 'wonderful properties' of plants, understood as teleologically self-organizing and self-maintaining beings, is Kant describing here? He highlights two phenomena that, according to him, are worth addressing through the epistemic lens of purposiveness: *plasticity* and *robustness* in plants. While vertebrates show these two features too, both occur in quite extraordinary forms in the plant kingdom. This is because plants are highly modular beings. That means that in plants various parts can become new wholes and that parts can be replaced by wholes to some extreme degrees.

Let us look at *robustness* and *regeneration* in plants first. Plants show various advanced forms of regeneration of their parts, like tissues and organs, but also, enhanced through tissue culture, regeneration of the whole organism (Bidabadi & Jain, 2020; Long et al., 2022). For example, when the shoot apical meristem of a plant is ablated, the wound gets regenerated. Also, various plant organs, like bark, young leaves, or the root apical meristem, can regenerate completely. Today, these capacities can be boosted to allow whole plant regeneration through biotechno-logical techniques like somatic embryogenesis (i.e., a plant is derived from a single somatic cell), which was first developed in the mid-twentieth century. During naturally occurring (or *in vitro*) regeneration, physiological and morphological abnormalities may occur in plants, as described by Kant. We may thus under-stand regeneration as a fragile process in which the parts in the whole plant are reorganized when certain components are lost or damaged and need to be replaced in order to maintain the whole plant. To understand this process, purposiveness is an important epistemic tool.

The other feature of plants Kant refers to is the *plasticity* and *totipotency* of their parts, for example, their ability to gain reproductive autonomy. Some plants reproduce through vegetative propagation (Swingle, 1940; Leakey, 2004). This is a form of asexual reproduction in which a bud grows and develops into a new genetically identical plant, usually through so-called runners (e.g., in strawberries), suckers (in elms), bulbs (in lilies), or tubers (in potatoes), among others. While many plants naturally reproduce in this way, it can also be induced artificially. Horticulturists have developed various techniques, like the method of grafting described by Kant (i.e., attaching a scion to the stem of another plant, a rootstock), which leads to the integration of a new organized being that shows multiple traits of two plants. An extreme example of this is the so-called Tree of 40 Fruit (Harlan, 2015), which is one of a series of fruit trees produced through grafting in which each single tree carries 40 different kinds of fruit of the genus *Prunus* on its branches, ripening sequentially between July and October.

For Kant, at least so it seems, grafting should exemplify that plants are things of natural purpose. Plants need to be studied as purposive beings, since they are cause and effect of themselves: Plants' parts reciprocally interact to produce a whole organism, and if placed into a new setting (via grafting), these parts—in plastic ways—interact with other new parts in an organized way to form and maintain a (somewhat mosaic-like) new plant whole. At the same time, we see him struggling with the high degree of autonomy of parts during plant development and reproduction. Parts are not only there to purposefully maintain the whole organism through reciprocal interaction. They may reproduce and become themselves new organisms. Kant hesitates to classify these new beings as parts or wholes, suggesting to see "each twig or leaf of the same tree as merely engrafted or inoculated into it, and so as an *independent tree* attached to another and parasitically nourished by it" (Kant, 1790/1914, AA 5, §64; emphasis added).

In sum, if plants' parts have totipotency and autonomy, how could we claim that only the whole plant is a purposefully unified individual? This problem to conciliate plants' special plasticity and modularity with a teleological view of their purposive features and organization would remain central in the nineteenth and twentieth centuries.

4.3   Plant Vitalism vs. Plant Mechanism

Until the early twentieth century, we find various authors who rejected the idea that plants are integrated individuals. As a consequence, many struggled with identifying them as units of purpose or agents, too. Similar to Kant, this was especially due to the problem that if many parts of plants are totipotent, highly plastic, and seemingly autonomous, how can we see the plant as a unified whole? Some, like Humboldt (1797), Erasmus Darwin (1800), Schleiden (1842-43/49), and Hegel (1986), argued that plants are merely aggregates or collectives, not individuals themselves. Others, like Carus (1846/51), Haeckel (1894-96, Vol. I), Driesch (1909), and Plessner (1928), highlighted the fact that plants, in contrast to animals, have an open organization.[1] They are not closed and—in line with what already Aristotle thought—have no core.

This poses a problem to any teleological conceptualization of the morphology, physiology, and behavior of plants. Obviously, parts of plants (cells, tissues, organs) are more than just a means for a purpose set by an integrated larger unit. Two lines of reasoning were advanced to discuss this kind of observations: a *vitalist tradition* and a *mechanistic tradition* (Allen, 2005; Nicholson & Gawne, 2015; Peterson, 2016; Baedke, 2019; Peterson & Hall, 2022; Donohue & Wolfe, 2023). Vitalists saw teleological phenomena as hinting toward immaterial forces working in organisms and bringing about purposive features.[2] Mechanists argued for reductive accounts of purposiveness (i.e., teleology can be ultimately replaced by mechanistic explanations) or—more commonly—they rejected the notion of teleology altogether (on epistemic or ontological grounds).

This early twentieth-century vitalism–mechanism debate (see also Fábregas-Tejeda, this volume) about purposeful features in nature was in substantial part driven by botanists and philosophers reasoning about plants. Their influence is testified by theoretical biologist Ludwig von Bertalanffy, who stated:

> A number of biologists try to fill the empty name of entelechy with concrete content, since they quite energetically call the organically functional principle *the soul*. The most important representatives of this *psychovitalism* are, in different shades, Pauly, Becher, Francé, Wagner and others. Strangely enough, *the majority of psychovitalists are botanists*, and so the doctrine of the *plant soul* stands in the center of this view, as first expressed by Fechner and Ed. v. Hartmann, and whose main modern representatives are Francé and A. Wagner.
> (Bertalanffy, 1927, p. 270; emphasis added; my translation)

The 'psychovitalists' Bertalanffy is referring to argue that while plant cells and structures may show totipotency, the whole plant still exerts a special teleological force on them and their functioning—a plant soul (*Pflanzenseele*).[3] In his *Nanna oder über das Seelenleben der Pflanzen* (Nanna or About the Psychic Life of Plants), Gustav Theodor Fechner (1848) described this soul as the core of a plant, which—albeit different than that of animals with central nervous systems—turns it into a conscious being with a life of feeling. By means of a socio-political analogy, he characterizes the internal organization of plants not as a centralized monarchy (like that of animals) but as a rather loosely centralized republic. Due to this special organization, the plant is a living and integrated agent perceiving its environment and acting on it.

The concept of plant soul gained new support among botanists around 1900, driven by new developments in plant physiology, especially on stimulus physiology, and plant behavior (see Stahlberg, 2006; Hiernaux, 2021). This included the discovery of the first chemical mechanism underlying plant behavior and the detection of electrical signals (without an existing nervous system) in plants. Against this background, the botanist and natural philosopher Raoul H. Francé called for a new research program that would build on the idea of 'plant psychology' as a working hypothesis (1909). Among others, between 1906 and 1910 he published four volumes on *Das Leben der Pflanze* (The Life of the Plant) and in 1920 the book *Die Pflanze als Erfinder* (The Plant as Inventor) in which he developed the idea of plants as sensitive, active, and creative cognitive agents (Francé, 1920). In 1909 he wrote:

> [T]he discovery [was made] that laws known from human psychology are also valid for plants [...]. If one adds that the whole structure of the plant as well as its activity is determined throughout by a similar purposefulness, as it is well known to us from our own activity and has virtually become the criterion of psychic activity, then the conclusion is also incontestable that enough *essential analogies are known between the structure and conscious activity of the human soul and the phenomena of the plant body and life* to be able to substantiate with facts the logical demand of the theory of evolution [*Entwicklungslehre*] that there must be psychic activity in the plant.
>
> (Francé, 1909, pp. 24–25; emphasis added; my translation)

Francé distinguished his empirically informed plant psychology from other new approaches to the 'plant soul' and plant agency, like the rather speculative-philosophical 'teleological plant psychology' of zoologist August Pauly (1905) who stressed the 'autoteleological causality' of the plant psyche in Lamarckian adaptation processes. Another influential author in this debate was botanist Adolf Wagner (1924), who in his *Die Vernunft der Pflanze* (The Reason of the Plant) argued for the existence of forms of rationality in plants.[4]

This vitalist movement and its interpretations of plants' purposiveness were strongly criticized by mechanistic botanists and plant physiologists. In the first half of the twentieth century, various new experimental findings were made on the mechanisms of plant growth, regulation, and development, as well as on photo-biology, the chemical processes of photosynthesis and plant bioenergetics, photo-synthetic CO_2 fixation, and plant pathology (for an overview, see Somerville, 2000; see also Nickelsen, 2017, 2022; Schürch, 2017). These developments were in line with those older concerns that the whole plant was just a loose aggregate, and the real driving forces determining its existence were located on lower levels of organ-ization. This new mechanistic program decomposed plants into their structural components. It questioned the usefulness of teleological reasoning about the whole plant organism in various ways. Some plant physiologists highlighted the numerous, seemingly non-purposeful traits and processes in plants (e.g., Küster, 1903). For example, Karl Goebel (1908, p. 184) argued that there is not only useless regen-eration but also useless movement in plants: "the view may be substantiated, [...] that the teleological view has often led astray [...] [and] that there are also useless movements" (Goebel, 1920, p. 5; my translation). Other scholars argued that many processes, like regeneration, lack a high level of perfection that would indicate purposiveness.[5] For instance, Bohumil Němec (1905) claimed that the regeneration of the root apical meristem exhibits a large waste of energy and material.

We come to see that, in contrast to the plant vitalists, mechanistic plant physiologists in the late nineteenth and early twentieth centuries argued that all processes in plants are, in fact, not teleological in nature, but only the result of a complex causal network of nutrients, environmental factors, various plant growth regulators (like auxin), and photosynthesis mechanisms. For them, only the chemical and physiological processes in and between plant cells and organs mattered, as they could be experimentally studied—the plant soul and its teleological urge could not.

4.4 Emil Ungerer: The Wholeness of Plants

The vitalism–mechanism controversy was in important ways driven by debates about the organization and teleological features of plants. However, botanists were not able to identify a clear winner, since both accounts got important things right about plants after all. The mechanists correctly highlighted the importance of plants' physiological structures for the whole organisms, and the vitalists rightly stressed that there must still exist 'something' that regulates and controls the seemingly autonomous parts according to higher-level goals and purposes set by the whole plant.

From the 1910s onwards, a serious attempt was made to integrate both positions and to find a middle ground. This middle ground was holism (*Ganzheitsbiologie*) and one of its most influential early advocates was botanist and theoretical biologist Emil Ungerer (1888–1976). Ungerer earned his PhD in 1918 from Heidelberg University with a theoretical work on plant physiology and development under the guidance of botanist Georg Albrecht Klebs (a pupil of Julius von Sachs and Wilhelm Pfeffer) and philosopher of biology Hans Driesch. Through working with his two supervisors, Ungerer was perfectly situated to integrate plant vitalism and mechanism. Driesch's vitalism made him aware of the importance of acknowledging a teleological perspective when studying plants. In addition, Klebs' experimental research on the development of form and its dependence on various environmental cues (e.g., in flower formation; see Ungerer, 1918) showed him the capacities located in plants' physiological structures on the cellular and subcellular level.

Ungerer's thesis led to the books *Die Regulationen der Pflanzen: Ein System der teleologischen Begriffe in der Botanik* (The Regulations of Plants: A System of Teleological Concepts in Botany; Ungerer, 1919, second expanded edition in 1926) and *Die Teleologie Kants und ihre Bedeutung für die Logik der Biologie* (Kant's Teleology and Its Meaning for the Logic of Biology; Ungerer, 1922).[6] In these works, Ungerer extensively delved into (neo-)Kantian positions on teleology, linking them to plant physiology, regulation, and regeneration, in an attempt to develop a new theory of organismal teleology. To do this, he consciously avoided metaphysical speculations and concepts, like Aristotle's final cause or Driesch's vitalist notion of entelechy, as well as cognitive analogies between plants and intentionally acting humans, like those endorsed by plant psychologists. Against Kant, Ungerer contended that it is possible to develop a perspective on teleology that differs from, what he called, 'real purpose analysis,' which draws on intentional teleology and human actions and agency. He called his approach the 'holistic view' (*Ganzheitsbetrachtung*). Ungerer picked up the concept of the *whole* or *wholeness* (*Ganzheit*) from Driesch (1911, 1924), who introduced it in order to avoid anthropomorphic biases linked to the concept of purposiveness (*Zweckmäßigkeit*). Then, for Driesch, the idea of 'entelechy' referred to the maintenance of a purposefully integrated 'whole.' Ungerer adopted the concept of whole and rejected purposiveness like Driesch but, against his supervisor, remained skeptical that a vitalist reading of wholeness through non-material entelechial forces can be integrated into a scientific worldview. Ungerer thus argued for a scientific approach to understanding wholeness, emphasizing its accessibility to empirical investigation. He conceptualized wholeness as the intrinsic ability of living systems to maintain their organization in a state of dynamic equilibrium.[7]

In order to develop this idea in detail, he turned to processes of regeneration and reproduction in plants. For example, like Kant in his '3rd Critique,' Ungerer (1919) discusses the problem of the totipotency of parts in plants. He investigates cases of grafting and vegetative propagation and their underlying physiological processes and mechanisms. He understands such cases as shedding light on the limits of 'holistic preservation' (*Ganzheitserhaltung*) or 'holistic production' (*Ganzheitsherstellung*). Where teleological wholeness ends, pathologies start. For Ungerer, in those cases where grafting one plant's part onto another plant is not successful and leads to anomalies (or to the later separation of scion and rootstock, and even the development of each component into individual organisms), we also come to the limits of teleological considerations. However, he highlights, not all abnormalities occurring during regeneration or symbiotic processes, like grafting, are indeed destroying wholeness: "The fact that certain abnormal findings co-occur with damage of the organism should not tempt us to regard them as necessarily destructive of the whole, as pathological" (Ungerer, 1919, p. 49; my translation). By drawing on the transplantation experiments by Vöchting (1892, 1908), he argues:

> If the combination of scion and rootstock succeeds, both form in a pronounced form a *functional whole* [*Funktionsganzheit*]; in terms of metabolism they constitute *one organism*. The situation is different with regard to the morphological characteristics, especially in the case of transplants of different plants and plant breeds or heterogeneous combinations of different organs of the same plant. Thus, according to Vöchting (1892, p. 23/24), the scion of *Beta vulgaris* 'has a peculiar, anatomically independent growth in the root.' 'By means of its own cambium, it forms a special secondary wood body, which is connected to the bundles of roots, but develops independently of them. The scion tries to shape and round itself according to its own growth law, even after it has entered into a firm tissue connection with the root; this is all the more striking when one considers how firm and closed the nutritional unit is that scion and root form here.' *The two organisms connected by grafting thus represent a functional wholeness, but by no means a form wholeness* [*Formganzheit*]. The formative processes through which the complete connection comes about—cambium activity, formation of the 'putty tissue' [*Kittgewebe*], and the vascular

connections, etc.—*are holistically preserving (or rather holistically creating) in relation to this functional wholeness.*

(Ungerer, 1919, p. 52; emphasis added; my translation)

Ungerer highlights here that plants qualify as organisms due to different criteria of individuality. In the above case (or, e.g., that of the 'Tree of 40 Fruits'), grafted symbionts of plants are physiological (or metabolic) individuals, but not morphological individuals. This is due to the fact that plants' parts are not only totipotent but also largely autonomous. Thus, a scion maintains its morphological autonomy when connected with the rootstock (e.g., a branch that still produces the same fruits). However, it loses its physiological individuality when the new plant actively forms a physiological whole. In this context, according to Ungerer, the complex regulatory processes that integrate the scion to a larger functional metabolic whole need to be conceptualized and studied in teleological terms. The morphological processes, however, are not part of this 'whole preserving process' and thus their investigation should not be guided by teleology.

In this analysis, Ungerer tries to give an answer to the problem that already puzzled Kant in his discussion of purposiveness: Are the parts of plants in fact parts or rather wholes themselves? Ungerer suggests that they are both, depending on how we individuate organisms. To make this point, he follows Driesch (1909) and his mentor Klebs (1903) in recognizing plants as openly organized beings, in contrast to animals: "The animal is 'finished' as a whole, just as all its typical parts are 'finished' individually. The plant as a whole is never 'finished'; only individual parts are 'finished'" (Ungerer, 1919, p. 54; my translation). In other words, the whole plant—as a physiological, morphological, or behavioral individual—has to integrate these 'finished' parts in its open life process in teleological ways to establish or maintain its dynamic equilibrium.

Based on this and related analyses, Ungerer (1919, pp. 47–48) distinguished three ways in which the plant organism can establish wholeness: (i) a plant restores its form as a morphological individual; (ii) it orders its metabolic processes to gain or maintain functional wholeness as a physiological individual (e.g., in the case of grafting); or (iii) it coordinates the activities of its motor apparatus as one integrated behavioral whole (e.g., in nastic movements like the coordinated periodic movements of many leaves on one plant). For Ungerer, all three conceptual facets allow accessing the internal teleological processes of the plant organisms through clear empirical methodologies—in contrast to entelechy or vital forces. Physiological studies should address these phenomena, he argued, to understand in which regulatory processes of plant physiology, development of form, reproduction, and behavior a teleological perspective is unavoidable, and in which it is not.

In sum, Ungerer's new teleological plant science aimed at defending the usefulness of teleological considerations against mechanists. But to do so, scientists should completely avoid the notion of purposiveness, introduced by Kant, and replace it with the idea of wholeness. In particular, Ungerer argued that the work of plant psychovitalists had shown that any anthropomorphic understanding of teleology needs to be avoided. For him, the concept of purposiveness ultimately leads to a speculative and biased science that draws on questionable analogies between human cognitive and intellectual abilities and plants' traits. Such an account cannot find an empirically fruitful research program—only holism can.

4.5 The End of Plant Holism

The non-vitalist shift from purposiveness to wholeness significantly influenced theoretical biology, both within the German-speaking community and internationally.[8] For example, the later founder of General Systems Theory, Ludwig von Bertalanffy, strongly aligned with Ungerer's view of teleology. By drawing on Ungerer, he proposed that empirical investigations into teleological phenomena are feasible by focusing on the dynamically maintained organization of organisms (Bertalanffy, 1929; see Baedke, 2019). Bertalanffy asserted that living organisms, operating as open systems far from thermodynamic equilibrium, exhibit metabolic self-organization. He emphasized that organisms sustain their states through constant changes in substances and energies, ensuring 'homeostasis' describable by natural laws (Bertalanffy, 1929, 1932, 1942). In this teleological perspective, living beings actively uphold their homeostatic internal organization as open systems, constrained by physical laws, especially by non-equilibrium thermodynamics. While organisms must adhere to these laws to survive, the specific pathways to reach a final homeostatic state are not determined by physical laws—a concept later termed 'equifinality' by Bertalanffy (1942). Consequently, teleology should be understood through its own biological laws.

Despite this positive reception of Ungerer's holistic approach, in general, the German-speaking community of *Ganzheitsbiologie* did not create a strong legacy when it comes to their views on teleology. Following World War II, holistic biology faced skepticism due to the association of key proponents, such as Ungerer and Bertalanffy, with National Socialist ideology in their careers or personal lives (e.g., Deichmann, 1992; Pouvreau, 2009; Dahn, 2019). While not all advocates of holism embraced Nazi biology (Deichmann, 1992), Emil Ungerer clearly did. In his 1936 report for a high school *Festschrift* in Karlsruhe, where he taught biology (besides working as an extraordinary professor at the University of Karlsruhe), he expressed enthusiasm for integrating Nazi race theory into school education (Ungerer, 1936). Utilizing analyses of physical and mental characteristics of races, along with practical exercises like phrenological skull measurements, he aimed to instill pride in German heritage in pupils and justify the eugenic policies of National Socialism.

Due to such involvements, also holistic theories of teleology faced increasing skepticism both by analytical philosophers and biologists, who questioned the empirical usefulness of the approach or simply equated it with vitalism (see Nicholson & Gawne, 2015; Baedke, 2019; Baedke & Fábregas-Tejeda, 2023). The late Ungerer (1965) himself rejected holism for not qualifying as a biological theory and being factually unjustified. Indeed, the main developments in plant biology after the war were not driven by a holistic-teleological approach, as envisioned by Ungerer, but by a further strengthened mechanistic agenda. For example, in botany, after holism had first replaced plant psychology, and then largely disappeared itself, new reductionist approaches filled the theoretical and methodological gap. Research questions that targeted parts of plants rather than wholes paved the way to a rapid increase in knowledge about plant genetics (e.g., the genetic mechanisms of plant pathologies and transposable elements), hormonal control of growth (e.g., the discovery of cytokinins), phytochromes and photosynthesis, and about the bio-chemical processes of metabolic control and CO_2 fixation, and this also stirred the development of various *in vitro* plant tissue culture techniques. For all of this, it seemed, the concept of organismal teleology was more of a hindrance than an aid.[9]

4.6 The Return of Vegetal Minds: A Dialectical Triad

Theoretical discussions of plants' agential abilities can be traced through the long history of philosophy and biology. This history is characterized by a continuing tension between refusing the crucial features of plants in contrast to animals, like being alive, experiencing their surroundings, or forming an integrated individual with a core, on the one hand, and recognizing their various agential abilities, like their plasticity, environmental responsiveness and perceptivity, regenerative abil-ities, behavioral repertoires, and reproductive versatility, on the other.

At the center of this tension lies the question of how to recognize the plant as a whole unit of purpose if its parts often show morphological or reproductive autonomy and can become wholes themselves. One may note that the solution to this problem offered by plant holism is only partially convincing: Ungerer's view on self-maintenance tries to highlight the potentials of the whole plant to persist despite the various changes that can happen to its parts. The plant actively 'uses' its (in contrast to animals) special loosely integrated form, for example, during regeneration, to maintain itself even though its internal organization or individual parts may change. However, in line with Kant's rationale, such holistic plant agency invariably pertains to preservation teleology. Its objective is always directed toward maintaining an already existing dynamic structure, rather than surpassing or transcending it. In fact, Ungerer tends to reduce the problem of totipotency of plants' components to a question about how to integrate these parts into self-maintaining processes in *another* existing being (e.g., by physiologically integrating a grafted scion into a new plant). He shies away from, for example, telling a story of autonomous parts as vegetal agents that retain developmental potency throughout the plant's lifetime to develop creatively into a new whole, rather than maintaining the old one. In addition, the experiential dimension of plant agency and phenomena of learning in plants, central for plant vitalists, is poorly covered in holistic plant theories, with their focus on physiological, mor-phological, or behavioral self-maintenance. In short, holism creates a view of plants as persisting conservative agents, but not as creative, learning agents. In this framework, plant agency is directed toward preserving the present self, not toward creating a novel future self.

Thus, the holistic account to plant agency is probably too narrow to address all of the various facets of agential actions we find in plants. As a consequence, in recent years, building especially on new empirical findings about plant learning, decision-making, plasticity, and behavior, another movement aiming at comprehending plant agency has emerged within botany, philosophy of mind, and philosophy of biology, as well as in embodied cognition research and cultural studies that aims to fill this explanatory gap. Interestingly, it does so by turning back the wheel of time. This movement ignores or actively rejects old holistic worries of anthropomorphizing, mentalizing, and intellectualizing plants as it tries to establish a new plant psychology. In recent discussions on plant neurobiology, plant intelligence, and plant cognition, various theoretical attempts are made to conceptualize the plant as an experiencing, environmentally responsive, learning, and acting subject (Calvo & Keijzer, 2011; Marder, 2011, 2012, 2013; Trewavas, 2015; Calvo, 2016; Maher, 2017; Gibson & Brits, 2018; Baluška et al., 2018; Sims, 2019; Baluška & Mancuso, 2021; Khattar et al., 2022; Segundo-Ortin & Calvo, 2022; Gilroy & Trewavas, 2023; Sims & Yilmaz, 2023; see also Hiernaux, 2021, 2023). Along these lines, various analogies to humans' or animals' agential capacities are explored and concepts traditionally reserved for non-vegetal beings with central nervous systems are widened or reframed to make them fit to plants—from plant souls, minds, intelligence, memory, learning, cognition, subjectivity, and consciousness, to intentionality.

What do we make of this recent trend? In his later work, holistic philosopher and historian of biology Adolf Meyer-Abich (1964, p. 163) suggested that the theoretical development of biology progresses in a 'dialectical triad' from holism to mechanism to vitalism, and again from holism to mechanism to vitalism, *ad infinitum*. In fact, this is what the history of theoretical debates about plant agency looks like: from proto-holistic Kant, to plant mechanism and plant 'psychovitalism' in the nineteenth and early twentieth centuries, to Ungerer's plant holism and then again to plant mechanism in the twentieth century, only to return to possibly another reincarnation of plant psychovitalism. Time will tell what this most recent theoretical development will contribute to our understanding of plant agency, and what its shortcomings will be.

Acknowledgments

I want to thank Alejandro Fábregas-Tejeda and Vera Straetmanns for constructive comments on an earlier version of this chapter. I gratefully acknowledge the financial support from the German Research Foundation (DFG; BA 5808/2–1).

Notes

1   On these positions, see also Toepfer (2011).
2   This refers to so-called 'substantival vitalism', rather than 'functional-epistemic vitalism' (see Wolfe, 2011).
3   On the history of the concept of plant soul since antiquity, see Ingensiep (2001).
4   For English-speaking advocates of plant psychology, see, e.g., Burroughs (1908), Gussow (1910), and Dixon (1914). For a critique against psychovitalism and replies to it, see Wagner (1923).

5 Among many plant physiologists around 1900, the widely held assumption was that plants, unlike animals, did not experience genuine regeneration—there was no direct 'restitution' of lost organs. Instead, some physiologists suggested that plants counteracted such losses by producing adventive shoots, which might emerge at some distance away from the site of injury (see Linsbauer, 1917, pp. 133–136).
6 For biographical notes and the impact of Emil Ungerer on early theoretical biology and philosophy of biology, see Baedke et al. (2024).
7 For other authors defending this standpoint in early twentieth-century biology, see Fábregas-Tejeda (this volume).
8 For a detailed discussion on these influences, see Baedke et al. (2024).
9 An exception to this trend is plant morphologist Agnes Arber, who developed a neo-Aristotelian view of plant teleology in the 1940s and 1950s that shows some interesting similarities to earlier holistic accounts (see, e.g., Arber, 1950).

References

Allen, G. E. (2005). Mechanism, vitalism and organicism in late nineteenth and twentieth-century biology: The importance of historical context. *Studies in History and Philosophy of Biological and Biomedical Sciences*, *36*(2), 261–283.

Arber, A. (1950). *The natural philosophy of plant form*. Cambridge University Press.

Aristotle. (2001). *On the parts of animals* (J. G. Lennox, Trans.). Clarendon Press.

Baedke, J. (2019). O organism, where art thou? Old and new challenges for organism-centered biology. *Journal of the History of Biology*, *52*(2), 293–324.

Baedke, J., Böhm, A., & Reiners-Selbach, S. (2024). From Kant to holism: The decline of neo-Kantianism and the rise of theoretical biology. In H. Pulte, J. Baedke, D. Koenig, & G. Nickel (Eds.), *New perspectives on neo-Kantianism and the sciences* (pp. 253–277). Routledge.

Baedke, J., & Fábregas-Tejeda, A. (2023). The organism in evolutionary explanation: From early twentieth century to the extended evolutionary synthesis. In T. E. Dickins & B. J. A. Dickins (Eds.), *Evolutionary biology: Contemporary and historical reflections upon core theory* (pp. 121–150). Springer.

Baluška, F., Gagliano, M., & Witzany, G. (Eds.) (2018). *Memory and learning in plants: Signaling and communication in plants*. Springer.

Baluška, F., & Mancuso, S. (2021). Individuality, self and sociality of vascular plants. *Philosophical Transactions of the Royal Society B: Biological Sciences*, *376*(1821), 20190760. https://doi.org/10.1098/rstb.2019.0760

Bertalanffy, L. v. (1927). Das Problem des Lebens. *Scientia*, *41*(180), 265–274.

Bertalanffy, L. v. (1929). Die Teleologie des Lebens. Eine kritische Erörterung. *Biologia Generalis*, *5*, 379–394.

Bertalanffy, L. v. (1932). *Theoretische Biologie* (Vol. I). Borntraeger.

Bertalanffy, L. v. (1942). *Theoretische Biologie* (Vol. II). Borntraeger.

Bidabadi, S. S., & Jain, S. M. (2020). Cellular, molecular, and physiological aspects of in vitro plant regeneration. *Plants*, *9*(6), 702. https://doi.org/10.3390/plants9060702

Burroughs, J. (1908). *Leaf and tendril*. Riverside Press.

Calvo, P. (2016). The philosophy of plant neurobiology: A manifesto. *Synthese*, *193*(5), 1323–1343.

Calvo, P., & Keijzer F. (2011). Plants: Adaptive behavior, root-brains, and minimal cognition. *Adaptive Behavior*, *19*(3), 155–171.

Carus, C. G. (1846/51). *Psyche: Zur Entwicklungsgeschicht der Seele*. Flammer & Hoffmann.

Coren, D. (2019). Aristotle on self-change in plants. *Rhizomata, 7*(1), 33–62.

Dahn, R. (2019). Big science, nazified? Pascual Jordan, Adolf Meyer-Abich, and the abortive scientific journal Physis. *Isis, 110*, 68–90.

Darwin, E. (1800). *Phytologia*. Johnson.

Deichmann, U. (1992). *Biologen unter Hitler: Vertreibung, Karrieren, Forschungsförderung.* Campus.

Dixon, R. (1914). *The human side of plants*. Frederick A. Stokes.

Donohue, C., & Wolfe, C. T. (Eds.). (2023). *Vitalism and its legacy in twentieth century life sciences and philosophy*. Springer.

Driesch, H. (1909). *Philosophie des Organischen*. Engelmann.

Driesch, H. (1911). Die Kategorie "Individualität" im Rahmen der Kategorienlehre Kants. *Kant Studien, 16*, 22–53.

Driesch, H. (1924). Kant und das Ganze. *Kant Studien, 29*, 365–376.

Fechner, G. T. (1848). *Nanna oder über das Seelenleben der Panzen*. Leopold Voß.

Francé, R. H. (1909). *Pflanzenpsychologie als Arbeitshypothese der Pflanzenphysiologie*. Franck.

Francé, R. H. (1920). *Die Pflanze als Erfinder*. Kosmos.

Gibson, P., & Brits, B. (Eds.) (2018). *Covert plants: Vegetal consciousness and agency in an anthropocentric world*. Punctum Books.

Gilroy, S., & Trewavas, T. (2023). Agency, teleonomy and signal transduction in plant systems. *Biological Journal of the Linnean Society, 139*(4), 514–529.

Goebel, K. (1908). *Einleitung in die experimentelle Morphologie der Pflanzen*. Teubner.

Goebel, K. (1920). *Die Entfaltungsbewegungen der Pflanzen und deren teleologische Deutung*. Fischer.

Gussow, H. T. (1910). Plant physiology versus psychology. *The Ottawa Naturalist, 24*(7), 113–116.

Haeckel, E. (1894-96). *Systematische Phylogenie* (3 Vols.). Reimer.

Harlan, B. (2015, July 31). What a 'tree of 40 fruit' tells us about agricultural evolution. *National Geographic*. www.nationalgeographic.com/culture/article/what-a-tree-of-40-fruit-tells-us-about-agricultural-evolution

Hegel, G. W. F. (1986). *Enzyklopädie der philosophischen Wissenschaften im Grundrisse* (Vols. 8–10; Part II). [(1817/30)]. Suhrkamp.

Hiernaux, Q. (2021). History and epistemology of plant behaviour: A pluralistic view? *Synthese, 198*(4), 3625–3650.

Hiernaux, Q. (2023). *From plant behavior to plant intelligence?* Éditions Quae.

Humboldt, A. v. (1797). *Versuche über die gereizte Muskel- und Nervenfaser* (2 Vols.). Rottmann.

Ingensiep, H. W. (2001). *Geschichte der Pflanzenseele philosophische und biologische Entwürfe von der Antike bis zur Gegenwart*. Kröner.

Kant, I. (1914). *Kant's critique of judgement* (J. H. Bernard, Trans.). Macmillan. (Original work published 1790)

Kant, I. (1936). *Opus Postumum*. In Königlich Preußische Akademie der Wissenschaften (Ed.), *Kant's gesammelte Schriften, Vol. XXI-XXII*. Reimer.

Khattar, J., Calvo, P., Vandebroek, I., Pandolfi, C., & Dahdouh-Guebas, F. (2022). Understanding interdisciplinary perspectives of plant intelligence: Is it a matter of science, language, or subjectivity? *Journal of Ethnobiology and Ethnomedicine, 18*(1), 41. https://doi.org/10.1186/s13002-022-00539-3

Klebs, G. (1903). *Willkürliche Entwickelungsänderungen bei Pflanzen*. Fischer.

Küster, E. (1903). *Pathologische Pflanzenanatomie*. Fischer.

Leakey, R. R. B. (2004). Physiology of vegetative reproduction. In J. Burley (Ed.), *Encyclopedia of forest sciences* (pp. 1655–1668). Elsevier.

Linsbauer, K. (1917). Studien über die Regeneration des Sproßvegetationspunktes. *Denkschriften der Akademie der Wissenschaften, 93*, 107–138.

Long, Y., Yang, Y., Pan, G., & Shen, Y. (2022). New insights into tissue culture plant-regeneration mechanisms. *Frontiers in Plant Science, 13*. www.frontiersin.org/articles/10.3389/fpls.2022.926752

Maher, C. (2017). *Plant minds: A philosophical defense*. Routledge.

Marder, M. (2011). Plant-soul: The elusive meanings of vegetative life. *Environmental Philosophy, 8*(1), 83–99.

Marder, M. (2012). Plant intentionality and the phenomenological framework of plant intelligence. *Plant Signaling & Behavior, 7*(11), 1365–1372.

Marder, M. (2013). *Plant-thinking: A philosophy of vegetal life*. Columbia University Press.

Meyer-Abich, A. (1964). *The historico-philosophical background of the modern evolution-biology* (Bibliotheca Biotheoretica 10). Brill.

Němec, B. (1905). *Studien über die Regeneration*. Borntraeger.

Nicholson, D. J., & Gawne, R. (2015). Neither logical empiricism nor vitalism, but organicism: What the philosophy of biology was. *History and Philosophy of the Life Sciences, 37*(4), 345–381.

Nickelsen, K. (2017). Growth, development and regeneration: Plant biology in Vienna around 1900. In G. Müller (Ed.), *Vivarium: Experimental, quantitative and theoretical biology at Vienna's Biologische Versuchsanstalt* (pp. 165–188). MIT Press.

Nickelsen, K. (2022). Physicochemical biology and knowledge transfer: The study of the mechanism of photosynthesis between the two world wars. *Journal of the History of Biology, 55*(2), 349–377.

Pauly, A. (1905). *Darwinismus und Lamarckismus: Entwurf einer psychophysischen Teleologie*. Reinhardt.

Peterson, E. L. (2016). *The life organic: The theoretical biology club and the roots of epigenetics*. University of Pittsburgh Press.

Peterson, E. L., & Hall, C. (2022). 'What is dead may not die': Locating marginalized concepts among ordinary biologists. *Journal of the History of Biology, 55*(2), 219–251.

Plessner, H. (1928). *Die Stufen des Organischen und der Mensch*. De Gruyter.

Pouvreau, D. (2009). *The dialectical tragedy of the concept of wholeness: Ludwig von Bertalanffy's biography revisited*. ISCE.

Schleiden, M. J. (1842-43/49). *Grundzüge der wissenschaftlichen Botanik nebst einer methodologischen Einleitung als Anleitung zum Studium der Pflanze*. Brockhaus.

Schürch, C. (2017). How mechanisms explain interfield cooperation: Biological–chemical study of plant growth hormones in Utrecht and Pasadena, 1930–1938. *History and Philosophy of the Life Sciences, 39*(3), 16. https://doi.org/10.1007/s40656-017-0144-3

Segundo-Ortin, M., & Calvo, P. (2022). Consciousness and cognition in plants. *WIREs Cognitive Science, 13*(2), e1578. https://doi.org/10.1002/wcs.1578

Sims, M. (2019). Minimal perception: Responding to the challenges of perceptual constancy and veridicality with plants. *Philosophical Psychology, 32*, 1024–1048.

Sims, R., & Yilmaz, Ö. (2023). Stigmergic coordination and minimal cognition in plants. *Adaptive Behavior, 31*(3), 265–280.

Somerville, C. (2000). The twentieth century trajectory of plant biology. *Cell, 100*, 13–25.

Stahlberg, R. (2006). Historical overview on plant neurobiology. *Plant Signaling & Behavior, 1*(1), 6–8. https://doi.org/10.4161/psb.1.1.2278

Swingle, C. F. (1940). Regeneration and vegetative propagation. *Botanical Review, 6*(7), 301–355.

Toepfer, G. (2011). Pflanze. In G. Toepfer (Ed.), *Historisches Wörterbuch der Biologie: Geschichte und Theorie der biologischen Grundbegriffe* (Vol. 3, pp. 11–33). Metzler.

Trewavas, A. J. (2015). *Plant behaviour and intelligence.* Oxford University Press.

Ungerer, E. (1918). Die Beherrschung der pflanzlichen Form. *Naturwissenschaften, 6,* 683–691.

Ungerer, E. (1919). *Die Regulationen der Pflanzen: Ein System der teleologischen Begriffe in der Botanik.* Springer.

Ungerer, E. (1922). *Die Teleologie Kants und ihre Bedeutung für die Logik der Biologie.* Brontraeger.

Ungerer, E. (1936). Vererbungs- und Rassefragen im biologischen Unterricht des Gymnasiums. In *350 Jahre Gymnasium Karlsruhe. Festschrift 1586–1936* (pp. 92–101). G. Braun.

Ungerer, E. (1965). Die Erkenntnisgrundlagen der Biologie: Ihre Geschichte und ihr gegenwärtiger Stand. In L. v. Bertalanffy & F. Gessner (Eds.), *Handbuch der Biologie* (Vol. I/1, pp. 1–94). Athenaion.

Vöchting, H. (1892). *Über Transplantation am Pflanzenkörper. Untersuchungen zur Physiologie und Pathologie.* H. Laupp.

Vöchting, H. (1908). *Untersuchungen zur experimentellen Anatomie und Pathologie des Pflanzenkörpers.* H. Laupp.

Wagner, A. (1923). *Das Zweckgesetz in der Natur: Grundlinien einer Meta-Mechanik des Lebens.* Rentsch.

Wagner, A. (1924). *Die Vernunft der Pflanze.* C. Reissner.

Wolfe, C. T. (2011). From substantival to functional vitalism and beyond: Animas, organisms and attitudes. *Eidos, 14,* 212–235.

5 Behavior, Purpose, and Teleology Revisited

Locating Cybernetic Teleology in Twentieth-Century Holism

Auguste Nahas

5.1 Introduction

In his foreword to a special issue on *Teleological Mechanisms*, published in the *Annals of the New York Academy of Sciences* in 1948, social scientist Lawrence K. Frank—speaking on behalf of a movement to be christened "cybernetics" that very year—attempted to offer some explanation for the topic of the issue. Teleology had been a subject of interest for members of the proto-cybernetic movement ever since the publication of Arturo Rosenblueth, Norbert Wiener, and Julian Bigelow's *Behavior, Purpose, and Teleology* (1943). In 1944, Howard Aiken, John von Neumann, and Wiener brought together a group of experts in engineering, mathematics, and neurophysiology and suggested along with Howard Aiken that they name themselves "The Teleological Society" and start a periodical named "Teleology" or "Teleologia" (Aiken et al., 1944). They were also looking to name their new science "devoted on the one hand to the study of *how purpose is realized* in human and animal conduct and on the other hand *how purpose can be imitated* by mechanical and electrical means" (Aiken et al., 1944; emphasis added). This emphasis on teleology carried over to the first post-war Macy conference in March 1946 (von Foerster et al., 1953, p. xix), though it virtually disappeared after the publication of Wiener's *Cybernetics* (Wiener, 1948/1985).[1] Nonetheless, teleology played a prominent role in the movement's infancy, and the cybernetic conception of purpose had an outsized influence on debates about teleology in molecular and evolutionary biology in the 1970s.

As Frank (1948) sees it, the cybernetic movement's interest in teleology came out of a "growing perplexity in biology and psychology," as researchers felt "further and further away from the organisms and the personalities they were trying to understand" (p. 190). This caused a search for new approaches "capable of dealing with the *larger wholes of organisms and personalities*" (Frank, 1948, p. 190; emphasis added). Frank (1948) locates the efforts of the proto-cybernetic group within such a broadly non-reductive, "holistic" line of thought that would render mechanism and teleology compatible: "the concept of teleological mechanisms" seeks to escape from older, inadequate mechanistic conception and instead seeks "*a conception of 'natural teleology'*" (p. 191; emphasis added). In making such claims, Frank posed two important questions: first, about the notion of "teleological

DOI: 10.4324/9781003413318-6

mechanisms" as the proto-cyberneticians' attempt to naturalize teleology with an updated notion of "mechanism" compatible with holism; and second about the movement's relationship to other anti-reductive movements of the interwar period. This chapter aims to provide answers to both questions.

I begin by revisiting the proto-cybernetic account of teleology (Section 5.2) to clarify Frank's claim that cybernetics was both mechanistic *and* holistic. In Section 5.3, I locate the cybernetic account of teleology in the thought of holistic thinkers in behavioristic psychology and philosophy in the decades prior to cybernetics. I will focus on the ties Wiener and others had with "purposive behaviorists" such as Edwin Holt, Ralph B. Perry, and Edward Tolman. I argue that problematic ambiguities about the distinction between intrinsic and extrinsic purposiveness permeated their accounts of purpose and that the same issues are found in cybernetics.

5.2 Cybernetic Teleology: Mechanistic, Holistic, or Both?

5.2.1 Behavior, Purpose, and Teleology

The paper, *Behavior, Purpose, and Teleology* by Rosenblueth, Wiener, and Bigelow aims (1) "to define the behavioristic study of natural events and to classify behavior," and (2) "to stress the importance of the concept of purpose" (Rosenblueth et al., 1943, p. 18). Their first aim is to extend the method of psychological behaviorism beyond its usual scope, on the basis that "no behaviorist has ever really understood the possibilities of behavior" (Wiener, 1942). The proposed "behavioristic" method could be used to study any event in a way which "omits the specific structure and the intrinsic organization of the object" and instead focuses on system–environment relationships (Rosenblueth et al. 1943, p. 18). This behavioristic method would operate in parallel with the contrasting "functional" method of study, which focuses on the internal workings of a system and brackets considerations pertaining to its relationship with the environment. While organisms and machines appear different from a functional perspective, a "uniform behavioristic analysis is applicable to both" (Rosenblueth et al. 1943, 22). For this reason, the authors aim to flesh out the usefulness of the behavioristic method as a unified approach to organisms and machines, along with the concepts proper to it, most notably *purpose* and *teleology*.

This second stated goal of the paper is to stress the importance of purpose as a concept for identifying certain *types of behavior* within broader classification. The authors distinguish active from passive behavior, which depend on whether the object is the source of the output energy. Active behavior is further subdivided into purposive behavior and non-purposive behavior. Purposive behavior is that which aims at "*a final condition* in which the *behaving object* reaches a definite correlation in time or in space with respect to another object or event" (Rosenblueth et al., 1943, p. 18; emphasis added). Such purposive behavior is subdivided—again, from a purely behavioristic standpoint—into *teleological (purposive) behavior*, where the behavior is continuously modulated on the basis of negative feedback, and *non-teleological (purposive) behavior*, where the action is too rapid to be modulated once initiated. Teleological behavior is of particular interest to the authors as they believe it to be the most fruitful for classifying "increasingly complex types of behavior of organisms" which involve multiple orders of prediction (Rosenblueth et al., 1943, p. 22). The behavioristic analysis thereby promises to span or even unify engineering, biology, and psychology (Abraham, 2012; Kline, 2020).

The paper concludes with the authors emphasizing the divide between the functional and behavioral mode of study, which proves crucial to avoiding the problem of backward causation: causality "initial or final" is a functional relationship, "whereas *teleology is concerned with behavior, not with functional relationships*" (Rosenblueth et al., 1943, p. 24; emphasis added). Such claims do not appear to preclude the possibility of studying systems that behave teleologically from a functional perspective and thereby uncover certain organizational preconditions for this kind of behavior, but they *do* imply that such a perspective would lose sight of the phenomenon of teleology *qua* behavior (see also Rosenblueth & Wiener, 1950, p. 320). That the authors so clearly thought of behavior and teleology as concepts belonging to the behavioristic rather than functional mode of study provides crucial clues to understanding Frank's claim that their account of teleology was both *mechanistic* and *holistic* at once.

5.2.2 The Meaning of "Mechanism" in Cybernetics

As Daniel Nicholson (2012) has argued, "mechanism" has at least three meanings: (1) *mechanicism*: a reductionist thesis about explaining systems in terms of their parts; (2) *causal mechanism*: a way of explaining the mode of operation of a causal process giving rise to a phenomenon of interest; and (3) *machine-mechanism*: a way of understanding organisms as being machine-like. This third sense is particularly open-ended, because its implications depend entirely on the particular *kinds* of machines one has in mind (Esposito & Baravalle, 2023). Indeed, whether "machine-mechanism" aligns with Nicholson's other two senses—particularly the reductionism of "mechanicism"—depends precisely on the particular machines in question.

I suggest we approach the "mechanism" of the proto-cyberneticians primarily in terms of machine-mechanism, which in this case meant a particular analogy between organisms and *servomechanisms* (i.e., machines regulated by negative feedback). This analogy was seen as fruitful because it had particular implications, limitations, and uses in advancing research that analogies with other machines (such as clocks) did not have. Furthermore, this analogy should be understood as methodological rather than metaphysical. Wiener among others repeatedly emphasized that servomechanisms were usefully interpreted as *imitations* or *models* of biological purposiveness (Rosenblueth et al., 1943; Aiken et al., 1944). This way of legitimating the organism–servomechanism analogy has its roots in the wartime origins of the major theses of the paper, namely, Wiener and Bigelow's research on anti-aircraft artillery and the behavior of airplanes, both of which were modeled on the basis of negative feedback (Galison, 1994). This "Manichean science" put a premium on successful control and prediction; questions of metaphysics were secondary (see Kay, 2000, pp. 83–84).

Also notable in this context is Norbert Wiener and Arturo Rosenblueth's common interest in general methodological questions in the philosophy of science (Montagnini, 2017, pp. 82–85). They wished to develop concepts and methods for understanding the *behavior* or "*performance*" of the system as an integrated whole interacting with an environment, i.e., from the *behavioral* mode of study, rather than *functional* one. In other words, bracketing functional considerations and focusing solely on behavior opened new possibilities for modeling (physically and mathematically) the behavior of servomechanisms and organisms in the same way. And just as organism-clock analogies might suggest a functional, reductive method of study, servomechanical machine-mechanism suggested approaching behavior primarily through black-box models of behaving systems as wholes. This, in turn, suggested novel scientific hypotheses to Wiener and Rosenblueth regarding the function of the cerebellum (Rosenblueth et al., 1943, p. 20; Wiener, 1948/1985, p. 8) which was the seed for "a statement of a program for a large body of experimental work," which they hoped could be carried out at a new interdisciplinary institute (Wiener, 1948/1985, p. 8).

These authors took the role of negative feedback in engendering complex behavior to be a crucial insight, one which brought together a functional perspective on the central nervous system as a complex organized system with a behavioristic perspective of it as a black-box system constantly coming to grips *with* and modulating its relationship *to* the environment through the feedback between sensation and action. For Wiener (1948/1985), this behavioristic conception of the nervous system marked "a new step in the study of that part of neurophysiology which concerns not solely the elementary processes of nerves and synapses but the performance of the nervous system *as an integrated whole*" (p. 8; emphasis added). Similarly, Ross Ashby called cybernetics "a 'theory of machines', [which] treats, not things but ways of behaving. It does not ask 'what is this thing?' but 'what does it do?'" (Ashby, 1957, p. 1).

The foregoing provides a tentative answer to the first of two questions raised by Frank's *Foreword*, namely, that the cyberneticians treated teleology and purposiveness as an empirically manifest feature of *behaviors* that can theorized in their own terms, without necessarily adverting to the way it is mechanically or "functionally" realized. In turn, this opened up new ways of thinking about organism-machine analogies that capture what is distinctive about purposive behavior, namely, the system-environment coupling. In both the organic and the machine, the nature of this coupling resists decompositional analysis; both systems must be treated as integrated, behaving wholes.

In the next section I turn to the second question concerning Frank's claim that cyberneticians shared the "growing perplexity of biologists and psychologists" regarding the disappearance of the whole "organisms or personalities" that they aimed to understand. Who were these researchers?

5.3 Cybernetic Teleology in Context

Many scholars have emphasized the parallels and possible lines of influence between the cybernetic account of teleology as negative feedback and Walter

Cannon's research on homeostasis, given that Rosenblueth worked in his laboratory at Harvard from 1930 and 1943 (Burbano & Reyes, 2022, pp. 1015–1016). However, Roberto Cordeschi (2002) has made a compelling case for broadening this narrative to include the various fields which contributed to the study of animal behavior. This included, among others, zoology, physiology, evolutionary biology, psychology, and philosophy. Indeed one sees in the 1943 proto-cybernetics paper reflections of many interrelated debates about teleology in the late nineteenth and early twentieth centuries which were discussed in prominent journals, such as *Philosophy of Science*, well into the 1950s (e.g., Krikorian, 1943; Lillie, 1943; see also Malaterre et al., 2019). Furthermore, an overemphasis on negative feedback and causal circularities obscures the crucial fact that this cybernetic definition of teleology was built on *a more basic notion of purposive behavior that did not invoke either* (see Section 5.2). Building on Cordeschi's work, I will focus on the close parallels between this notion of purpose and that of a lesser-known strand of thinking in behaviorist philosophy and psychology which I will dub "purposive behaviorism." Its defenders sought to go beyond reductive "Watsonian" behaviorism and advance a conception of behavior that could accommodate purpose. In so doing, they embodied precisely those concerns over holism and naturalizing teleology in psychology that Frank discussed in his *Foreword*.

Though the remarkable resemblance between the behaviorist and cybernetic accounts of purpose was remarked shortly after 1943 (Churchman & Ackoff, 1950; Taylor, 1950a), this has now largely been forgotten. The resemblance is no coincidence, however. Wiener came out of his PhD in philosophy at Harvard well versed in debates about mechanism, purpose, and teleology (see, e.g., Wiener, 1924a; 1924b; 1924c). These were topics that interested the pragmatist and neo-realist philosophers whom he knew personally and whose seminars he attended: Josiah Royce, George Santayana, Ralph Barton Perry, and Edwin Holt (Montagnini, 2017, p. 15). As we shall see, Perry and Holt were particularly keen on defending a purposive account of behavior. Wiener corresponded with Perry during his post-doc at the University of Cambridge (1913–1914) and would speak very highly of Royce's seminar on scientific method (Montagnini, 2017, p. 15; Wiener, 2018, p. 130). The seminar was attended by L. J. Henderson who had recently defended an account of teleology (1913) which Royce (1914), Perry (1917a), and Holt (1920) engaged with and critiqued. These connections that Wiener had to debates about teleology in philosophy and psychology provide crucial context to the cybernetic account of purpose.

5.3.1 Purposive Behaviorism

Scholars have long challenged the received view of behaviorism as a monolithic movement inaugurated by John B. Watson (1913). Many of its supposedly distinctive tenets—such as its opposition to "unscientific" introspectionism in contrast to the "objective" study of behavior—were already defended by psychologists and philosophers in decades prior (Costall, 2006; Klein, 2020). Furthermore, behaviorism was a heterogeneous movement that comprised many different perspectives on the place of mental concepts and their relationship to behavior (Hatfield, 2003). Rather than eliminating the concepts of purpose and teleology, some scholars aimed to find a place for them within a broadly functionalist, and later behavioristic, view of the mind.

Support for a purposive conception of organismal behavior was shared by philosophers, such as Edgar Singer, Arthur Ernest Davies, Ralph Barton Perry; psychologists such as Howard Warren, Leonard Hobhouse, Edward Tolman, William McDougall; and scholars who lived in both worlds, such as Edwin Holt. Similarly, articles on the topic were published in philosophical venues (*The Monist, The Philosophical Review, The Journal of Philosophy*), psychological ones (*Psychological Review, American Journal of Psychology*), and hybrid journals (*Psychology and Scientific Methods, Philosophy of Science*). Though not a true "school" in its own right, purposive behaviorism can be characterized by a shared set of philosophical commitments regarding the inherent purposiveness of behavior, along with strong institutional ties to Harvard, and shared influences from American Pragmatists such as Peirce and James who had defended naturalistic views of teleology (Short, 1981; James, 1983; Kuklick, 2001; see also Pearce, 2020). Following Peirce's pragmatist maxim which urges us to consider the "practical effects" of our beliefs, purposive behaviorism saw the mind and its associated concepts—such as intention, belief, and purpose—as being intimately tied to behavior in the world, rather than something that resides "in the head" (Perry, 1909; Singer, 1911). On this view, rejecting introspectionism did not mean eliminating mental concepts but rather meant properly resituating them as a feature of behavior (Carroll, 2017, p. 33; Kuklick, 2001, Chapter 11).

To varying degrees, purposive behaviorists pitted themselves against reductive views of behaviorism. The British psychologist William McDougall—who took up the William James chair in psychology at Harvard in 1920—emphasized the shortcomings of the "mechanistic" behaviorism of Watson (influenced by his teacher, the reflexologist Jacques Loeb), which interprets behavior solely through the lens of its atomistic components such as the reflex arc.[2] In Watson's reductive mechanistic view, the brain "appear[s] as nothing but a vast automatic Telephone exchange" and humans as a vastly complex "penny-in-the-slot machine" (McDougall, 1923, p. 278). Such an overly simplistic view of organisms forces the mechanistic psychologist to have recourse to other concepts, such as " 'motor sets,' 'trends,' 'drives,' and 'determining tendencies,' all of which are terms of the Purposive Psychology, thinly disguised" (McDougall, 1923, p. 288). Ralph B. Perry was more conciliatory: Watson's view, while limited, was compatible with the purposiveness of behavior as they defined it (Perry, 1918, p. 3).

The distinction between reductive and holistic forms of behaviorism also appeared under the terms "molar" and "molecular" behaviorism, coined by the British emergentist C. D. Broad (1925). The former referred to behavior as gross "overt action" of the whole organism, while the latter referred to behavior inside the organism, such as change in blood pressure (Broad, 1925, p. 616). Broad used this distinction to argue against the behaviorist project: "a man starts as a Molar Behaviourist and is then pushed back by criticism into Molecular Behaviourism, at which stage his theory has lost most of its interest" (Broad, 1925, p. 617). But purposive behaviorists thought that there was a genuine distinction to be had here. Tolman, for instance, critiqued Watsonian "molecular" behaviorism as being a theory of "muscle twitchism" while insisting that molar behavior has certain "emergent" properties, including purposiveness. Perry (1918) said that purpose requires "an epistemological construction that lies beyond the scope of a strictly physiological [viz., molecular] behaviorism," without necessarily contradicting the basic theses of behaviorism (p. 7).

Similar arguments had been made by Edwin B. Holt who, alongside Perry, taught courses at Harvard which were followed by both Tolman and Wiener between 1911 and 1913 (Montagnini 2017, p. 15; Carroll, 2017, pp. 21, 184). As Holt had it, "nerves and muscles, sense-organs, reflex arcs, stimulation, and muscular response [...] these things do not reach the core of the matter, and that they never can [...]." The biologist has "overlooked the *form of organization* of these reflex arcs" (Holt, 1915b, pp. 49–50, 76; emphasis added). For Holt, it is out of this larger "form of organization" that emerges a capacity for *behavior* that makes psychology irreducible to physiology. Holt, and much like Tolman after him, is careful to specify that the organized whole is not necessarily "more" than the sum of the parts "thus organized" but can *do* something which the parts on their own cannot do. As such, "new phenomena, new laws and functions have been developed" (Holt, 1915a, p. 365). Perry presented the divergence between reductive and holistic views in slogan form: "in so far as psychology divides the organism it approaches physiology, and in so far as physiology integrates the organism it approaches psychology" (Perry, 1921, p. 85). Tolman emphasized the fact that reactions must be understood in terms of the broader purposes and functions they serve (Carroll, 2017, p. 76).

Holt illustrated the way purposive behavior emerges through the organized integration of multiple reflexes with a thought experiment. We are asked to imagine a phototropic creature—much like the servomechanical robots that cyberneticians would build decades later (see Riskin, 2016, Chapter 9)—which has a light receptor on one side of its body that stimulates the action of a fin on the opposite side, causing to rotate in a circle (Holt, 1915b, pp. 52–57). With this single reflex arc, the creature is capable of motion, but not *bona fide* behavior. With a second light receptor and fin on the opposite side, however, it can now swim *toward* the light. As light hits one eye, it will rotate toward it, and light will hit its opposite eye, causing its opposite fin to flap and redress its motion. It now has the capacity to respond to a specific feature of its environment. With the addition of a third reflex that senses heat, it would be able to move toward a bright and hot source without burning itself (Holt, 1915b, pp. 52–57).

For Holt, and later Tolman, purposiveness was an emergent property of the whole behaving organism in its environment, which cannot be explained in terms of individual reflex actions caused by stimuli impinging on sense organs. Such an explanation would fail to account for the behavior emerging from the organized integration of these reflexes (Holt, 1915a, p. 366).[3] Furthermore, whereas a single reflex acts like an ignited rocket that happens to fly in a direction, full-blown *behavior* has reference to a particular object of the environment. Put differently, while we can fully understand the rocket based on its internal conditions, we cannot understand purposive behavior by looking inside the organism, because the object in the environment toward which it behaves is part of the description of behavior (c.f. the behavioral/functional distinction in Rosenblueth et al., 1943). Purpose, in this view, is therefore always a feature of behavior directed toward objects in an environment, an idea that would later strongly influence one of Holt's students, James J. Gibson (Holt, 1915b, pp. 54–55; see Shaw, 2011).

In this way, purpose became a central topic of interest for scholars interested in a middle ground between vitalistic and reductive-mechanistic accounts of animal behavior (see also Peterson, 2016). In an article published a year after Holt's, Princeton psychology professor Howard Warren claimed that

> Notwithstanding the vigorous controversy that has recently waged between vitalists and mechanists [...] neither side has so far given an adequate account of the nature of purpose itself. A thorough examination of its fundamental character might reveal a middle ground on which the two opposing parties could unite.
>
> (Warren, 1916a, p. 6; emphasis added)

It appears that the very possibility of finding a suitable middle position was, to some extent, contingent on the kind of "machine-mechanism" which I discussed in the previous section—though this is at times hard to see due to ambiguities in the meaning of causality and mechanism. Those who subscribed to Clark Hull's "robot approach" to simulating purposive behavior did so in part as a way of validating their naturalist (i.e., non-vitalistic) credentials (Cordeschi, 2002, p. 144). In a similar vein, Warren defended a conception of teleology as anticipatory behavior while emphasizing the way in which it was analogous to the governor regulating the speed of a wheel. This indicated to him that "mechanistic processes may fall into the purposive type" and that there is therefore no need to ascribe "a resident entelechy to the regulator" (Warren, 1916b, p. 61). The use of such machine analogies were part of an extended effort on the part of purposive behaviorists to insist on the existence of purpose, without overemphasizing its distinctiveness and risking the charge of dualism or vitalism.

Holt's thinking on this matter is worth detailing. He insisted on the emergent novelty of behavior, while also arguing that it falls within a broader view of scientific laws which applies equally to solar systems and organisms (Holt, 1915a). He justified this based on a distinction between two kinds of causation. According to the outdated "bead theory," the previous state of a system "causes" the next, which implies that there is an entelechy or occult power residing *within* the system in question. According to the modern "function theory," one describes the behavior of the system as a constant "function" in relation to something else (not to be confused with Rosenblueth et al.'s quasi-*opposite* use of the term "functional"). For Holt, just as the motion of a falling object is a function of its mass and the mass of the attracting object, purposive behavior is a function of the organism and the environmental object toward which it behaves purposefully (Holt, 1915ap. 367; c.f. Rosenblueth & Wiener, 1950, p. 322). This view allows behavior to be described in terms of physical laws. As Holt (1915a) sees it, every law is "in the last analysis the statement of a constant function between one process or thing and some other process or thing" (pp. 367–368). Here too one finds hints of the precarious balance that had to be struck by middle way approaches: for while Holt (1915a) emphasized such continuities between physics and psychology, he simultaneously used the parallel to admonish us to keep the organism whole "if it is behavior that we are studying" (p. 369).

In sum, one finds in the work of Holt and other purposive behaviorists a need to simultaneously highlight the continuities and discontinuities between purpose and mechanism; psychology and physiology; the life and the physical sciences. It is in this context that one finds what may be the earliest expression of the cybernetic view that servomechanisms like thermostats and regulators may behave purposively (Warren, 1916b, p. 61; Perry, 1917b, p. 488). In fact, one notices that this was already a straightforward consequence of Holt's definition of purpose, which is remarkably similar to the cybernetic account: on this view, any entity capable of behaving with respect to a specific feature of the environment is purposive. But as mentioned above, there were some such as Warren who argued that teleology required anticipation as an additional ingredient. Perry took this further, arguing that true purposiveness required the capacity for learning. In what follows, I shall argue that underlying these disagreements was a more fundamental ambiguity about the very meaning of purpose and the limits of what a behavioristic account could accommodate. I shall conclude by showing how these same difficulties are to be found in the cybernetic account of teleology.

5.3.2 *Intrinsic and Extrinsic Teleology*

In Perry's 1917 article, *Purpose as Tendency and Adaptation*, he argued that various ways of conceiving teleology behavioristically—including Holt, Warren, and even the organicist Ralph S. Lillie's (for discussion, see also Fábregas-Tejeda, this volume)—were fundamentally the same. In all cases, purposiveness is constituted by an adjustment which a system makes to a change in conditions. These come in three types: (1) "compensatory adjustment": the tendency to maintain equilibrium, as typified by a thermostat; (2) "progressive adjustment": the creation and maintenance of new equilibria, such as in growth; and (3) "preparatory adjustment": the anticipatory maintenance of an equilibrium (Warren's view; or what Rosenblueth et al. called "predictive behavior"). Perry deemed all three cases to be fundamentally the same in so far as

> the independent variable is the environmental action, the dependent variable is the organic response, while the outcome or resultant is constant [...] the organism behaves in a variety of situations in such a way as to produce a constant result.

(Perry, 1917a, p. 493)

For his part, Perry remained agnostic over whether equilibration processes amounted to purposive behavior. On the one hand, he deemed the three kinds of adjustment to be worth distinguishing as a distinctive class of processes and recognized that one may find it "useful"—as Rosenblueth et al. later claimed—to call these purposive. However, he also argued that there were good reasons to think that purposiveness requires more complex behavior, because talk of "means" and "ends" otherwise drops out of the picture: in all cases of equilibration, there is

> a give and take between environment and organism, in which the environment makes the first move, and in which the exchange of actions obeys a constant law [...]. The response *is* complementary; but it means nothing to say that it occurs *in order to* be complementary, or *for the sake of* the complementary outcome.
>
> (Perry, 1918, p. 1; emphasis in original)

For us to meaningfully speak of behavior that occurs for the *sake of something*, we must look to learning behavior. In this case, making sense of the way an animal selects and learns to deploy a particular action in a given context requires speaking of two "propensities": a "higher" or "selective" propensity and a "lower" propensity (Perry, 1918, 1921). The former "higher" propensity refers to that which motivates the animal's effortful trying and which determines what counts as success. "In the maxim 'if at first you don't succeed, try, try again' the higher propensity both accounts for the repeated trials and defines what shall constitute success" (Perry, 1918, p. 3). The lower propensity is the specific actions which are adopted as a means toward reaching that end. These can rightly be called "means" precisely because they have been *selected* in virtue of their suitability to reach the goal.

On the surface, such disagreements appear to be over how basic and omni-present teleology is in organismal behavior, i.e., whether teleology rightly applies only to the level of learning (Perry) or to more basic tropistic and equilibration processes (Holt). Tolman and McDougall defended positions in between these (Cordeschi, 2002, 136). But underlying these disagreements was a more funda-mental ambiguity about the very concept of teleology, namely, whether it can be defined without having recourse to claims about the nature of, or features intrinsic to, the behaving object (c.f. Taylor, 1950b, pp. 327–329). As discussed in the pre-vious section, for Holt the very scientific status of purpose—and the possibility of extracting laws of behavior—hung on it not appealing to causal powers internal to the system in question. This parallels Rosenblueth and Wiener's insistence that purpose "be recognizable from the nature of the act, not from the study of or from any speculation on the structure and nature of the acting object" (Rosenblueth & Wiener, 1950, p. 323). According to these views, goal-directedness is a "functional relationship" (in the Holtian sense) between the behaving object and a feature of the environment. But worries about backward causation and so-called Aristotelian teleology led the more conservative version of purposive behaviorism defended by Holt and the cyberneticians to insist that goals are always observable features of the environment, such that one can speak of "signals" emanating "from the goal" and guiding behavior (Rosenblueth et al., 1943, p. 19).

These worries also emerged in the context of a disagreement over whether pur-posiveness was truly being *identified* as the behavior or if it was merely *inferred* from the behavior. Highly influenced by Holt and Perry, Tolman first insisted on the former and critiqued McDougall for holding the latter (Tolman, 1925, p. 37). But this was easier said than done: Tolman, along with other purposive behaviorists, was accused by his student Zing-Yang Kuo for implicitly doing the same thing (Kuo, 1928). Tolman conceded, but turned around to defend, the inferentialist position on the grounds that it was still objective: "our concept of purpose, or drive, is not, *even though inferred, therefore mentalistic or introspectionistic*. We do not describe or define these purposes in terms of their introspected 'feels'" (Tolman, 1928, p. 524, as cited in Carroll, 2017, pp. 75–80; emphasis added; see also Cordeschi, 2002, pp. 135–136). Tolman, in effect, aimed for a difficult compromise.

The need for such a middle ground was also driven by the need for accounting for goals that were not located in the environment—an issue was raised decades later by Richard Taylor (1950a) in his critique of cybernetics. Holt (1915b) had attempted to circumvent this by stating that more complex behaviors occurred with respect to "a larger and more comprehensive situation" (p. 197). Perry, in turn, later made the case that the complementarity between behavior and object could be hypothetical rather than actual. In

> the case of hunting for a pin, the organism is not, strictly speaking, responding to an object or fact of its environment. [...] The finding of any particular pin is the hypothetical complement to its present response. It is related to it as a hypo-thetical key is related to some lock which it would fit if it did exist.
>
> (Perry, 1918, p. 7)

Here too, it appears that we are forced to admit that the goal—since it is no longer an objective feature of the environment—is inferred from the behavior to the internal "propensity." The problem is that it was never quite clear what this "pro-pensity" really is. It all too often appears as a stand-in for many more problematic concepts, many of which Perry suggested to describe it, such as "trying," "agency," or "impulse," and even at times seemed to resemble what Tolman later called a "cognitive map"—a precursor to "mental representation" (Carroll, 2017).

Such debates about the nature of goals—whether they are internal to the pur-posive system in question, merely to the behavior itself, or to a "goal-object" in the environment—closely paralleled another set of concerns over whether servomechanisms are intrinsically teleological or whether their goal-directedness is merely a projection of the observer. Since at least the time of Aristotle, philosophers have recognized a fundamental difference between the purposiveness *intrinsic* to a human mind and the purpose—in the sense of the "function"—of an artifact which is *extrinsically* bestowed onto it by its maker or user.[4] The major problem here is making sense of what makes teleology *intrinsic* in the first place. It is no coinci-dence that the rise of servomechanical robots, combined with the redefinition of purpose in behavioristic terms that saw goals as empirically observable objects in the world, radically reconfigured the scope of the organism–artifact analogy to now comprise intrinsic purpose. Though a torpedo or thermostat's "goal" appears to be bestowed onto it by its user or maker, and not intrinsic to it, it is not clear on what grounds a behaviorist and cybernetic account of teleology could make such distinctions—since teleology "just is" any behavior that reaches "a definite correlation in time or in space with respect to another object or event" (Rosenblueth et al., 1943, p. 18).

Many supporters of purposive behaviorism recognized the significance of this issue in the decades prior to cybernetics. Warren (1916) argued that self-regulating systems are purposive only insofar as that purpose is bestowed onto them. While Perry did not explicitly make the same claim, his concern that "means–ends" talk does not apply to self-regulating systems seems to drive toward the same idea. He quoted an article by William James, in which he argues that the mind "not only *serves* a final purpose, but *brings* a final purpose—posits, declares it" (James, 1878, p. 2, as cited in Perry, 1921, p. 101). As Perry sees it,

> No one would now be disposed to dispute the essential soundness of this position. The human individual does not merely do things that are useful as judged by an external observer, but by its own activity adopts and seeks that result in relation to which its deeds are useful.
>
> (Perry, 1921 p. 101)

Perry had addressed this topic in an earlier article, where he rejected a definition of teleology in terms of "systematic unity" (i.e., appearance of order and design) that we find in both artifacts and organisms, because teleology is about the existence of entities that "not only exhibit unity, but must be 'for the sake of' that unity" (Perry, 1917a, p. 477). In other words, the mere fact of being a systematically unified whole, as artifacts are, is insufficient to define teleology because it does not tell us the purpose or function that such a unity serves. The essential question for Perry was whether making sense of intrinsic purposiveness requires a dualistic sundering of mind and body. His answer was negative: "The better the organism is understood, the more does it assume just those characters which James insists upon as the prerogatives of mind" (Perry, 1921, p. 102).

But Perry also recognized the difficulty of accounting for intrinsic purpose (and mindedness more generally) in behavioristic terms in so far as it required going beyond a conception of behavior and goal-objects, defended by the like of Holt.[5] Perry's alternative was to insist that purposive behavior is "spontaneous" and "internally conditioned." He hoped that such a view would "not [...] contradict the fundamental thesis of behaviorism," though it did appear to "forbid any hasty or contemptuous dismissal of the traditional association of purpose with nonphysical or 'ideal' entities" (Perry, 1918, p. 7). In other words, while Holt defined purpose in terms of features external to the animal, Perry and later Tolman emphasized the need to refer to something *internal* to it: a propensity, impulse, or curiosity, which allowed the organism to learn its behavior in the first place, and governs this behavior continuously (Perry, 1918, p. 13). For Perry (1917a), distinguishing automatic action (in the sense of behavior as tendency and adaptation) from intelligent action (in which one can speak of means and ends) requires "over and above the external action of the stimulus, […] some *additional factor* [...] which decisively determines the direction which the discharge [viz., action] takes" (p. 359; emphasis added). This "third and differential factor [...] constitutes their purposive aspect" (Perry, 1917a, p. 360). It is this "third factor" which Perry appeared to be describing when he talked of the "general propensity" that drives an animal's behavior in scenarios of trial and error.

The importance that Perry saw in learning for demarcating intrinsic and extrinsic teleology did not carry over to cybernetics. For their part Rosenblueth, Wiener, and Bigelow—who defined purposiveness in a broadly Holtian way as "a final condition in which the behaving object reaches a definite correlation in time or in space with respect to another object or event"—attempted to re-inscribe the distinction between intrinsic and extrinsic purposiveness within such a definition, to strange effect. As Rosenblueth et al. (1943) state, "the term servomechanisms has been coined precisely to designate machines with intrinsic purposeful behavior," as opposed to systems that perform in an orderly but non-purposeful way, such as a clock, since "there is no specific final condition toward which the movement of the clock strives" (p. 19). But such conceptions of "intrinsic" purposiveness seem to stray very far from their original meaning, given that it is no longer clear what the contrast with "extrinsic" teleology would even amount to. Consider, for instance, a roulette machine which clearly plays a functional role (an extrinsic teleology) for the purposes of gambling. Such devices, on their view, are "designed precisely for purposelessness" (Rosenblueth et al., 1943, p. 19). Taylor (1950a) was quick to show the odd consequences of this view: if one were to add a weight to a roulette, or if a clock happened to break down at midnight on the 1st of January, it would suddenly become intrinsically purposive. Furthermore, it appears that all the critiques which Perry leveled at Holt's account could also be extended to the cybernetic account, most notably the fact that there is no room for a distinction between "means" and "ends."

It appears, then, that the various attempts to naturalize purpose from a behavioristic standpoint in the decades prior to cybernetics struggled mightily to make sense of the distinction between intrinsic and extrinsic teleology. It seemed that more needed to be said about the nature of the organism itself to demarcate its *intrinsic* purposiveness from the *extrinsic* purposiveness of artifacts. As the philosopher Hans Jonas famously put it, such a conception of teleology confused "serving a purpose" with "carrying out" a purpose (Jonas, 1966/2001, p. 303).

These shortcomings can be traced to Rosenblueth and Wiener's particular sensitivity to the need for purpose to be *useful* for science in practice. In this sense, their understanding of naturalization was closely tied to the scientific operationalization of teleology rather than any kind of deeper metaphysical claim about its nature. Their view rendered moot the question of whether servomechanisms *really are* intrinsically teleological or whether this was a projection of the observer. Based on their distinction between functional and behavior modes of study, they saw it as essential that purpose "be recognizable from the nature of the act, not from the study of or from any speculation on the structure and nature of the acting object"

(Rosenblueth & Wiener, 1945, p. 323). As I argued in Section 5.2, this was also what was required for purpose to be both applicable to non-living machines and to be holistic, according to Frank. But methodological holism of this variety was not enough, in so far as it remained omitted to keeping at arm's length the deeper metaphysical questions that inevitably reappeared at every corner. As Rosenblueth and Wiener saw it, one's operational commitments did not necessarily commit one to a metaphysical position (Rosenblueth & Wiener, 1945). But their constant attempt at seeking refuge in the heuristic "as if" stance put them at constant risk of perniciously reifying their own models and abstractions (see Abraham, 2012; Dupuy, 2009, p. 138; Chirimuuta, 2020). By the end of the 1940s, the phrase "teleological mechanisms" was replaced by a new term, "cybernetics," and the movement swiftly moved on from the topic of teleology, leaving the purposive behaviorist project behind and incomplete.

5.4 Conclusions

This chapter attempted to re-evaluate the cybernetic account of teleology on two fronts. First, I aimed to understand Frank's claim that their account can be legitimately deemed both non-reductive or holistic and mechanistic at once. This argument centered around the distinction between the behavioristic and functional mode of study proposed by Rosenblueth et al., which, despite being essential to their account of teleology, has been largely ignored by those who have since made references to cybernetic teleology. Second, I aimed to situate this conception of purpose within a broader family of non-reductive behaviorist views. I showed that a common problem permeated all these accounts, namely, a difficulty in accounting for the distinction between intrinsic and extrinsic purposiveness, and that this problem also haunted the cybernetic account of teleology.

Notes

1 Perhaps because focus had turned to a newer and more exciting concept: "information." Like teleology, the concept appeared capable of bringing together otherwise disparate phenomena and methods, and in doing so rhetorically capture the movement's transdisciplinary ambitions (Abraham, 2012; Galison, 1998).
2 McDougall's position is difficult to characterize. Some might argue that the incompatibility he saw between mechanism and teleology, and his self-labeling as a dualist and animist, sets him apart from other purposive behaviorists. However, what he meant by "mechanism" is unclear, and he asserted that his dualism was not a "metaphysical Dualism, or indeed any metaphysical or ontological doctrine" (McDougall in Boden, 1965, p. 4; c.f. Lawrence & Weiss, 1998, p. 12; Rose, 2016).
3 Though whether such an integrated view of reflexes, famously defended by Charles Sherrington, was sufficiently "holistic" was itself a matter of debate (Goldstein, 1995).
4 In this way my account departs from Cordeschi's (2002), who sees the issues raised by the intrinsic/extrinsic distinction as restricted to the specific subset of views which Perry identified in his (1917b) article. In contrast, I see these to have permeated the entire debate about purpose within the behavioristic framework.

5   Holt would have deemed appeals to features intrinsic to the behaving as a regression to the unscientific "bead theory" (see previous sub-section).

References

Abraham, T. H. (2012). Transcending disciplines: Scientific styles in studies of the brain in mid twentieth century America. *Studies in History and Philosophy of Science Part C: Studies in History and Philosophy of Biological and Biomedical Sciences, 43*(2), 552–568.

Aiken, H., Von Neumann, J., & Wiener, N. (1944, December 28). Letter to H. H. Goldstine (Box 4, folder 66). *Norbert Wiener Papers, MIT Archive.*

Ashby, W. R. (1957). *An introduction to cybernetics* (2nd ed.). Chapman & Hall Ltd.

Boden, M. A. (1965). McDougall revisited. *Journal of Personality, 33*(1), 1–19.

Broad, C. D. (1925). *The mind and its place in nature.* Kegan Paul, Trench, Trubner & Co., Ltd.

Burbano, A., & Reyes, E. (2022). Capsaicin and cybernetics: Mexican intellectual networks in the foundation of cybernetics. *AI & Society, 37*, 1013–1025.

Carroll, D. W. (2017). *Purpose and cognition: Edward Tolman and the transformation of American psychology.* Cambridge University Press.

Chirimuuta, M. (2020). The reflex machine and the cybernetic brain: The critique of abstraction and its application to computationalism. *Perspectives on Science, 28*(3), 421–457.

Churchman, C. W., & Ackoff, R. L. (1950). Purposive behavior and cybernetics. *Social Forces, 29*(1), 32–39.

Cordeschi, R. (2002). *The discovery of the artificial* (Vol. 28). Springer Netherlands.

Costall, A. (2006). 'Introspectionism' and the mythical origins of scientific psychology. *Consciousness and Cognition, 15*(4), 634–654.

Dupuy, J.-P. (2009). *On the origins of cognitive science: The mechanization of the mind.* MIT Press.

Esposito, M., & Baravalle, L. (2023). The machine-organism relation revisited. *History and Philosophy of the Life Sciences, 45*(3), 34. https://doi.org/10.1007/s40656-023-00587-2

Frank, L. K. (1948). Foreword. *Annals of the New York Academy of Sciences, 50*(4), 189–196.

Galison, P. (1994). The ontology of the enemy: Norbert Wiener and the cybernetic vision. *Critical Inquiry, 21*(1), 228–266.

Galison, P. (1998). The Americanization of unity. *Daedalus, 127*(1), 45–71.

Goldstein, K. (1995). *The organism: A holistic approach to biology derived from pathological data in man.* Zone Books.

Hatfield, G. (2003). Behaviourism and psychology. In T. Baldwin (Ed.), *The Cambridge history of philosophy 1870–1945* (pp. 640–648). Cambridge University Press.

Henderson, L. J. (1913). The fitness of the environment, an inquiry into the biological significance of the properties of matter. *The American Naturalist, 47*(554), 105–115.

Holt, E. B. (1915a). Response and cognition I: The specific-response relation. *The Journal of Philosophy, Psychology and Scientific Methods, 12*(14), 365–373.

Holt, E. B. (1915b). *The Freudian wish and its place in ethics.* Henry Holt and Company.

Holt, E. B. (1920). Professor Henderson's "Fitness" and the locus of concepts. *The Journal of Philosophy, Psychology and Scientific Methods, 17*(14), 365–381.

James, W. (1878). Remarks on Spencer's definition of mind as correspondence. *The Journal of Speculative Philosophy, 12*(1), 1–18.

James, W. (1983). *The principles of psychology.* Harvard University Press.

Jonas, H. (2001). *The phenomenon of life: Toward a philosophical biology*. Northwestern University Press (Original work published 1966).

Kay, L. E. (2000). *Who wrote the book of life?: A history of the genetic code*. Stanford University Press.

Klein, A. (2020). The death of consciousness? James's case against psychological unobservables. *Journal of the History of Philosophy, 58*(2), 293–323.

Kline, R. (2020). How disunity matters to the history of cybernetics in the human sciences in the United States, 1940–80. *History of the Human Sciences, 33*(1), 12–35.

Krikorian, Y. H. (1943). Life, mechanism and purpose. *Philosophy of Science, 10*(3), 184–190.

Kuklick, B. (2001). *A history of philosophy in America, 1720–2000*. Clarendon Press.

Kuo, Z. Y. (1928). The fundamental error of the concept of purpose and the trial and error fallacy. *Psychological Review, 35*(5), 414–433.

Lawrence, C., & Weisz, G. (1998). *Greater than the parts: Holism in biomedicine, 1920-1950*. Oxford University Press.

Lillie, R. S. (1943). The psychic factor in living organisms. *Philosophy of Science, 10*(4), 262–270.

Malaterre, C., Chartier, J.-F., & Pulizzotto, D. (2019). What is this thing called philosophy of science? A computational topic-modeling perspective, 1934–2015. *HOPOS: The Journal of the International Society for the History of Philosophy of Science, 9*(2), 215–249.

McDougall, W. (1923). Purposive or mechanical psychology? *Psychological Review, 30*(4), 273–288.

Montagnini, L. (2017). *Harmonies of disorder- Norbert Wiener: A mathematician-philosopher of our time*. Springer.

Nicholson, D. J. (2012). The concept of mechanism in biology. *Studies in History and Philosophy of Science Part C: Studies in History and Philosophy of Biological and Biomedical Sciences, 43*(1), 152–163.

Pearce, T. (2020). *Pragmatism's Evolution: Organism and Environment in American Philosophy*. The University of Chicago Press.

Perry, R. B. (1909). The hiddenness of the mind. *The Journal of Philosophy, Psychology and Scientific Methods, 6*(2), 29–36.

Perry, R. B. (1917a). Purpose as systematic unity. *The Monist, 27*(3), 352–375.

Perry, R. B. (1917b). Purpose as tendency and adaptation. *The Philosophical Review, 26*(5), 477–495.

Perry, R. B. (1918). Docility and purposiveness. *Psychological Review, 25*(1), 1–20.

Perry, R. B. (1921). A behavioristic view of purpose. *The Journal of Philosophy, 18*(4), 85–105.

Peterson, E. L. (2016). *The life organic: The theoretical biology club and the roots of epigenetics*. University of Pittsburgh Press.

Riskin, J. (2016). *The restless clock: A history of the centuries-long argument over what makes living things tick*. University of Chicago Press.

Rose, A. C. (2016). William McDougall, American psychologist: A reconsideration of nature-nurture debates in the interwar United States. *Journal of the History of the Behavioral Sciences, 52*(4), 325–348.

Rosenblueth, A., & Wiener, N. (1945). The role of models in science. *Philosophy of Science, 12*(4), 316–321.

Rosenblueth, A., & Wiener, N. (1950). Purposeful and non-purposeful behavior. *Philosophy of Science, 17*(4), 318–326.

Rosenblueth, A., Wiener, N., & Bigelow, J. (1943). Behavior, purpose and teleology. *Philosophy of Science, 10*(1), 18–24.

Royce, J. (1914). The mechanical, the historical and the statistical. *Science, 39*(1007), 551–566.

Shaw, R. (2011). Ecological realism as a reaction to new realism: Holt's legacy to Gibson. In *A new look at new realism: The psychology and philosophy of EB Holt* (pp. 157–190). Transaction Publishers.

Short, T. L. (1981). Peirce's concept of final causation. *Transactions of the Charles S. Peirce Society, 17*(4), 369–382.

Singer, E. A. (1911). Mind as an observable object. *The Journal of Philosophy, Psychology and Scientific Methods, 8*(7), 180–186.

Taylor, R. (1950a). Comments on a mechanistic conception of purposefulness. *Philosophy of Science, 17*(4), 310–317.

Taylor, R. (1950b). Purposeful and non-purposeful behavior: A rejoinder. *Philosophy of Science, 17*(4), 327–332.

Tolman, E. C. (1925). Behaviorism and purpose. *The Journal of Philosophy, 22*(2), 36–41.

Tolman, E. C. (1928). Purposive behavior. *Psychological Review, 35*, 524–530.

von Foerster, H., Mead, M., & Teuber, H. L. (1953). A note by the editors. In H. von Foerster, M. Mead, & H. L. Teuber (Eds.), *Cybernetics. Circular causal and feedback mechanisms in biological and social systems. Transactions of the ninth conference* (pp. xi–xx). Josiah Macy, Jr. Foundation.

Warren, H. C. (1916a). A study of purpose I. *The Journal of Philosophy, Psychology and Scientific Methods, 13*(1), 5–26.

Warren, H. C. (1916b). A study of purpose. III: The role of purpose in nature. *The Journal of Philosophy, Psychology and Scientific Methods, 13*(3), 57–72.

Watson, J. B. (1913). Psychology as the behaviorist views it. *Psychological Review, 20*(2), 158–177.

Wiener, N. (1924a). Dualism. In *The Encyclopedia Americana* (Vol. 9). The Encyclopedia Americana Corporation.

Wiener, N. (1924b). Mechanism and vitalism. In *The Encyclopedia Americana* (Vol. 18). The Encyclopedia Americana Corporation.

Wiener, N. (1924c). Metaphysics. In *The Encyclopedia Americana* (Vol. 18). The Encyclopedia Americana Corporation.

Wiener, N. (1942, June 22). *Letter to J. B. S. Haldane* (box 4, folder 62). Norbert Wiener papers, MIT Archive.

Wiener, N. (1985). *Cybernetics or control and communication in the animal and the machine.* MIT Press (Original work published 1948).

Wiener, N. (2018). *Norbert Wiener—A life in cybernetics: Ex-prodigy: My childhood and youth and I am a mathematician: The later life of a prodigy.* MIT Press.

Part II
Evolutionary Perspectives on Agency

6 The Baldwin Effect and the Potentialities for Thoughtful Darwinism around 1900

Gregory Radick

6.1 1900 and All That

For anyone concerned with heredity and evolution, 1900 marks, first and foremost, the year of the "rediscovery of Mendel." Independently of each other, three European botanists engaged in experimental and quantitative crossbreeding research—Hugo de Vries in the Netherlands, Carl Correns in Germany, and Erich von Tschermak in Austria—published papers which converged strikingly not only on each other's work but on that of a long-dead and previously little-known Augustinian monk from Moravia, Gregor Mendel (see, e.g., Olby, 1990, pp. 528–530). As enthusiasm for Mendel's original paper began to spread, first under the banner of "Mendelism," then under "genetics," so too did the remarkable story of its rediscovery. Soon that story took its place with similar ones from the history of science—Newton and Leibniz converging on the calculus, Darwin and Wallace converging on natural selection—as an illustration of a curious general truth: when a scientific discovery's time has come, it becomes irresistible, inevitable. In the Mendelian case, had there not been three rediscoverers in 1900, there would have been six in 1901: so, at any rate, judged the American anthropologist Alfred Kroeber, looking back from 1917 (Kroeber, 1917, p. 199). The standing of the Mendelian rediscovery went on to climb still higher over the next decades, as defenders of Darwin's theory of natural selection succeeded in showing that the Mendelian gene concept was what natural selection had needed all along. From the perspective of this new "Modern Synthesis," 1900 was a pivotal year, when the revival in the fortunes of Darwinism, then at a nadir, first became a possibility.

Less well remembered is another triple convergence, no less relevant to heredity and evolution, and taking place just a few years earlier—yet about as different as could be from the Mendelian triple as a clue to the energies animating the era's biological debates. In 1896, the American psychologist James Mark Baldwin, the American palaeontologist Henry Fairfield Osborn, and the English comparative psychologist Conwy Lloyd Morgan independently proposed a form of natural selection that could result in acquired characters becoming hereditary. Called "organic selection" by Baldwin, and a number of other names since then, it was enduringly renamed "the Baldwin effect" in a 1953 essay in *Evolution*, the house journal of the Modern Synthesis, by its major palaeontological contributor, George

DOI: 10.4324/9781003413318-8

Gaylord Simpson. In Simpson's view, the Baldwin effect is best thought of as involving three distinct (but partly simultaneous) steps:

(1) Individual organisms interact with the environment in such a way as systematically to produce in them behavioral, physiological, or structural modifications that are not hereditary as such but that are advantageous for survival, *i.e.*, are adaptive for the individuals having them.
(2) There occur in the population genetic factors producing hereditary characteristics similar to the individual modifications referred to in (1), or having the same sorts of adaptive advantages.
(3) The genetic factors of (2) are favored by natural selection and tend to spread in the population over the course of generations. The net result is that adaptation originally individual and non-hereditary becomes hereditary.

(Simpson, 1953, p. 112)

Simpson devoted most of his essay to evaluating the status of the Baldwin effect in *circa* 1950 biology, including those precincts of it—notably in France—where debate about the evolutionary significance of purposive activity by individual organisms remained lively. But near the start he reflected on what it revealed about *circa* 1900 biology:

That three workers independently thought of the Baldwin effect at the same time demonstrates that the idea was in the air, that it was the inevitable outgrowth of the intellectual atmosphere of the time. *That time was at the height of the neo-Darwinian versus neo-Lamarckian controversy and shortly before the rediscovery of Mendelism gave a radically different turn to biological thought.* There was a sharp issue, still familiar to all of us. Organism and environment obviously interact and obviously are closely fitted, that is, adapted to each other. Yet, as was already clear in the 1890's, it is improbable (to say the least) that the effects of the interaction can become heritable directly and in the same form. The Baldwin effect ostensibly provides a reconciliation between neo-Darwinism and neo-Lamarckism. To the extent that it may really occur, it provides a mechanism that is capable of making acquired characters hereditary—or of seeming to do so. Baldwin, Lloyd Morgan, and Osborn all explicitly postulated the Baldwin effect as a way out of the neo-Darwinian–neo-Lamarckian dilemma.

(Simpson, 1953, p. 110; emphasis added)

Elsewhere (Radick, 2023, pp. 301–302, 365, 381) I have recommended Simpson's juxtaposing of the Baldwinian and Mendelian triples as a corrective to a picture of *circa* 1900 biology that many of us inherited along with elementary Mendelism—a picture that tends to make the Mendelian turn in biology look like a foregone conclusion.[1] Yes, for some biological workers, curious to know what could be gleaned via crossing and counting about the transmission patterns of all-or-nothing unit characters, the Mendelian ideas of dominance, recessiveness, the three-to-one ratio, and so on were in the air. But for other workers, attuned to questions about how developing organisms and their descendants adapt to—and sometimes act on—environments, and how variation and selection bring about evolutionarily consequential change, what was in the air was the Baldwin effect.

Here I want to expand the frame around the Baldwinian triple to exhibit something of the creative resourcefulness of Darwinian discussion in the 1890s and 1900s. As we shall see, it was a thoughtful discussion twice over: in its recognition of complexity, and attention to the drawing of distinctions and articulating of options; and in its openness to what Conwy Lloyd Morgan called "mental factors in evolution." I begin with Morgan's survey under that title for a 1909 volume commemorating a hundred years since Charles Darwin's birth and 50 years since the publication of the *Origin*. Next, I consider in a little more detail three topics that Morgan touched upon: female choice in sexual selection, the Baldwin effect ("organic selection" for Morgan), and the English comparative anatomist Edwin Ray Lankester's proposal about what he called "educability." Finally, I turn to look at the thoughtful Darwinism of an instructive figure for all interested in that era in developmental plasticity, adaptive evolution, and inheritance, the English zoologist and biometrician Walter Frank Raphael Weldon. In conclusion I offer some brief remarks on how differently the period between the rediscovery of Mendel and the Modern Synthesis looks once we give the Baldwinian triple its due.

6.2 A Centennial Celebration

Near the start of "Mental Factors in Evolution," Morgan quoted Darwin counterfactually imagining that our world had been one where "no organic being excepting man had possessed any mental power," or where "his powers had been of a wholly different nature from those of the lower animals." In such a world, Darwin had guessed, "we should never have been able to convince ourselves that our high faculties had been gradually developed" (Darwin, 1871, vol. 1, pp. 34–35; Morgan, 1909, p. 425). In writing as extensively as he had on what Morgan called "mental evolution," Darwin had thus taken full advantage of the opportunities that the actual world presented for pressing home his general case for "organic evolution." What was more, he had brought to the task what Morgan judged to be a winning combination of considerable skill as an observer, a straightforwardly naturalistic view of the mind as a product of nervous physiology, and confidence that the puzzle of the origin of consciousness was no more his worry than the puzzle of the origin of life. "Mental Factors in Evolution" was Morgan's appreciation and updating of his brilliant predecessor's treatments of mind and behaviour in four key works: the *Origin of Species* (1859, though Morgan cited the sixth edition of 1872); the *Expression of the Emotions in Man and Animals* (1872, in the posthumous 1890 second edition); the sexual-selection chapters—really a book within a book—of the *Descent of Man, and Selection in Relation to Sex* (1871, citing the second edition of 1874); and the evolution-of-mind-and-morals chapters of the *Descent*.

For Morgan, the most important updating was to do with the role that Darwin had assigned to Lamarckian inheritance in the shaping of animal instincts and emotional expressions. In the chapter on instinct in the *Origin*, for example,

Darwin had represented complex instincts as primarily due to natural selection, yet allowed for Lamarckian supplementation as individual organisms coping with new environments formed new adaptive habits which, over time, became hereditary (Darwin, 1859, p. 209). Morgan's own experimental work in comparative psychology had been devoted to the disentangling of instinctive actions from learned ones (see, e.g., Boakes 1984, pp. 32–44), and he wrote admiringly of Darwin's emphasis on the distinction. But, after the experimental and theoretical assault on Lamarckian inheritance of the German zoologist August Weismann in the late 1880s, the idea of the learned becoming hereditary was no longer widely accepted, and Morgan instead drew attention to a couple of more recent proposals that, in strictly Darwinian fashion, honoured the more general insight about the biological significance of the interaction of learning and inheritance. One was Lankester's notion that, in Morgan's phrase, "'educability,' not less than instinct, is hereditary" (Morgan, 1909, p. 427). The other was organic selection, in which adaptive modifications in behaviour and even structure acquired during an individual's lifetime are not, as they were for Lamarckians, "the parents of inherited variations" in the same direction, but merely their "foster-parents or nurses," on the view that the acquired modifications, in helping to adapt the organism to its environment, contribute to the survival of whatever coinciding variations the organism chanced to be born with (Morgan, 1909, pp. 428–429). As for Darwin's work in the comprehensively Lamarckian *Expression*, Morgan suggested that Darwinian reform lay in taking seriously how emotions and expressions alike can invigorate action in ways that promote survival and so be subject to natural selection (Morgan, 1909, p. 435).

Emotions and their expression are nowhere more extreme than during courtship—for Darwin, the realm of sexual selection, in the forms of male combat and female choice (Darwin, 1871, vol. 1, p. 232). From the start, the prominence that Darwin gave to the latter had been controversial, not least with Wallace. As Wallace saw it, males attempting to outcompete each other, with victory going to the ones that chanced to be born with a competitive edge, and the passing on of their advantageous variations to their offspring as the prize, was just natural selection. But females exercising choice in line with their sense of beauty was something else, with, in Wallace's disapproving words (quoted by Morgan), "none of that character of constancy and of inevitable result that attaches to natural selection, including male rivalry" (Wallace, 1889, p. 283; Morgan, 1909, pp. 436–437). Morgan did his best to suggest that in the details Darwin and Wallace were not really all that far apart and that in any case their disagreement should not distract attention from the undoubted importance of pairing situations in driving mental evolution. In the male, the stimulus of seeing, smelling, or otherwise sensing a female unleashes instinctive, often highly baroque sequences of emotionally charged display: the legacy of eons of selection among ancestors whose mental and behavioural repertoires were thus advanced beyond the point where natural struggle alone could have taken them. And in the female, what mattered in a complementary way, psychologically and evolutionarily, was the perceptual discrimination as well as emotional energy that she brought to the business of accepting only the "most vigorous, defiant and mettlesome male" as a partner (Morgan, 1909, p. 438).

Even in courtship, however, there is learning from experience as well as instinctive action. For Morgan, the *Descent* overall had laid the foundations for burgeoning psychological studies into the complex interactions between the hereditary and the acquired. From play in animals to progress in human civilization, the teasing apart of the contributions of instinct, emotion, and intelligence, and the understanding of each of these elements as a product of natural selection, was the work of investigators building on Darwin's precedents in comparative psychology, developmental psychology (then called "genetic psychology"), and other sciences. There was growing awareness, for example, that with natural selection having long ago ceased to operate among civilized peoples, the instinctive bases of the social behaviour underpinning morality were, in Morgan's words, "somewhat out of date" (Morgan, 1909, p. 445). In closing he returned to educability, as a key notion for evaluating both our current evolutionary position and Darwin's achievement in elucidating it:

> The history of human progress has been mainly the history of man's higher educability, the products of which he has projected on to his environment. This educability remains on the average what it was a dozen generations ago; but the thought-woven tapestry of his surroundings is refashioned and improved by each succeeding generation. Few men have in greater measure enriched the thought-environment with which it is the aim of education to bring educable human beings into vital contact, than has Charles Darwin. His special field of work was the wide province of biology; but he did much to help us to realise that mental factors have contributed to organic evolution and that in man, the highest product of Evolution, they have reached a position of unquestioned supremacy.
>
> (Morgan, 1909, p. 445)

6.3 A (Deflationist) Defence of Choice-Making in Females

I want from here to concentrate on the years 1894–1906 in order to examine more closely four expressions of the period's thoughtful Darwinism touched upon in Morgan's essay: the defence of female choice; the discovery of organic selection; the Darwinizing of educability; and what, in his discussion of sexual selection, he described as a near future when "Mendelism and mutation […] have been more fully correlated with the basal principles of selection" (Morgan, 1909, p. 437). Although all are relevant to the biological understanding of organismal agency, female choice bears on it so directly that we should pause to note both how much emphasis Darwin gave to female choice in the *Descent* and, in consequence, how much debate that emphasis provoked, within the scientific community and well beyond it. Could the subtle shadings in the patterns on the wing feathers of male Argus pheasants really have come into being incrementally for no other reason than that female Argus pheasants found such ornamentation beautiful and chose their mates accordingly? Darwin revelled in the idea, but his critics balked

(Darwin, 1871, vol. 2, pp. 400–401; Cronin, 1991, pp. 165–181; Richards, 2017, pp. 466–516; Milam, 2010, pp. 9–28). As one reviewer of the *Descent* wrote, "We must attribute to the hen Argus Pheasant the aesthetic powers of a Raphael in order to account for the decorations of her mate" (quoted in Richards, 2017, p. 466). By the 1890s, there was even a political edge; in her 1894 book *The Evolution of Woman*, the American suffragist Eliza Burt Gamble argued that the lesson for the human world was surely that women needed to be economically emancipated, for only then would they be free to choose the fittest men and so keep the evolution of the species on track—a message that resonated with other socialist and feminist writers (Erskine, 1995, pp. 112–113; Hamlin, 2014, pp. 128–165).

That year saw the publication of Morgan's *An Introduction to Comparative Psychology*, where he first put in canonical form the deflationist methodological rule that came to be known as "Morgan's canon." In essence, it commands the comparative psychologist not to attribute to animal minds more than is needed psychologically in order to explain observed behaviour. If you see a dog opening a gate, do not assume that the dog understands the principle of the latch and then reasoned its way to the action. Before you are entitled to that reason-attributing interpretation, you need to rule out the possibility that what you saw was the result of something psychologically humbler—that the dog was just imitating another dog, say, or learned to open the gate through a prior history of blind trial-and-error. In an earlier statement of his canon, Morgan had justified it on the view that since language and reason go together, and non-human animals seem to lack language, they probably lack reason too. Strikingly, in his *Introduction*, he offered an entirely different, more elaborate, more Darwinian rationale.[2] Consider, he suggested, three hypothetical relationships between a dog's mind and your mind. It could be that the dog's mind is like your mind except that it is missing one or more higher psychological "levels" that you enjoy, like reason. It could be that the dog's mind has all of the levels your mind has, but to a lesser degree. Or it could be that the dog's mind is fitted to the struggle for existence as the dog's wild ancestors knew it—with the result that the dog's mind could well be very different indeed from your mind. In Morgan's view, the third possibility had to command assent from the evolutionist. But it was also the most demanding to put into practice, since the least anthropomorphic (Morgan, 1894, pp. 53–59).

With the Darwinian credentials of his anti-anthropomorphic "canon of interpretation" in place, Morgan proceeded to recast comparative psychology by its light through the rest of the book, coming at the end to the problem of comparing animal and human minds, not least in their powers of aesthetic judgement. Characteristically, Morgan now introduced a distinction along canonical lines, between holding a standard in the mind, and judging that one thing comes closer to the standard than another, and merely preferring one sense-experience over another. For Morgan, since choosiness among female birds during courtship could be satisfactorily explained as an instance of the latter, it should be:

> Many biologists [...] believe that birds select their mates from among numerous suitors because of their song or because of their bright plumage. Suppose a bird

has two males before it, both of which are endeavouring by display of plumage, and by love-antics to win her choice. She selects the brighter, and more graceful performer. Does not this, it may be asked, imply that she has a standard of excellence, and selects that mate which she perceives as the nearer of the two to such standard? But admitting, for the purpose in hand, the correctness of the biological interpretation, that there is an exercise of choice on the part of the hen-bird, it does not necessarily follow that [...] she compares the two competing males to an ideal standard, or even the one with the other. It is quite enough to suppose that A evokes a stronger emotion and a stronger appetence than B, and that she is therefore drawn to A rather than to B. There is no necessary [...] framing of an ideal of excellence. And if the facts, supposing them to be biologically well founded, can be explained on the hypothesis of sense-experience, the greater appetence prevailing, we are bound by our canon of interpretation not to assume the higher faculty [of judgment against a standard].

(Morgan, 1894, p. 366)

When, in his centennial essay, Morgan recommended this psychologically abstemious view of the mind of the choosy female, he suggested that Darwin himself in certain passages of the *Descent* came close enough to it that it "seems to have Darwin's own sanction" (Morgan, 1894, pp. 438–439).

6.4 Learning, Inheritance, Evolution, and the Baldwin Effect

To have agency is to exercise choice, which requires some consciousness of options, some power to select between them, and some ability to learn from experience. In Morgan's next book, *Habit and Instinct* (1896), he continued to emphasize how psychologically modest a thing choosiness could be, but also how dependent it was on inherited capacities, and how fundamental it was to intelligence and all that it made possible. He illustrated with an example from his own experimental studies of newborn chicks. Offered a mix of two sorts of caterpillar to eat, a chick learns the hard way, via instinctive pecking, that one sort is tasty and the other not. When offered the same mix again, it selects only the tasty sort, shrinking away from the nasty sort. Plainly the learning was associative; and though exactly what had happened in the chick's brain when the associations formed was, Morgan reckoned, a task for future science to elucidate, the broad outlines were clear: the stimulus experienced in the chick's consciousness as unpleasant had resulted, via some sort of cortical disturbance, in the inhibiting of the movements that would bring about repetition of the stimulus, while the stimulus experienced as pleasant had resulted, by similar means, in the enhancement of the stimulus-repeating movements. No less clear, Morgan stressed, was that the capacity of the chick's brain to form these associations, and ultimately the habits that grew from them, was as much a part of its hereditary, evolutionarily bequeathed endowment as the instinct to peck. It was thanks to that endowment that at the chick's first encounter with the different larvae, indiscriminate pecking was automatic, but at the second encounter, the sight of the nasty sort summoned up a memory of their unpleasant taste and the chick avoided them. In Morgan's view, such choice-making, guided by experience-informed consciousness in a brain hereditarily and so evolutionarily organized to enable the requisite associations, was the basis on which all of mental evolution was founded (Morgan, 1896, pp. 147–152).

This impressive thoughtfulness about thoughtfulness, as arising where adaptive habits meet adaptive instincts and so where learning meets inheritance and evolution, was the backdrop for Morgan's proposal of his version of organic selection. He set it out in the book's penultimate chapter, entitled "Modification and Variation," and using "modification" to refer to a bodily difference acquired in the course of an animal's experience and "variation" to refer to a bodily difference "of germinal origin" (Morgan, 1896, p. 309). The chapter begins with a review of the Weismannian case for doubt about the Lamarckian inheritance of modifications, and Weismann's own recent attempt, in his Romanes lecture at Oxford in 1894, to suggest how the post-Lamarckian Darwinian, theorizing only with variations, might re-interpret what the Lamarckian holdouts regarded as their best cases. Could the Darwinian convincingly explain how, say, the variations which made deer antlers adaptively heavier happened to coincide with exactly the variations needed in skull, neck, musculature, and so on to support the heavier antlers? In response, Weismann sketched out the possibility that the bodies of individual deer born with slightly heavier antlers grew to accommodate that heaviness, and that this developmental accommodation, recurring in descendant after descendant, sufficed to support the heavier antlers until eventually the supporting anatomy acquired its own variational basis—after which, a deer with even heavier antlers was born, and the whole process repeated, and repeated, with modification assisting variation every step of the way upward (Weismann, 1894; Morgan, 1896, pp. 312–315).

Morgan offered his proposal as an advance on Weismann's, in showing how something like the reverse process could work too, with modification driving the adaptive change, and variation taking the role of assistant. "Modification would lead; variation follow in its wake," as Morgan put it. He supposed that modification-led adaptation would come into its own at times of rapid environmental change, when the need to adapt to new circumstances would come on too fast for new variations to do the adapting work—with the prospect of an even longer delay if, as Morgan suspected, a lineage's long survival under constant conditions actually damped down tendencies to vary. Under the altered circumstances, less plastic races would go extinct, leaving behind the more plastic ones, whose individual members could develop adaptively. Thanks to this developmental plasticity, the race would survive long enough for congenital variations in the same adaptive directions to begin appearing and be selectively preserved. In Morgan's words, "persistent modification through many generations, though not transmitted to the germ, nevertheless affords the opportunity for germinal variation of like nature"— an opportunity for natural selection to build a new instinct, afforded only thanks to innate plasticity which is no less the product of natural selection. The upshot for habit and instinct was that the connection between them could be evolutionarily consequential, but it was "indirect and permissive," not "direct and transmissive" (Morgan, 1896, pp. 315–322, quotations on pp. 319, 322).

The book had started as lectures that Morgan delivered in the United States in the winter of 1895–1896. He presented his ideas on modification-leading, variation-following adaptation at a meeting of the New York Academy of Science, only to find that another speaker at the same meeting, James Mark Baldwin, presented what both recognized as the same proposal (Richards, 1987, pp. 398–399). Baldwin, but not Morgan, called it "organic selection" (Baldwin, 1896). A few months later, Henry Fairfield Osborn, independently of the other two, promulgated the same idea again, later adopting the name "organic selection" (Bristol, 1896; Osborn, 1897). How to account for the convergence? The short answer is roughly the one that Simpson gave: with Lamarckian inheritance on the backfoot, in an evolutionist scientific community whose members still understood the intellectual appeal of Lamarckian inheritance for certain adaptive characters (indeed, in Osborn's case, were still Lamarckians), there was creative ferment in thinking about how apparently Lamarckian adaptation could be Darwinized, and more generally in thinking afresh about how to put together learning, habit, instinct, development, environment, and natural selection. For the longer answer, we are indebted to the historian of science Robert J. Richards, whose authoritative history of the episode supplies the details that bring out features common to other "in the air" convergences. In the case of Morgan and Baldwin, for example, there seems to have been a shared inheritance at work, in their reading of a posthumous volume from the Lamarckism-accepting comparative psychologist George John Romanes, who, without fanfare, described what in retrospect can be identified as the Baldwin effect (Romanes, 1895; Richards, 1987, p. 402, and more generally pp. 398–404, pp. 480–495).[3]

6.5 Natural Selection and the Organ of Educability

If, in *circa* 1900 Baldwinian spirit, we grant that learning in particular and plasticity in general are products of natural selection, no less than instincts are, and furthermore that the direction of evolutionary travel for a lineage under severe adaptive pressure will be from habits to instincts, then what about at larger scales? First in 1899 in a French scientific volume, and then again in 1900 in the pages of *Nature*, the distinguished English comparative anatomist Edwin Ray Lankester argued that, over the Darwinian-evolutionary long run, the trajectory went the other way, with instincts tending to give way to habit or, in Lankester's attractive term, "educability" (see Lester, 1995, pp. 172–173). He began by pointing out, that on the whole, the cerebrums of extinct mammals were much smaller than the cerebrums of their living counterparts, even when their bodies were about the same size, or when the extinct mammals were larger. What was more, that generalization seemed to hold more widely—for the reptiles, for example—and it certainly held for humans and the anthropoid apes compared with fossil pithecoids. Assuming that natural selection lay behind the trend, what, precisely, was adaptive about larger brains? Lankester's answer was that the greater the mass of cerebral tissue, the greater the ability of the individual to respond flexibly to a given environmental situation, instead of having to rely on inflexible, one-size-fits-all instincts.

The advanced position of humans could be assessed in terms of our comparative paucity of instincts. Man, according to Lankester,

> has a greater capacity for "learning" and storing his *individual* experience, so as to take the place of the more *general* inherited brain-mechanisms of lower mammals. Obviously such brain mechanisms as the individual thus develops (habits, judgments, &c.) are of greater value in the struggle for existence than are the less specially-fitted instinctive in-born mechanisms of a race, species or genus. The power of being educated—"educability" as we may term it—is what man possesses in excess as compared with the apes. I think we are justified in forming the hypothesis that it is this "educability" which is the correlative of the increased size of the cerebrum. If this hypothesis be correct—then we may conclude in all classes of Vertebrata and even in many Invertebrata—there is and has been a continual tendency to substitute "educability" for mere inherited brain-mechanisms or instincts, and that this requires increased volume of cerebral substance. [...] The ancient forms with small brains though excellent "automata" had to give place, by natural selection in the struggle for existence, to the gradually increasing brains with their greater power of mental adaptation in the changing and varied conditions life; until in man a creature has been developed which, though differing but little in bodily structure from the monkey, has an amount of cerebral tissue and a capacity for education which indicates an enormous period of gradual development during which, not the general structure, but the organ of "educability," the cerebrum, was almost solely the objective of selection.
>
> (Lankester, 1900, p. 625; emphases in original)

Lankester went on to spell out two consequences of this view. The first was that the concentration of selection on the brain as the organ of educability, especially in humans but in any group where it happened, probably came at the expense of selection on bodily structure, thus putting even more of a survival premium on the ability to adapt to new circumstances by learning. The second was that the old Lamarckian view of instincts as "lapsed intelligence" was even more wrong than hitherto understood; for not only were the results of education un-transmissible biologically, but brain tissue could be devoted to learning only so far as it ceased to be devoted to instinct. Lankester (1900) said, "To the educable animal—the less there is of specialised mechanism transmitted by heredity, the better. The loss of instinct is what permits and necessitates the education of the receptive brain" (p. 625).[4]

Baldwin appreciated the resonances with his own views, as he noted in his 1902 book defending and elaborating organic selection, *Development and Evolution* (Baldwin, 1902, p. 35). As for Morgan, in his 1909 essay he heaped praise on Lankester's linking of behavioural plasticity, natural selection, and cerebral anatomy, declaring him to have thus laid "the biological foundations for a further development of genetic psychology" (Morgan, 1909, p. 441). Later in the essay Morgan indicated something of how that development was already underway.

What was increasingly becoming clear, he wrote, was that there are two orders of educable intelligence, a lower perceptual order and a higher conceptual order. The former is, in fact, connected with instinct. But the latter, involving not just the greater cerebral mass that occupied Lankester but greater surface area through convolutions, is much as Lankester described. "It is through educability of this order," wrote Morgan,

> that the human child is brought intellectually and affectively into touch with the ideal constructions by means of which man has endeavoured, with more or less success, to reach an interpretation of nature, and to guide the course of the further evolution of his race—ideal constructions which form part of man's environment.
>
> (Morgan, 1909, p. 443)[5]

6.6 Natural Selection and the Dependence of Development on Environments

Elementary Mendelism invites no curiosity about how the hereditary and the environmental interact, let alone about how that interaction might develop as an organism develops. In a basic Mendelian cross, the environment is background, either fully under control (in which case you get those lovely patterns) or not fully under control (in which case—grrr—you need to fix it, in both senses, in order to get those lovely patterns), while development is a time-consuming, potentially pattern-wrecking impediment between you and the all-or-nothing unit characters you want to count. If, in asking what was up in biology *circa* 1900, we have only the Mendelian triple to think with, then it is hard not to suppose that the convergence of de Vries, Correns, and Tschermak reflects a very general orientation, and to see Mendelism itself as an inevitability, so perfect was its apparent fit with its times. To be sure, there *were* workers—notably William Bateson, the English zoologist and lead Mendelian—who came through the debates over Lamarckism convinced that the whole question of gradual adaptation to environments, whether Lamarckian or Darwinian, should be shunned in favour of a concentration on the shuffling of unit characters and the shifting of forms between internally stable states. But other workers came through those debates more engaged than ever by the question, and more adventuresome in the Darwinian thinking they brought to answering it, as the Baldwinian triple reminds us. Among their number was Bateson's most formidable opponent in the controversy over Mendelism, W. F. R. Weldon.

In the spring of 1902, shortly after publishing his controversy-sparking critique of Mendel's original paper on crossbred peas, Weldon published a second critique in the same journal (*Biometrika*, founded not long before by Weldon with his mathematical-biological allies Francis Galton and Karl Pearson) of a new book by de Vries on his anti-selectionist "mutation theory." As Baldwin wrote later that year in *Development and Evolution*, de Vries' theory, "which holds that species originate in abrupt or 'sport' variation, called 'mutation,' strikes at the very foundations of the Darwinian conception." Baldwin commended Weldon's "able, negative criticism" (Baldwin, 1902, p. 33), as well as an embryological experiment that Weldon reported in the article. He had made a hole in a hen's egg and then artificially replaced the water that ordinarily evaporates from the egg as the chick develops. The effect, he discovered, was to disrupt or even suppress entirely the normal development of the fluid-filled sac (amnion) around the embryo (Weldon, 1902, pp. 367–368). For Baldwin, what the experiment illustrated was how an eggshell normally functions to protect an environment-within-an-environment, so that an embryonic chick's immediate surroundings approximate to the environment which its free-living ancestors (which the embryo resembles) were adapted to thrive in (Baldwin, 1902, p. 193). For Weldon, the point was a complementary but rather different one. In Weldon's view, the tests that de Vries had conducted and then declared selection to have failed were bad tests, because de Vries had not considered the possibility that when he imposed new conditions on some wheat plants, the changes that ensued were not due to the selective elimination of the less fit individuals, as he had assumed, but instead were due to altered development in the new conditions. So when de Vries re-imposed the old conditions, and the plants took on the old characters, although he took himself to have shown that selection was not capable of producing permanent change, it was entirely possible that the plants had never undergone selection in the first place.

For Weldon, the chick experiment dramatized how easy it can be to underestimate the extent to which normal development depends on normal conditions, and more generally to underestimate—as de Vries had done—the sensitivity of development to environmental changes. The visible form of an organism always had to be understood as the result of the hereditary and the environmental. As Weldon put it,

> Now it cannot be too strongly insisted upon that every character of an animal or of a plant, as we see it, depends upon two sets of conditions; one a set of structural or other conditions inherited by the organism from its ancestors, the other a set of environmental conditions. There is probably no race of plants or of animals which cannot be directly modified, during the life of a single generation, by a suitable change in some group of environmental conditions.
>
> (Weldon, 1902, p. 367)

In support he cited the work of French and German experimental embryologists whose recent research had gone a long way towards showing "that some of the most normal and universal phenomena of animal development are each directly dependent for their occurrence upon a certain group of external conditions" (Weldon, 1902, p. 367). But the conviction that, thanks to natural selection, developing forms were adapted to particular environments, and could change as those environments changed, was an old and deep one for Weldon, going back to his student days. In the 1890s research for which he was best known, demonstrating natural selection at work in modifying the shore crabs of Plymouth, he had actually done the (hard) work that he was now scolding de Vries for not having done—monitoring the growth of huge numbers of crabs in bottles to be sure that the statistical changes he was detecting in wild crabs were due to selective elimination and not developmental convergence. And when, in late summer 1902, he lectured a popular audience on "Inheritance" at the British Association for the Advancement of Science meeting in Belfast, he told them early on about the chick experiment and also experiments done by his student Ernest Warren on the waterflea *Daphnia*, whose spines got shorter and shorter as their water got more polluted, but whose offspring if born in pure water would grow a full-length spine. "Now clearly the condition of the spine [...] is not exclusively acquired; and it is not exclusively inherited. It belongs to both categories" (Radick, 2023, p. 74, pp. 117–120, pp. 312–313, p. 484n32).

That characters are not either acquired or hereditary but always, and complexly, the product of interaction between both categories of cause was something of a motto for Weldon (who blamed Weismann for spreading the misleading idea to the contrary). A version of it appeared in the manuscript "Theory of Inheritance" that he was working on at his death in 1906, and most actively in 1904–1905. There Weldon attempted to set out his alternative to Mendelism—an alternative centred on a conception of character expression not as all-or-nothing, and dependent only on the presence or not of a dominant factor or of two recessive factors, but as variable depending on contexts, from the chromosomal-ancestral to the physico-chemical (Radick, 2023, p. 232, pp. 245–254, p. 466n39).

6.7 Concluding Remarks

With our field of vision of biology *circa* 1900 now expanded beyond the Mendelian triple to include the Baldwinian triple, flanked by the post-Lamarckian debate over mental evolution, the post-Morgan's-canon debate over female choice, Lankester's Darwinian case for educability, and Weldon's Darwinian dissolution of the "acquired character," we depart 1900 with a much richer sense of the options available then for being thoughtfully Darwinian, not least about thoughtfulness, including the riddle of organismal agency. In closing, let's now look forward from 1900 and, returning to Simpson's (1953) essay, notice two ways in which awareness of the potentialities for thoughtful Darwinism also alters our perception of what followed.

The first is to do with that "radically different turn to biological thought" as twentieth-century biology Mendelized. By 1909, as we have seen, Morgan sensed not only the turn but the new job of theoretical work it had generated for biologists as committed to the reality of selection as they were to the reality of the Mendelian gene (so named in that year). The evolutionary biology duly forged in the decades that followed recast selection as principally a matter of changing genotype frequencies, and organisms as principally bundles of gene variants. Ever since, biologists and others wanting to give organisms in their developmental, environmentally situated, choice-making complexity their evolutionary due have been critics, protesting against an orthodoxy that finds it easy to brush them aside (Radick, 2017, p. 56). They should take courage from knowing that the persistent marginalizing of their concerns is an accident of history. Had Weldon lived to complete and publish his synthesis of selection theory with chromosomal physiology, experimental embryology, Galtonian biometry, and data from Mendel-style crossing (illuminating as long one remembered its limitations), the thoughtful Darwinism of Weldon's era—Baldwin effect included—would have framed the advances that came after (see Radick, 2023, esp. pp. 317, 365, 399–400).

My second observation is a counterpart to one I have made elsewhere in relation to the Baldwin effect's belonging to the Modern Synthesis. Recalling the prominence that Julian Huxley gave to that agency-friendly process can help us keep our treatment of organismal agency from lapsing into caricature when discussing the Modern Synthesis (Huxley, 1942; Radick, 2017, p. 56). Likewise, for all that Simpson's essay has a deserved reputation for being downbeat about the importance of the effect he named, his closing section offers a tour-de-force reinterpretation of the genetics of Synthetic Darwinism in terms of reaction norms, by way of highlighting the good that the effect could do in directing research attention that way. "Genetical systems," he begins, "do not directly and rigidly determine the characteristics of organisms but set up reaction ranges within which those characteristics develop" (Simpson, 1953, p. 111). By the lights of *circa* 1900 biology, it is much more Weldonian than Batesonian. And by the lights of 2020s history and philosophy of biology, it is closer to the genetic-determinism-rejecting Extended Evolutionary Synthesis than to conventional ideas of what the Modern Synthesis stood for. Thoughtful Darwinism, it turns out, has unrealized potentialities all over the timeline.

So what? A more accurate image of the biological past is worth having for its own sake, of course. When we enlarge our picture of biology around 1900 to include the Baldwinian triple as an outgrowth of far-reaching debates on environments, heredity, and adaptive change in developing individuals and evolving lineages—debates encompassing everything from the plastic morphologies of waterfleas and wheat, to the emotionally charged nervous systems of feeding and courting birds, to human culture as an endlessly improvable medium for our species' self-creation—we see, as Simpson did, how misleading is the standard emphasis on the Mendelian triple, in which absolutely none of that matters. And whatever the general value of getting the biological past right, the payoff in this particular case may be greater still. An impoverished historical account is impoverishing, leaving us less able to identify the full range of paths for thought and action now, partly because we look to history for guidance on what was and is possible, partly because our world is so much the product of thoughts and actions taken in line with that same narrowed vision of the past. A better account cannot in itself undo the accumulated limitations. But it can fruitfully shine a light on their origins and, in so doing, raise doubts about their taken-for-granted wisdom, in ways that may ultimately suggest new lines of inquiry. It can, in other words, boost agency in the present. For the thoughtful-Darwinian history reconstructed here, no fate could be more fitting.

Notes

1 To quote from the Biology 101 textbook that I used: "It was not until the year 1900 that biology finally caught up with Gregor Mendel" (Campbell, 1993, p. 280).

2 For a reconstruction of the events that led to this shift in rationales, see Radick (2000).

3 Likewise, the Mendelian rediscoverers were all aware of Mendel's paper (Olby, 1990, pp. 528–529), and Wallace and Darwin were both devoted Lyellians (Radick, 2009, p. 154). And just as apparent "sameness" dissolves the closer one looks at the individual writings in the Mendelian and Darwinian convergence cases, so too with the Baldwinian one, with David Depew going so far as to suggest that the term "Baldwin effect" does not pick out a single process (Depew, 2003). The most extensive collection of primary sources relating to the Baldwinian triple is in Baldwin's *Development and Evolution* (Baldwin, 1902, Appendix A).

4 A note on Lankester, acquired characters, and Lamarckism. As discussed in the next section, a trademark *kvetch* of W. F. R. Weldon's was that the term "acquired character" was lousy, since every character was to some extent acquired and to some extent ger-minally based. In a 1912 popular science book, Weldon's second-in-command at Oxford, the Darwinian comparative anatomist E. S. Goodrich, included a passage that sounds just like Weldon on this point (Goodrich, 1912, pp. 32–38, quotation on p. 37). So good is the passage that it appears *verbatim* in an obituary notice of Goodrich by an English Darwinian comparative anatomist of the next generation, Gavin de Beer. However, de Beer celebrated the passage as a brilliant development of insights that Goodrich had picked up not from Weldon but from an earlier mentor, Lankester (de Beer, 1947, p. 484). Why did de Beer think that? My guess is that, reading Goodrich's obituary notice for Lankester, de Beer misinterpreted Goodrich's praise for 1894 *Nature* letters from Lankester on Lamarckism and acquired characters (Goodrich, 1931, p. 379) in the light of the Weldonian passage in Goodrich's book. Be that as it may, anyone who goes back to Lankester's *Nature* letters will find that he came nowhere near to finding fault with the distinction between inherited and acquired characters (Lankester, 1894a & 1894b, affirming the argument of Lankester, 1890, affirmed again in Lankester, 1906, pp. 29–30). Alas de Beer went on to compound his misattribution in the august *Dictionary of Scientific Biography*, where he credited the view that every character is both inherited and acquired, and so the distinction between them meaningless, to Lankester in an article on Lankester himself (de Beer, 1973, p. 27).

5 "Educability" went on to surface here and there in the twentieth century (Poulton, 1937, p. 402; Dobzhansky & Montagu, 1947).

References

Baldwin, J. M. (1896). A New Factor in Evolution. *American Naturalist, 30*, 441–451.

Baldwin, J. M. (1902). *Development and Evolution*. Macmillan.

Boakes, R. (1984). *From Darwin to Behaviourism: Psychology and the Minds of Animals*. Cambridge University Press.

Bristol, C. L. (1896). [Report from the 9 March Meeting of the Biological Section of the] New York Academy of Sciences. *Science, 3*, 529–530.

Campbell, N. (1993). *Biology* (3rd ed.). Benjamin/Cummings.

Cronin, H. (1991). *The Ant and the Peacock: Altruism and Sexual Selection from Darwin to Today*. Cambridge University Press.

Darwin, C. (1859). *On the Origin of Species*. John Murray.

Darwin, C. (1871). *The Descent of Man, and Selection in Relation to Sex*. John Murray.

Darwin, C. (1872). *The Expression of the Emotions in Man and Animals*. John Murray.

de Beer, G. (1947). Edwin Stephen Goodrich 1868–1946. *Biographical Memoirs of Fellows of the Royal Society, 15*, 477–490.

de Beer, G. (1973). Lankester, Edwin Ray. *Dictionary of Scientific Biography, 8*, 26–27.

Depew, D. J. (2003). Baldwin and His Many Effects. In B. H. Weber & D. J. Depew (Eds.), *Evolution and Learning: The Baldwin Effect Reconsidered* (pp. 3–31). MIT Press.

Dobzhansky, T., & Montagu, M. F. A. (1947). Natural Selection and the Mental Capacities of Mankind. *Science, 105*, 587–590.

Erskine, F. (1995). The *Origin of Species* and the Science of Female Inferiority. In D. Amigoni & J. Wallace (Eds.), *Charles Darwin's The Origin of Species: New Interdisciplinary Essays* (pp. 95–121). Manchester University Press.

Gamble, E. B. (1894). *The Evolution of Woman: An Inquiry into the Dogma of Her Inferiority to Man*. G. P. Putnam's.

Goodrich, E. S. (1912). *The Evolution of Living Organisms*. T.C. & E.C. Jack.

Goodrich, E. S. (1931). The Scientific Work of Edwin Ray Lankester. *Quarterly Journal of Microscopical Science, 74*, 363–381.

Hamlin, K. A. (2014). *From Eve to Evolution: Darwin, Science, and Women's Rights in Gilded Age America*. University of Chicago Press.

Huxley, J. (1942). *Evolution: The Modern Synthesis*. George Allen & Unwin.

Kroeber, A. L. (1917). The Superorganic. *American Anthropologist, n.s. 19*, 163–213.

Lankester, E. R. (1890). The Inheritance of Acquired Characters. *Nature, 41*, 415–416.

Lankester, E. R. (1894a). Acquired Characters. *Nature, 51*, 54.

Lankester, E. R. (1894b). Acquired Characters. *Nature, 51*, 102–103.

Lankester, E. R. (1900). The Significance of the Increased Size of the Cerebrum in Recent as Compared with Extinct Mammalia. *Nature, 61*, 624–625.

Lankester, E. R. (1906). *President's Address. Report of the BAAS* (pp. 3–42). John Murray.

Lester, J. (1995). *E. Ray Lankester and the Making of Modern British Biology*. P. J. Bowler (Ed.), *BSHS Monographs, 9*.

Milam, E. L. (2010). *Looking for a Few Good Males: Female Choice in Evolutionary Biology*. Johns Hopkins University Press.

Morgan, C. L. (1894). *An Introduction to Comparative Psychology*. Walter Scott.

Morgan, C. L. (1896). *Habit and Instinct*. Edward Arnold

Morgan, C. L. (1909). Mental Factors in Evolution. In A. C. Seward (Ed.), *Darwin and Modern Science* (pp. 424–445). Cambridge University Press.

Olby, R. C. (1990). The Emergence of Genetics. In R. C. Olby, G. N. Cantor, J. R. R. Christie, & M. J. S. Hodge (Eds.), *Companion to the History of Modern Science* (pp. 521–536). Routledge.

Osborn, H. F. (1897). Organic Selection. *Science, 4*, 583–587.

Poulton, E. B. (1937). The History of Evolutionary Thought at the BAAS. *Nature, 140*, 395–407.

Radick, G. (2000). Morgan's Canon, Garner's Phonograph, and the Evolutionary Origins of Language and Reason. *British Journal for the History of Science, 33*, 3–23.

Radick, G. (2009). Is the Theory of Natural Selection Independent of its History? In M. J. S. Hodge & G. Radick (Eds.), *The Cambridge Companion to Darwin* (pp. 147–172). Cambridge University Press.

Radick, G. (2017). Animal Agency in the Age of the Modern Synthesis: W. H. Thorpe's Example. *BJHS Themes, 2*, 35–56.

Radick, G. (2023). *Disputed Inheritance: The Battle over Mendel and the Future of Biology.* University of Chicago Press.

Richards, E. (2017). *Darwin and the Making of Sexual Selection.* University of Chicago Press.

Richards, R. J. (1987). *Darwin and the Emergence of Evolutionary Theories of Mind and Behavior.* University of Chicago Press.

Romanes, G. J. (1895). *Darwin and After Darwin. Vol. 2: Post-Darwinian Questions— Heredity and Utility.* C. L. Morgan (Ed.). Open Court.

Simpson, G. G. (1953). The Baldwin Effect. *Evolution, 7,* 110–117.

Wallace, A. R. (1889). *Darwinism: An Exposition of the Theory of Natural Selection, with Some of its Applications.* Cambridge University Press.

Weismann, A. (1894). *The Romanes Lecture, 1894: The Effects of External Influences Upon Development.* Clarendon Press.

Weldon, W. F. R. (1902). Professor de Vries on the Origin of Species. *Biometrika, 1,* 365–372.

7 The Higher-Order Norm of Reaction

Biological Agency and Adaptive Phenotypic
Response

Denis M. Walsh and Sonia E. Sultan

7.1 Introduction

In recent years, the prevailing gene-centered model of evolution has encountered acute challenges (Pigliucci & Müller, 2010; Laland et al., 2014; Huneman & Walsh, 2017). One such challenge, the agency perspective, charges that this framework misidentifies the canonical unit of adaptive evolutionary explanation. Rather than explaining the dynamics of evolution by appeal to the capacities of genes, we should think of evolution as a consequence of how organisms work and what they do. In that light, the evolutionary process reflects a spiralling feedback between organismic properties as its products and its subsequent causes. In making this explanatory shift, organisms are construed as natural purposive agents: entities that build themselves; respond to environmental, genetic, and developmental perturbations; and actively regulate and orchestrate their constituent parts and processes (Walsh, 2015; Sultan et al., 2022; Nadolski & Moczek, 2023; Uller, 2023).

What difference might an agency perspective make to our understanding of evolution and our approaches to studying it? Here we attempt to make the case for a specific empirical and explanatory difference. We canvass a familiar biological phenomenon—the norm of reaction—as a test case for the agency perspective. We argue that the agency perspective is required in order to properly interpret the norm of reaction and to account for its role in evolutionary dynamics.[1]

The norm of reaction is a common way of conceptualizing organism–environment relations (e.g., Stearns, 1989; Scheiner, 1993; Sarkar & Fuller, 2003; Sultan & Stearns, 2005; Sultan, 2007). At its most simple, it is an empirically derived description of the way that phenotype varies as a function of genes and environment. As such, it is used to illustrate and explain a variety of biological phenomena: the environmental sensitivity of development, the causal interactions between genes and environment, the often (but not always) adaptive developmental plasticity of organisms (Sultan, 2021; see also Gilbert & Epel, 2015; Sultan, 2015; Pfennig, 2021 and references therein). The norm of reaction is usually depicted by means of a two-dimensional graph, with the value of a specified aspect of the phenotype as the ordinate (Y axis) and the value of some environmental parameter along the abscissa (X axis). Lines plotted on the graph represent the way in which, for a given genotype, phenotype varies as a function of environment. Different

DOI: 10.4324/9781003413318-9

lines on the same graph depict the ways in which genotypic and environmental variations are reflected in phenotypic differences. But representation is one thing; interpretation is another. How should we interpret the norm of reaction?

It is usually interpreted as a property of a genotype. In an attenuated but trivial sense, this is clearly true, inasmuch as a norm of reaction graph plots phenotype as a function of environment (*E*) *for a given genotype* (*G*). But if the norm of reaction is to be explanatory, we should further ask what *accounts for* the fact that, for a given genotype, the phenotype varies as a function of environment. We argue that a proper understanding of the explanatory role of the norm of reaction requires an agential interpretation. Here, then, is our exploratory test case for the difference made by the agency perspective. We proceed in the following way. We briefly outline the gene-centered and agential conceptions of evolution. They differ in many ways, but two of these differences will be salient for our purposes. One difference is to be found in their respective conceptions of organism–environment relations. The other crucial difference is to be found in the kind of explanation each deploys in accounting for adaptively plastic responses to environmental conditions. After differentiating the gene-centered and agential views on evolution, we survey some recent empirical work on the norm of reaction, including cases where individuals that encounter some kind of environmental challenge are capable of transmitting to their offspring a functionally adaptive capacity to respond to *their* environments.

We argue that the phenomenon of transgenerationally adaptive response has two implications. The first is that the norm of reaction must be a property of an adaptive individual, an organismal agent.[2] The second is that the individual possesses a *higher-order* norm of reaction—a suite of possible norms of reaction rather than a single, genetically pre-determined one. The capacity of an organism to, effectively, transmit an adaptive norm of reaction to its offspring from among these many possibilities makes a difference to evolutionary dynamics. This difference cannot be adequately characterized without citing the *adaptiveness* of the norm of reaction. We argue that whenever the *adaptiveness* of an organism is invoked to explain the dynamics of evolution, one is implicitly invoking agency.

7.2 Agential Evolutionary Biology

The central tenet of contemporary gene-centered evolutionary biology is that evolution is a consequence of *the capacities of genes*. For evolution to happen there must be novelty (variation), development, inheritance, and adaptively biased change over generations. Genes are understood to be intimately involved in each of these component processes of evolution. Genes are units of evolutionary novelty in that ultimately new evolutionary traits arise through their random mutation and recombination. They are units of phenotypic control in that they comprise a coded program for the production of phenotypes. They are units of phenotypic inheritance as the materials copied and passed to offspring. Furthermore, genes are units of adaptive evolutionary change inasmuch as they are differentially retained in a population according to the aptness of the phenotypes they build for the environments they are in. When genes participate in novelty, development, inheritance, and biased change, the result is adaptive evolution.[3]

According to the agential perspective, by contrast, evolution occurs as a consequence of *the activities of organisms*. Organisms are—distinctively—self-building, self-regulating, adaptive systems. They synthesize the materials out of which they are constructed; they organize and integrate the collective workings of their parts, regulate their genes, and construct their environments, in pursuit of the functional goals that constitute their way of life. On this view, organisms and their constituent processes are agents, naturally purposive systems. The guiding insight of the agential perspective is that through all this self-sustainingly purposive activity organisms enact evolution (Thompson, 2007; Walsh, 2012, 2023). By "enacting" evolution we mean that individual organisms have the capacity to substantially adapt themselves to their conditions of existence, by marshaling their genetic, extragenetic, environmental, and behavioral resources via evolved systems of regulation and response. As Richard Lewontin put it, "the organism cannot be regarded as simply the passive object of autonomous internal and external forces; it is also the subject of its own evolution" (Lewontin, 1985, p. 89). It is the purposiveness of organisms that shapes the component processes of evolution—novelty, development, inheritance, and adaptive bias—and it further integrates these processes in a way that realizes adaptive evolution. The adaptive, purposive coping, adjusting, and innovating that is the very nature of organisms is also the driver of adaptive evolution (Walsh & Rupik, 2023).

Any evolutionary biology seeks to explain the fit and diversity of organisms. One prominent feature of agential biology is the license it provides biologists to explain fit and diversity by appeal to the adaptive purposiveness of organisms. Agents have repertoires of alternate possible responses to their circumstances. When an agent responds adaptively to its conditions it does so by implementing one element of its repertoire rather than any of the other possible responses. The agency perspective allows one to appeal to the fact that the response is adaptive to explain why it, rather than any of the possible alternatives, occurred. This feature of agential explanation, it will turn out, is crucial for understanding the role of the norm of reaction in evolution.

7.3 Interpreting the Norm of Reaction

The norm of reaction graph depicts the way that phenotype varies as a function of environment for a given genotype. The general point is that both genotypes and environments affect phenotypes; a population's total phenotypic variance can be attributed to variation due to different genotypes, G, and variation due to different environments, E, with the non-additive portion of these effects expressed as a statistical interaction between genes and environment ($G \times E$).

On its introduction by Woltereck in 1909, the norm of reaction (*Reaktionsnorm*) was posited as a general feature of organisms: the inherited capacity of an organism to mount a multifarious plastic response to its environmental circumstances (Sarkar, 2004).[4] In the intervening century, with the acceptance of the late "Modern

Synthesis" view of strictly gene-based variation as the substrate for evolution, it has become commonplace to interpret the norm of reaction more narrowly as a deterministic "property of a genotype" (Nager et al., 2000); "the set of phenotypes produced by the gene in different environments" (Maynard Smith, 1994, p. 95), that is, as "an environmental response program in the genes" (de Jong, 1999). On this interpretation, the norm of reaction arises as a genetically determined outcome of prior selection (usually in quantitative genetics terms, as well explained in Chevin et al., [2013]). As such, the norm of reaction graph represents the capacity of a genotype to produce different phenotypes across a specified range of environments. In view of the recent recognition that gene expression is inherently context-dependent (Wray et al., 2014), perhaps the "phenotypic repertoire" of a genotype is an apt phrase (Sultan, 2019).

One important role for the norm of reaction is in the explanation of adaptive phenotypic responses—that is to say, when the norm consists of phenotypes that are appropriate to the environmental states that elicit them. In the simplest case, such plasticity occurs when a given genetic individual produces alternative, specifically adaptive phenotypes in response to contrasting developmental conditions; familiar cases include animals and plants that produce metabolically costly defense structures and chemicals only in the presence of predators, allocate proportionately more tissue to resource-collecting structures in nutrient-poor conditions, or alter life-cycle transitions to match the timing of favorable conditions (see Gilbert & Epel, 2015; Sultan, 2015; Pfennig, 2021 and references therein). Such adaptive phenotypic response patterns are not ubiquitous (Scheiner, 1993), but where they occur they require an explanation. Henceforth, we focus on adaptive norms of reaction.

Viewing the norm of reaction as a property of genotypes—i.e., "an environmental response program in the genes"—justifies the key operational step of allowing the analytical and predictive tools of conventional evolutionary theory to be extended to individual plasticity (DeWitt & Scheiner, 2004; e.g., Via & Lande, 1985; Gomulkiewicz & Kirkpatrick, 1992; Gavrilets & Scheiner, 1993). Sophisticated applications of such theory are being increasingly drawn upon to examine evolution of plasticity as a potential source of adaptive rescue for plants and animals confronting climate warming and other environmental challenges (e.g., Chevin et al., 2010, 2013). Although in some cases key phenotypic response patterns may be broadly consistent for particular genotypes and hence could be treated as such for practical purposes, the urgent need to incorporate realized phenotypic outcomes into our understanding of the organism–environment relation suggests that the time is right to more fully interrogate norm of reaction causation.

The agency approach offers a distinctive interpretation of the norm of reaction. On this view, it is a property not of the genotype but rather of the organism as agent. The adaptive phenotypic variation between individuals across different environments is a manifestation of the purposive engagement of organisms with their environments in drawing on and guiding their systemic developmental (including genetic, epigenetic, behavioral) processes.

One question, then, is whether the norm of reaction is better suited to explaining occurrences of adaptive phenotypic response when considered as a property of a genotype or a property of an agential organism. Here it seems that *neither* the traditional gene-centered approach nor the agency perspective enjoys a particular explanatory advantage. The traditional gene-centered approach explains these adaptive norms of reaction as a manifestation of selection in the past on genes that produce the appropriate phenotype for a given environment. In this view, the norm of reaction is a response of a genotype to alternative environments that has been selected in the past. In this respect, adaptive plasticity is no different from any other evolutionary adaptation. According to the agency interpretation, on the other hand, the norm of reaction explains adaptive plastic response as a manifestation of the capacity of organisms to produce a particular phenotype *because it is adaptively advantageous*, by drawing on their evolved systems of regulatory adjustment in real time. Both approaches appear to explain the adaptive responses of organisms to their developmental environments equally well, so it seems that neither interpretation of the norm of reaction claims any particular explanatory advantage over the other.

However, the agential interpretation of the norm of reaction has two distinct advantages, to be developed below (cf. Potter & Mitchell, this volume). The first is that it is particularly congenial to the emerging understanding of the complex relations between genes, organisms, and environments. The second, and more definitive, advantage is that it offers a more satisfactory explanation of transgenerational adaptive plasticity.

7.4 Organism and Environment

A significant point of departure between gene-centered and agential conceptions of evolution can be seen in their respective conceptions of the relation between organism and environment. Because the norm of reaction is essentially a phenomenon of this fundamental relation, these divergent conceptions are of critical importance. According to gene-centered evolutionary thinking, the organism is essentially the passive interface between the internal workings of genes and the external environment. Organisms are built by genes and selected by external environments that determine whether they succeed or fail. This separation of internal developmental causes from external "selection pressures" frames the environment as distinct from—and largely independent of—the organism, despite their evident mutual influence (see Lewontin, 1978, 1985; Odling-Smee et al., 2003; Laland et al., 2014; Sultan, 2015). The gene-centered view can only account for the norm of reaction by ascribing the environment's influence on the phenotype to the genes (de Jong, 1999; DeWitt & Scheiner, 2004).[5]

However, this reassignment fails to capture the nature of development as the joint and inseparable outcome of how a given set of genes are expressed in a particular environment:

> Developmental outcomes are shaped by multiple types of information and not by DNA sequence alone [...]: phenotypes emerge from the real-time regulatory interactions of the evolved genotype with the transient environmental and/

or [environmentally induced or stochastic] epigenetic influences that occur at timescales from within a generation to several or many generations. These factors cannot be pulled apart as individual causes, because they contribute interactively to phenotypic expression [...].

(Sultan, 2019, p. 117)

Recognizing the inherently context-dependent, responsive nature of development calls for replacing the organism as passively determined interface with the organism as evolutionary agent (Walsh, 2015; Nadolski & Moczek, 2023). Organisms marshal their phenotypic resources—including genetic, epigenetic, microbial, behavioral, and environmental—in ways that respond to and mitigate the consequences of deleterious conditions and exploit beneficial ones.[6]

An agential perspective also involves recognizing the reciprocal effects of the organism's body and behavior on its environment as evolutionary factors. This is not a new point; indeed, it is a venerable one: "The organism is thus no more determined by the surroundings than it at the same time determines them. The two stand to one another, not in the relation of cause and effect, but in that of reciprocity" (Haldane, 1884, pp. 32–33, as cited in Baedke et al., 2021, p. 4). But this recognition is only beginning to find a comprehensive revival in the agential view of evolution (see also Odling-Smee et al., 2003; Laland et al., 2019).

Organisms respond to their environments, including effecting changes to their environments which in turn alter the way that genes affect phenotypes. Environments thus have an influence on the effects that genes (via organisms) have on their environments, and genes have an influence (via organisms) on the effect that environments have on gene action. The expression of genes involves "intricate networks of cause and effect that are mediated by an organism's physiology, behavior, and interactions with the environment" (Rockman, 2008, pp. 738–744). The causal relations between genes, organisms, and environments are complex and imbricating, a "triple helix" (Lewontin, 2000).

One implication of this causal intertwining of gene, organism, and environment is that one cannot decompose differences in phenotype into that portion exclusively caused by genes and that caused by environments. The prevailing interpretation of the norm of reaction—as a capacity of genes to produce different phenotypes in different environments—is predicated on the supposition that the principal causes of phenotypic variation across environments can be located in the DNA sequence. Because this decomposition cannot be made, we cannot interpret the environmentally contingent phenotypic response pattern as an "environmental response program in the genes." In short, the interpretation of the norm of reaction as a property of the genotype is incapable of explaining the complex way in which genes, organisms, and environments jointly contribute to variable phenotypes.

7.5 Transgenerational Plasticity and the Higher-Order Norm of Reaction

Alongside the well-known adaptive responses of organisms to their environments, it is now well documented that adaptive responses to environmental conditions can also be transmitted to offspring (Mousseau & Fox, 1998; Uller, 2008; Donohue, 2009; Herman & Sultan, 2011, Bonduriansky, 2021): "[…] Individuals in a number of plant and animal taxa have the ability to adaptively alter their offspring's development in response to environmental stresses, such that the offspring show increased tolerance to the stress in question" (Herman et al., 2012, p. 78).

Transgenerationally robust adaptive responses have been observed in a wide range of organisms (Salinas et al., 2013). In one set of experiments with a common annual plant, Sultan and colleagues subjected genetically uniform lineages to naturalistic environmental stresses including drought, shade, and neighbor competition. These developmental conditions are known in this and many other plant species to elicit adaptive norms of reaction that mitigate these stresses, such as increased investment in root tissue by plants grown in dry soil and in leaf tissue by shaded individuals.[7] Remarkably, these experiments showed that adaptive responses by resource-stressed individuals were transmitted to their offspring and in some cases their grand-offspring, resulting in the expression of adaptively enhanced phenotypes by seedlings developing under similar stress conditions. For example, plants grown in dry soil developed relatively larger, thinner root systems; seedlings whose parents had been grown in dry soil developed even larger root systems, and seedlings whose parents and grandparents had both experienced drought stress produced cumulatively enlarged root systems that enabled them to survive significantly longer in dry soil than genetically identical seedlings lacking drought exposure in previous generations.

Adaptive transgenerational effects on offspring phenotypes have also been extensively documented in animal species. Many studies show that maternal encounters with predators result in offspring with predator avoidance traits, such as longer wings in birds (great tits) and greater dispersal in *Zootoca* lizards, or protective behaviors, such as tighter shoaling in stickleback fish (Bestion, 2014 and references therein). Miller et al. (2012) found that when parental anemonefish had been exposed to elevated levels of dissolved carbon dioxide, their juvenile offspring developed normally in this ordinarily damaging environmental treatment; in another coral reef fish (*Acanthochromis polyacanthus*), Ryu et al. (2018) found that acclimation to elevated temperature was transmitted to progeny, showing the adaptive potential for such transgenerational responses to predicted climate change. Similarly, exposure of grandparental and parental tropical damselfish to higher water temperature resulted in physiologically adapted juveniles with improved aerobic capacity (Bernal et al., 2022). In the well-studied nematode *Caenorhabditis elegans*, adaptive transgenerational plasticity has been documented in response to resource availability, osmotic stress, and pathogens (Baugh & Day, 2020 and references therein). For example, learned avoidance of a bacterial pathogen by this tiny animal is epigenetically transmitted, resulting in gene expression changes in sensory neurons associated with avoidance behavior that provides a survival advantage across four generations (Moore et al., 2019).

How are these transgenerational environmental effects on progeny and descendent phenotypes brought about? In some cases, they result from the maternal and/or paternal individual's epigenetic modulation of the expression of its genes, for instance, through the activity of methyltransferases that add or remove DNA methyl groups (e.g., Herman & Sultan, 2016; Baker et al., 2018; reviewed by Quadrana & Colot, 2016; Perez & Lehner, 2019). In other cases, parental conditions interact with complex physiological or hormonal pathways in ways that alter developmental conditions for embryos within a seed or egg or, for mammals, *in utero*. Such systemic adjustments underlie the transmission of adaptive dispersal and aggression traits in western bluebirds in response to maternal social and ecological environments (Potticary & Duckworth, 2020). In wild populations of North American red squirrels, exposing mothers to territorial calls indicating competitive density led to a stress hormone change that resulted in offspring with significantly higher growth rates, which are strongly related to fitness in dense populations due to greater probability of surviving the first winter (Dantzer et al., 2013). The age of a parent can lead to adaptive life-history changes to progeny, as in freshwater shrimp of the genus *Daphnia*. In *Daphnia*, as well as many insects and birds, clutches produced later in a mother or grandmother's lifetime consisted of larger neonates with an accelerated reproductive schedule, counteracting the expected effects of maternal age on progeny fitness but resulting in shorter offspring lifetimes (Plaistow et al., 2015 and references therein). In general, then, parent organisms that adaptively adjust their offspring do so by drawing on the same epigenetic, physiological, and developmental systems that are notably flexible and self-regulating during their own lifetimes.

The examples above illustrate possible ways that the environment during a parent's or grandparent's lifetime can result in changes to the phenotypes expressed by its offspring. In other words, the offspring's norm of reaction will be different if its parent or further ancestors encountered one environment rather than another: offspring and grand-offspring of drought-stressed plants respond differently to a range of soil moisture conditions than do offspring and grand-offspring of non-stressed plants. Juvenile anemonefish respond differently to a range of CO_2 levels depending on whether their parents experienced high concentrations.

The standard interpretation of the norm of reaction as a consequence of the developmental interaction of an individual's genes and its environment (e.g., Haldane, 1946; Gupta & Lewontin, 1982; Scheiner, 1993) does not take into account the influence of previous-generation environments on an organism's phenotypic responses. In effect, such transgenerational effects demonstrate that there is no such thing as *the* norm of reaction for a genotype. Whether inherited influences of previous environments are adaptive or maladaptive (Uller et al., 2013, Heckwolf et al., 2018), each $G \times E$ relation must be conditioned on previous environments. Studies implementing alternative environments across several successive generations reveal that progeny phenotypes are influenced by the precise number, sequence, and combination of environments (e.g., Herman et al., 2012; Alvarez et al., 2021; see also Sultan, 2019). Accordingly, for a given genotype there may be any number of norms of reaction depending on the precise maternal and/or paternal conditions during one or more previous generations. Once this more complex basis of an individual's phenotypic response to its environment is recognized, it becomes clear that the norm of reaction cannot be understood as a genotypically determinate response pattern. Across the range of current environments, the response of genotypes to current environments, $G \times E_{\text{current}}$, is affected by the parental and even grandparental environments.[8] The norm of reaction, then, should be represented by $[(G \times E_{\text{current}}) \times E_{\text{parental}}]$ or $[(G \times E_{\text{current}}) \times E_{\text{parental}} \times E_{\text{grandparental}}]$, and so on (Sultan, 2019). This transgenerational adaptive plasticity has two general implications for the interpretation of the norm of reaction, and consequently the explanation of adaptive plasticity, that appear to weigh strongly in favor of the agential view. They are as follows: (1) the norm of reaction is a property not of genotypes but of organisms embedded in their current and ancestral environments; and (2) transgenerational adaptive plasticity requires an agential interpretation. We take these in turn.

7.5.1 A Property of Organisms

We argued above that the norm of reaction should be considered to be a property of an individual organism. Transgenerational adaptive responses offer a further, definitive reason why. One major implication of the transgenerational adaptive response experiments is that norm of reaction is *underdetermined by genotype*. For a given genotype, there may be any number of norms of reaction. There is, for example, the norm of reaction manifested in the response to variable moisture conditions during the individual's lifetime ($G \times E_{\text{current}}$). There is, further, a norm of reaction expressed by offspring of parents who encounter drought-stress environments, denote these as $E_{\text{parent},d}$. So individuals in the current environment whose parents were drought stressed may have the norm of reaction $[(G \times E_{\text{current}}) \times E_{\text{parent},d}]$. A different norm of reaction is expressed by offspring whose parents encounter moist, non-stressful soil conditions, $E_{\text{parent},m}$—$[(G \times E_{\text{current}}) \times E_{\text{parent},m}]$—or plants exposed to the stress of anoxic soil due to flooding (denoted by "f") instead $E_{\text{parent},f}$—$[(G \times E_{\text{current}}) \times E_{\text{parent},f}]$. Moreover, there are the norms of reaction expressed by the grand-offspring of individuals with drought-stressed parents and grandparents—$[(G \times E_{\text{current}}) \times E_{\text{parent},d} \times E_{\text{grandparent},d}]$—which differ from the norm of reaction of individuals with flood-stressed parents and grandparents $[(G \times E_{\text{current}}) \times E_{\text{parent},f} \times E_{\text{grandparent},f}]$. These norms of reaction explain the different phenotypic responses of genetically identical individuals to a range of environments encountered over several generations.

It appears, then, that there is no such thing as *the* norm of reaction of a genotype. Depending on its ancestral environmental (and consequently epigenetic) history, a genotype has access to a latent suite of norms of reaction: *a higher-order norm of reaction*.[9] Moreover, an individual of a given genotype and environmental history will respond differently to a range of conditions for one environmental factor, such as soil moisture, depending on its previous and current conditions for *other* environmental factors—for instance, light and nutrient availability, air temperature, microbial symbionts, and competition from neighbors—further expanding the dimensionality of its norm of reaction. We note that this concept is closer to Woltereck's original recognition of an organism's multifaceted repertoire of contingent responses, "[…] the complete reaction norm with all its innumerable specific relations" (Woltereck, as cited in Stearns, 1989, p. 438).

An individual's realized phenotypic responses will reflect its particular multi-generation environmental history as well as its highly multifactorial environmental conditions. This idiosyncratic causal complex suggests that the norm of reaction, and the higher-order norm of reaction underlying it, can be understood as a property of an individual organism with its particular history and circumstances, rather than as a constitutive property of a genotype. This idiosyncratic complexity inheres uniquely in the individual organism, suggesting that the norm of reaction might more accurately be captured by the complex, imbricated interaction of gene, organism, and environment we discussed above. In this model, the organism with its self-regulating epigenetic, developmental, physiological, and behavioral capacities integrates the engagement of the current environment with pathways of gene expression. These systemic capacities are all determinants of an organism's norm of reaction and of the specific modifications to the norm of reaction that a parent passes to its offspring.

The agential view nicely accommodates the phenomenon of transgenerational environmental inheritance. What an offspring inherits, on the agential view, is not simply a genome. Rather, the offspring inherits an entire suite of developmental resources—genetic, epigenetic, cytoplasmic, environmental—required to produce an organism. The higher-order norm of reaction is simply this suite of organismal resources for response. The actual, realized phenotype reflects the fact that the organism draws upon one disposition out of many, partly in consequence of the transgenerational disposition drawn on by its parents; the organism as agent has the capacity to exploit and integrate these resources in the production of a viable individual, more or less aptly suited to its conditions of existence.

7.5.2 *Explaining Transgenerational Adaptive Responses*

The agential interpretation of the higher-order norm of reaction suggests for it a role in the dynamics of adaptive evolution that is not countenanced by gene-centered evolutionary biology. In cases of adaptive transgenerational effects in response to parental conditions, an organism influences its offspring's norm of reaction in ways that will be adaptively advantageous if the offspring encounter a similar environment. For example, the parent *Polygonum* plant responds adaptively to drought stress by altering its phenotype in response to the immediate conditions and by specifically altering the norms of reaction of its offspring such that, in dry soil, its seedlings develop drought-adaptive phenotypes by rapidly producing especially deep, extensive root systems, yet if instead they encounter moist soil, they produce a phenotype without this costly allocational shift (Sultan et al., 2009). It is not the fact of transgenerational environmental effects that has been so surprising, but rather their often specifically adaptive nature. In this case, the environmentally conditioned parental influence disposes the offspring to respond appropriately to their soil moisture conditions. This account of how the norm of reaction figures in transgenerational adaptive responses appears to have the form of an agential explanation. It explains an outcome, the transmission of influence from parent to offspring, by appeal to the fact that it is adaptive.

One tempting rejoinder on behalf of gene-centered evolutionary biology is that agency is not required to explain the capacity of organisms to adaptively influence their offspring's phenotype (e.g., Haig, 2011). Mutations that, for example, enabled a plant in a dry soil environment to implement and transmit to its offspring a facultative capacity for enhanced response to drought would surely be selectively advantageous and would lead to the differential retention in the population of that organism's genes. Further mutations that permitted these individuals also to implement and transmit, say, an adaptive norm of reaction in response to flooded soil, or to an entirely different environmental variable, would be additionally selectively advantageous. In this way, the full capacity to transgenerationally match functionally appropriate norms of reaction could evolve through iterated episodes of mutation and selection (depending on spatial and temporal environmental distributions, fitness trade-offs, and other factors in selective evolution).

We do not dispute that selective evolution can account for complex, self-regulating developmental and transgenerational response capacities that have adaptive consequences for individuals and their progeny, including those underlying the higher-order norm of reaction. Indeed, we take Sewall Wright's point that "individual adaptability [...] is itself perhaps the chief object of selection" (Wright, 1931, p. 147). The agential interpretation may not be required to account for the origin of the higher-order norm of reaction, but it is required to account for the difference it makes to subsequent evolutionary dynamics via its effects on offspring fitness. In order for offspring to express an adaptive norm of reaction, that norm of reaction must be actuated by the parent from amongst an array of possible alternative norms of reaction that might have been implemented. If the adaptiveness of the higher-order norm of reaction makes any difference to the *tempo* and *mode* of evolution, then its *adaptiveness* will be required to explain that difference. The point generalizes. If the adaptiveness of the adaptive responses of organisms makes a difference to evolution, then there is an indispensable role for agential explanation in evolution (cf. Potter & Mitchell, this volume).

Of course, in other cases an organism's response to its own environment, or the influence on response that it sends to its offspring, is maladaptive rather than adaptive. Maladaptive phenotypes are explained in selective terms by appeal to genetic or other constraints on the evolution of adaptive phenotypes. In the case of maladaptive plasticity within or across generations, one might evoke disrupted environmental cues as well as similar constraints on the extent of response capacities to environmental challenges. Adaptations are neither perfect nor unlimited. Our question is how we can most attentively interpret the distinctive features and capacities of organisms.

An adequate understanding of the higher-order norm of reaction requires that we interpret it as a complex capacity of individual organisms, grounded in their adaptive agency. It consists in the ability of organisms to respond to conditions of existence. This capacity is grounded in the complete suite of an organism's genetic, epigenetic, developmental, behavioral, and environmental resources. Both the organism's immediate phenotypic response and its adaptive influences on offspring are agential capacities. Only agents have the capacity to implement one element of a repertoire preferentially over other elements on the grounds that under the circumstances the actuated response is advantageous. We explain why one element of the organismal repertoire was actuated rather than some other, by appeal to the fact that it is adaptive. Consequently, the higher-order norm of reaction, together with the phenomenon of adaptive transgenerational plasticity, points to a view of the norm of reaction as a manifestation of organismal agency.

7.6 Conclusion

Gene-centered evolutionary biology and agential evolutionary biology differ crucially on the explanatory role that each accords to organismal adaptiveness. According to gene-centered evolutionary biology, evolution happens because of what genes do. Along these lines, the adaptive purposiveness of organisms is at best a phenomenon to be explained. According to the agential perspective, organisms are goal-directed systems. They have the capacity to respond to their circumstances in adaptively biased ways. As such, organisms are the paradigm cases of natural agents. The agential perspective holds that this adaptive, intrinsic purposiveness of organisms is centrally implicated in evolution. In order to understand the dynamics of evolution we must understand the contributory role played by agency. It should be expected, then, that evolutionary dynamics look very different from the gene-centered and organismal perspectives.

We have used the norm of reaction as a test case to contrast these two approaches. On either approach we should expect variation in both genes and environments to contribute to variations in phenotype. The phenomenon of transgenerational adaptive response, however, illuminates a difference between the two approaches. We have argued that transgenerational adaptive responses require the concept of a "higher-order norm of reaction." The higher-order norm of reaction is a property of an organism that consists in two organismal capacities. The first is the capacity of the organism to respond to its conditions of existence. This capacity is grounded in the complete suite of an organism's genetic, epigenetic, developmental, behavioral, and environmental resources. The second is the capacity of an organism to bias the norm of reaction of its offspring. These are agential capacities because in each of the cases, the organism typically possesses a range of possible responses to its circumstances. We explain why one element of the organismal repertoire was actuated rather than some other, by appeal to the fact that it is *adaptive*.

The higher-order norm of reaction is simply not explained by its standard interpretation as a self-contained genotypic program for environmental response. Here, then, is an empirical difference between the gene-centered and agential approaches to understanding evolution, a difference that weighs in favor of the agential approach.

Biologists are increasingly inclined to invoke the adaptive bias, the responsiveness of organisms, and the capacity of organisms to marshal their developmental repertoires in ways that affect the process of evolution (Moczek et al., 2011; Pfennig et al., 2010; Hu et al., 2020; Feiner et al., 2021). Our contention is that whenever these capacities of organisms are being invoked, their agency is tacitly being appealed to. Understanding the role of agency can have significant implications for the questions we ask about evolutionary phenomena and the answers we offer.[10]

> Agential concepts thus direct attention to "internal" or "agential" sources of consistent bias in evolution that may account for the evolution of particular adaptations, diversification or evolvability […]. In so doing, those concepts set an alternative explanatory agenda; they structure scientific investigation of evolution according to criteria of explanatory adequacy that are different to those of the neo-Darwinian representation of evolution by natural selection.
>
> (Uller, 2023, p. 332)

We hope to have demonstrated here that the norm of reaction is a biological phenomenon that calls for such an "alternative explanatory agenda." We expect that there will be many more.

Notes

1 The concept of agency in use here has been worked out in detail in Walsh (2015). See also Sultan et al. (2022). It *does not* impute cognitive or intentional states to organisms.
2 Agency in this sense can inhere at any of several levels of functional organization: an individual organism, its constituent regulatory systems and processes, and in some cases a social colony.
3 Ågren (2021, 2023) offers a comprehensive update on the precepts of gene-centred evolutionary biology.
4 We realize that there are terminological nuances here. As Sarkar (2004) points out, Woltereck considered the "norm of reaction" to be a property of a genotype, but in a very different sense of the latter concept. See also Nicoglou (2015) for a definitive historical discussion of plasticity.
5 Sultan (2019) provides a discussion of this argument.
6 See Jablonka and Raz (2009), Gilbert and Epel (2015), Sultan (2015) and references therein.
7 For experimental details and references to similar studies, see Herman et al. (2012), Baker et al. (2018, 2019), and Waterman and Sultan (2021).
8 A note on notation. Environmental variable, E, with the subscript, $E_{current}$, denotes the range of occurrent environments. Environmental variable, $E_{parental}$, denotes the range of parental environments (and so on). Environmental variables with double subscripts, $E_{x,n}$, take generations—e.g., "current", "parental", "grandparental" as values of x—and their conditions—e.g., drought stressed, moist, flooded—as values of n.
9 Each norm of reaction may also be a higher-order norm of reaction inasmuch as any inherited norm of reaction could be transmitted to offspring as any of an array of possible norms of reaction.
10 For recent surveys, see Walsh and Rupik (2023) and Nadolski and Moczek (2023).

References

Ågren, A. (2021). *The gene's-eye view of evolution*. Oxford University Press.

Ågren, A. (2023). Genes and organisms in the legacy of the modern synthesis. In T. E. Dickins & J. A. Dickins (Eds.), *Evolutionary biology: Contemporary and historical reflections upon core theory* (pp. 555–568). Springer.

Alvarez, M., Bleich, A., & Donohue, K. (2021). Genetic differences in the temporal and environmental stability of transgenerational environmental effects. *Evolution, 75*(11), 2773–2790.

Baedke, J., Fábregas-Tejeda, A., & Prieto, G. I. (2021). Unknotting reciprocal causation between organism and environment. *Biology & Philosophy, 36*(5), 5. https://doi.org/10.1007/s10539-021-09815-0

Baker, B. H., Berg, L. J., & Sultan, S. E. (2018). Context-dependent developmental effects of parental shade versus sun are mediated by DNA methylation. *Frontiers in Plant Science, 9.* https://doi.org/10.3389/fpls.2018.01251

Baker, B. H., Sultan, S. E., Lopez-Ichikawa, M., & Waterman, R. (2019). Transgenerational effects of parental light environment on progeny competitive performance and lifetime fitness. *Philosophical Transactions of the Royal Society B, 374*(1768), 20180–20182.

Baugh, L. R., & Day, T. (2020). Nongenetic inheritance and multigenerational plasticity in the nematode *C. elegans. Elife, 9,* 584–598.

Bernal, M. A., Ravasi, T., Rodgers, G. G., Munday, P. L., & Donelson, J. M. (2022). Plasticity to ocean warming is influenced by transgenerational, reproductive, and developmental exposure in a coral reef fish. *Evolutionary Applications, 15*(2), 249–261.

Bestion, E., Teyssier, A., Aubret, F., Clobert, J., & Cote, J. (2014). Maternal exposure to predator scents: Offspring phenotypic adjustment and dispersal. *Proceedings of the Royal Society B: Biological Sciences, 281*(1792), 20140701.

Bonduriansky, R. (2021). Plasticity across generations. In D. W. Pfennig (Ed.), *Phenotypic plasticity & evolution: Causes, consequences, controversies* (pp. 327–348). CRC Press.

Chevin, L. M., Collins, S., & Lefèvre, F. (2013). Phenotypic plasticity and evolutionary demographic responses to climate change: Taking theory out to the field. *Functional Ecology, 27*(4), 967–979.

Chevin, L. M., Lande, R., & Mace, G. M. (2010). Adaptation, plasticity, and extinction in a changing environment: Towards a predictive theory. *PLoS Biology, 8*(4), e1000357.

Dantzer, B., Newman, A. E., Boonstra, R., Palme, R., Boutin, S., Humphries, M. M., & McAdam, A. G. (2013). Density triggers maternal hormones that increase adaptive off-spring growth in a wild mammal. *Science, 340*(6137), 1215–1217.

de Jong, G. (1999). Unpredictable selection in a structured population leads to local genetic differentiation in evolved reaction norms. *Journal of Evolutionary Biology, 12,* 839–851.

DeWitt, T. J., & Scheiner, S. M. (2004). Phenotypic variation from single genotypes, a primer. In J. DeWitt & S. M. Scheiner (Eds.), *Phenotypic plasticity: Functional and conceptual approaches* (pp. 1–9). Oxford University Press.

Donohue, K. (2009). Completing the cycle: Maternal effects as the missing link in plant life histories. *Philosophical Transactions of the Royal Society of London B, 364,* 1059–1074.

Feiner, N., Brun-Usan M., & Uller, T. (2021). Evolvability and evolutionary rescue. *Evolution & Development, 23*(4), 308–319.

Gavrilets, S., & Scheiner, S. M. (1993). The genetics of phenotypic plasticity. VI. Theoretical predictions for directional selection. *Journal of Evolutionary Biology, 6,* 49–68.

Gilbert, S. F., & Epel D. (2015). *Ecological developmental biology: Integrating epigenetics, medicine, and evolution* (2nd ed.). Sinauer Associates.

Gomulkiewicz, R., & Kirkpatrick, M. (1992). Quantitative genetics and the evolution of reaction norms. *Evolution, 46,* 390–411.

Gupta, A. P., & Lewontin, R. C. (1982). A study of reaction norms in natural populations of *Drosophila pseudoobscura. Evolution, 36*, 934–948.

Haig, D. (2011). Lamarck ascending! *Philosophy, Theory and Practice in Biology, 3*(4). http://hdl.handle.net/2027/spo.6959004.0003.004

Haldane, J. B. S. (1946). The interaction of nature and nurture. *Annals of Eugenics, 13*, 197–205.

Haldane, J. S. (1884). Life and mechanism. *Mind, 9*(33), 27–47.

Heckwolf, M. J., Meyer, B. S., Döring, T., Eizaguirre, C., & Reusch, T. B. (2018). Transgenerational plasticity and selection shape the adaptive potential of sticklebacks to salinity change. *Evolutionary Applications, 11*(10), 1873–1885.

Herman, J. J., & Sultan, S. E. (2011). Adaptive transgenerational plasticity: Case studies, mechanisms, and implications for natural populations. *Frontiers in Plant Genetics and Genomics, 2*, 102.

Herman, J. J., & Sultan, S. E. (2016). DNA methylation mediates genetic variation for adaptive transgenerational plasticity. *Proceedings of the Royal Society B: Biological Sciences, 283*(1838), 20160988.

Herman, J. J., Sultan, S. E., Horgan-Kobelski, T., & Riggs, C. E. (2012). Adaptive transgenerational plasticity in an annual plant: Grandparental and parental drought stress enhance performance of seedlings in dry soil. *Integrative & Comparative Biology, 52*, 1–12.

Hu, T., Linz, D., Parker, E. S., & Moczek, A. (2020). Developmental bias in horned dung beetles and its contributions to innovation, adaptation, and resilience. *Evolution & Development, 22*(1–2), 165–180.

Huneman, P., & Walsh, D. M. (Eds.). (2017). *Challenging the modern synthesis: Adaptation, inheritance, development.* Oxford University Press.

Jablonka, E., & Raz, G. (2009). Transgenerational epigenetic inheritance: Prevalence, mechanisms, and implications for the study of heredity and evolution. *Quarterly Review of Biology, 84*, 1331–1176.

Laland, K., Odling-Smee, J., & Feldman, M. W. (2019). Understanding niche construction as an evolutionary process. In T. Uller & K. Laland (Eds.), *Evolutionary Causation* (pp. 127–152). MIT Press.

Laland, K., Uller, T., Feldman, M., Sterelny, K., Müller, G., Moczek, A., Jablonka, E., & Odling Smee, J. (2014). Does evolutionary theory need a rethink? *Nature, 514*, 161–164.

Lewontin, R. C. (1978). Adaptation. *Scientific American, 239*, 212–230.

Lewontin, R. C. (1985). The organism as the subject and object of evolution. In R. Levins & R. C. Lewontin (Eds.), *The dialectical biologist* (pp. 85–106). Harvard University Press.

Lewontin, R. C. (2000). *The triple helix: Gene, organism and environment.* Harvard University Press.

Maynard Smith, J. (1994). *Evolutionary genetics.* Oxford University Press.

Miller, G. M., Watson, S.-A., Donelson, J. M., McCormick, M. I., & Munday, P. L. (2012). Parental environment mediates impacts of increased carbon dioxide on a coral reef fish. *Nature Climate Change, 2*, 858–861.

Moczek, A. P., Sultan, S. E., Foster, S., Ledón-Rettig, C., Dworkin, I., Nijhout, D. H., Abouheif, E., & Pfennig, D. W. (2011). The role of developmental plasticity in evolutionary innovation. *Proceedings of the Royal Society B, 278*, 2705–2713.

Moore, R. S., Kaletsky, R., & Murphy, C. T. (2019). Piwi/PRG-1 argonaute and TGF-β mediate transgenerational learned pathogenic avoidance. *Cell, 177*(7), 1827–1841.

Mousseau, T. A., & Fox, C. (1998). *Maternal effects as adaptations.* Oxford University Press.

Nadolski, E., & Moczek, A. (2023). Promises and limits of an agency perspective in evolutionary developmental biology. *Evolution & Development, 25*(6), 371–392.

Nager, R. G., Keller, L. F., & Van Noordwijk, A. J. (2000). Understanding natural selection on traits that are influenced by environmental conditions. In T. Mousseau, B. Sinervo, & J. A. Endler (Eds.), *Adaptive genetic variation in the wild* (pp. 5–115). Oxford University Press.

Nicoglou, A. (2015). The evolution of phenotypic plasticity: Genealogy of a debate in genetics. *Studies in History and Philosophy of Biological and Biomedical Sciences, 50*, 67–76.

Odling Smee, J., Laland, K., & Feldman, M. W. (2003). *Niche construction: The neglected process in evolution.* Princeton University Press.

Perez, M. F., & Lehner, B. (2019). Intergenerational and transgenerational epigenetic inheritance in animals. *Nature Cell Biology, 21*, 143–151.

Pfennig, D. W. (Ed.). (2021). *Phenotypic plasticity & evolution: Causes, consequences, controversies.* CRC Press.

Pfennig, D. W., Wund, M. A., Schlichting, C., Snell-Rood, C. E., Cruikshank, T., Schlichting, C., & Moczek, A. (2010). Phenotypic plasticity's impacts on diversification and speciation. *Trends in Ecology and Evolution, 25*, 459–467.

Pigliucci, M., & Müller G. (Eds.) (2010). *The extended evolutionary synthesis.* MIT Press.

Plaistow, S. J., Shirley, C., Collin, H., Cornell, S. J., & Harney, E. D. (2015). Offspring provisioning explains clone-specific maternal age effects on life history and life span in the water flea, *Daphnia pulex. The American Naturalist, 186*(3), 376–389.

Potticary, A. L., & Duckworth, R. A. (2020). Multiple environmental stressors induce an adaptive maternal effect. *The American Naturalist, 196*(4), 487–500.

Quadrana, L., & Colot, V. (2016). Plant transgenerational epigenetics. *Annual Review of Genetics, 50*, 467–491.

Rockman, M. V. (2008). Reverse engineering the genotype – phenotype map with natural genetic variation. *Nature, 456*, 738–744.

Ryu, T., Veilleux, H. D., Donelson, J. M., Munday, P. L., & Ravasi, T. (2018). The epigenetic landscape of transgenerational acclimation to ocean warming. *Nature Climate Change, 8*(6), 504–509.

Salinas, S., Brown, S. C., Mangel, M., & Munch, S. B. (2013). Non-genetic inheritance and changing environments. *Non-Genetic Inheritance, 1*, 38–50.

Sarkar, S. (2004). From the *Reaktionsnorm* to the evolution of adaptive plasticity: A historical sketch, 1909–1999. In T. J. DeWitt & S. M. Scheiner (Eds.), *Phenotypic plasticity: Functional and conceptual approaches* (pp. 10–30). Oxford University Press.

Sarkar, S., & Fuller, T. (2003). Generalized norms of reaction for ecological developmental biology. *Evolution & Development, 5*(1), 106–115.

Scheiner, S. M. (1993). Genetics and evolution of phenotypic plasticity. *Annual Review of Ecology and Systematics, 24*(1), 35–68.

Stearns, S. C. (1989). The evolutionary significance of phenotypic plasticity. *Bioscience, 39*(7), 436–445.

Sultan, S. E. (2007). Development in context: The timely emergence of eco-devo. *Trends in Ecology & Evolution, 22*(11), 575–582.

Sultan, S. E. (2015). *Organism and environment: Ecological development, niche construction and adaptation.* Oxford University Press.

Sultan, S. E. (2019). Genotype-environment interaction and the unscripted reaction norm. In T. Uller & K. Laland (Eds.), *Evolutionary causation* (pp. 99–126). MIT Press.

Sultan, S. E. (2021). Phenotypic plasticity as an intrinsic property of organisms. In D. W. Pfennig (Ed.), *Phenotypic plasticity & evolution: Causes, consequences, controversies* (pp. 3–24). CRC Press.

Sultan, S. E., Barton, K., & Wilczek, A. M. (2009). Contrasting patterns of transgenerational plasticity in ecologically distinct congeners. *Ecology, 90*(7), 1831–1839.

Sultan, S., Moczek, A. P., & Walsh, D. M. (2022). Bridging the explanatory gaps: What can we learn from a biological agency perspective? *BioEssays, 44*(1). https://doi.org/10.1002/bies.202100185

Sultan, S. E., & Stearns, S. C. (2005). Environmentally contingent variation: Phenotypic plasticity and norms of reaction. In B. Hall & B. Hallgrimsson (Eds.), *Variation: A central concept in biology* (pp. 303–332). Academic Press.

Thompson, E. (2007). *Life and mind*. Harvard University Press.

Uller, T. (2008). Developmental plasticity and the evolution of parental effects. *Trends in Ecology & Evolution, 23*(8), 432–438.

Uller, T. (2023). Agency, goal-orientation and evolutionary explanations. In P. Corning, S. Kauffman, D. Noble, J. Shapiro, & R. Vane-Wright (Eds.), *Evolution 'on purpose': Teleonomy in living systems* (pp. 325–339). MIT Press.

Uller, T., Nakagawa, S., & English, S. (2013). Weak evidence for anticipatory parental effects in plants and animals. *Journal of Evolutionary Biology, 26*(10), 2161–2170.

Via, S., & Lande, R. (1985). Genotype-environment interaction and the evolution of phenotypic plasticity. *Evolution, 39*, 505–522.

Walsh, D. M. (2012). Situated adaptationism. In W. P. Kabesanche, M. O'Rourke, & M. Slater (Eds.), *The environment* (pp. 89–116). MIT Press.

Walsh, D. M. (2015). *Organisms, agency, and evolution*. Cambridge University Press.

Walsh, D. M. (2023). Evolutionary foundationalism and the myth of the chemical given. In P. Corning, S. Kauffman, D. Noble, J. Shapiro, & R. Vane-Wright (Eds.), *Evolution 'on purpose': Teleonomy in living systems* (pp. 341–362). MIT Press.

Walsh, D. M., & Rupik, G. (2023). The agential perspective: Countermapping the modern synthesis. *Evolution & Development, 25*(6), 335–352.

Waterman, R., & Sultan, S. E. (2021). Transgenerational effects of parent plant competition on offspring development in contrasting conditions. *Ecology, 102*(12), 03531.

Wray, G. A., Hoekstra, H. E., Futuyma, D. J., Lenski, R. E., Mackay, T. F., Schluter, D., & Strassmann, J. E. (2014). Does evolutionary theory need a rethink? No, all is well. *Nature, 514*(7521), 7521. https://doi.org/10.1038/514161a

Wright, S. (1931). Evolution in Mendelian populations. *Genetics, 16*, 97–159.

8 A Critique of the Agential Stance in Development and Evolution

Henry D. Potter and Kevin J. Mitchell

8.1 Introduction

The agential perspective is an emerging conceptual framework within theoretical and philosophical biology, which seeks to foreground the view that organisms are agents and then explore the consequences of this insight for key topics in biology (Walsh, 2006, 2015, 2018; Walsh & Rupik, 2023; Sultan et al., 2022; Jaeger, 2022; Uller, 2022; Fábregas-Tejeda & Baedke, 2023; Nadolski & Moczek, 2023; Snell-Rood & Ehlman, 2023; Fulda, 2023). We have previously argued along these lines for the position that organisms *themselves* are the agents of their own behaviour (Potter & Mitchell, 2022; Mitchell, 2023; see also Walsh, 2015). Organisms are not mere automata, driven around by complicated genetic or neuronal happenings, nor by the conditions of their immediate or historical environment. On the contrary, over ontogenesis and maturation they develop the capacity to *act* in the world, with holistic, integrative, purposive, and goal-directed behaviours that are genuinely 'up to them' *qua* agents. Consequently, organismal behaviours are not generally amenable to a completely reductive analysis that abstracts away from the agent itself to find the 'real' causes of behaviour in neuronal, genetic, or atomic activity.

This agential view of behaviour has some important implications for how we conceptualise and understand evolution. First, it brings into focus the fact that what organisms within an ecosystem collectively *do* is fundamentally what shapes the evolutionary trajectories of lineages. Their actions alter or even primarily create the selective pressures and ecological opportunities to which the organisms within that niche adapt. In this sense, organisms are the entities that *enact* natural selection through their choices and actions (Walsh, 2015).

Second, and more particularly, the agential view identifies processes of niche construction and cultural inheritance as clear instances in which organisms play an active role in directing evolutionary change by purposefully modifying their own environments. Taking both these points, it becomes clear that "[a]daptive evolution does not unfold as populations migrate along fixed adaptive landscapes" within fixed environments (Nadolski & Moczek, 2023, p. 12). Rather, the ongoing interplay between organisms and their environments, where individuals actively and continuously modify (and are modified by) their surroundings, exerts a large influence over how lineages end up evolving. The agency of behaving organisms

DOI: 10.4324/9781003413318-10

is therefore an essential factor to include in any model of evolution—a view that should not be considered controversial.

More controversial is the view that an agential perspective is needed in the developmental domain, too. This is the argument that organisms are the agents of their own development, and that the exercise of this developmental form of organismal agency can actively shape the trajectories of evolution (Walsh, 2006, 2015, 2018; Walsh & Rupik, 2023; Walsh & Sultan, this volume; Sultan et al., 2022; Nadolski & Moczek, 2023; Snell-Rood & Ehlman, 2023; Fulda, 2023). We refer to this view as the 'agential stance on development' (sometimes 'agential stance' for short), in order to differentiate it from the wider agential perspective of which it is a part.

Our focus in this chapter is to offer a critical analysis of the agential stance on development and of its proposed implications for evolutionary theory.[1] We ask: what do the claims of the agential stance consist of? What is the evidence that is taken to support these claims? Are they to be interpreted literally or, more metaphorically, as serving a heuristic function? What is taken to follow from adopting the agential stance on development? And are these implications valid, justified, or useful?

We conclude that if the agential stance on development is interpreted literally, then its core claims are not supported by the empirical evidence. If the agential stance is interpreted in a heuristic sense, which its proponents sometimes endorse, then it does not offer the sort of novel or distinctive explanatory insights it is commonly claimed to.

8.2 The Agential Stance on Development

8.2.1 *Claims of the Agential Stance*

The agential stance is a particular way of conceptualising biological development. Its central thesis is that "the development of phenotypes is under the active control of the developing organism" (Sultan et al., 2022, p. 9). On this view, morphogenesis is not a passive process. Instead, it is posited to be the process by which developing organisms *actively* "direct their own development" towards particular outcomes, in a purposive and goal-seeking manner (Walsh, 2015, p. 84), by dynamically controlling their own internal structure and function—much like "clay modelling itself" (Russell, 1924, p. 61, as cited in Baedke, 2021, p. 83).

In other words, the agential stance is the view that development is an agential process; morphogenesis is something the embryo 'does,' as opposed to something that merely 'happens' to it. As philosopher Denis Walsh, one of the leading proponents of the view, puts it, the organism *itself* is the "unit that exerts executive control over development," such that "[p]roper development depends upon the capacity of organisms to assimilate, integrate and orchestrate the causal contributions from genes, epigenetic structures, tissues, organs, behaviour and the physical, ecological and cultural setting" (Walsh, 2015, p. 157).

A number of radical conceptual and empirical implications are argued to follow from this view. For the purposes of this chapter, we have grouped these into three categories:

(i)   *Organism-level control*: The agential stance is often characterised by the causal and theoretical privilege it ascribes to the organism, *as a whole*, in answering the question 'how is organismal form generated during development?' In comparing the agential stance with other systems-focused approaches in developmental biology, such as Developmental Systems Theory (DST; Oyama, 2000), Nadolski and Moczek (2023) explicitly highlight organism-level control as one of the agential view's primary distinguishing features. While other approaches may recognise a multitude of causal factors and levels, the agential stance is unique in privileging the "ordering influence from the system as a whole" (p. 5) and thereby positioning the embryo as the 'executive control' unit of its own development: "Insofar as any single entity can be said to 'control', 'regulate' or 'orchestrate' this widely distributed plexus of causes [in development], it is the organism as a whole" (Walsh, 2015, p. 18).

(ii) *Agential adaptation:* The second implication that is suggested to follow from adopting an agential stance relates to the problem of how to explain biological adaptation. Standard explanations of the fittedness of organisms to their environments invoke adaptation by natural selection—the statistical tendency for (combinations of) heritable variants that predispose to better adapted phenotypes to increase in frequency in the population. Adopting the agential stance is suggested to offer an alternative or distinctive mode of explanation: the purposive agency of the developing organism. From this perspective, the observable adaptedness of developmental outcomes is taken to be the consequence of the organism's purposive 'pursuit of its goals' during development (i.e., its agency). Importantly, this is taken to be a complementary, but still *distinctive*, explanatory strategy for explaining organismal adaptation to that of adaptation by natural selection (Walsh, 2015; Baedke & Fabregas-Tejeda, 2023); one that is altogether more active and organism-centred, wherein "development is the manifestation of the purposiveness of organisms" (Walsh, 2015, p. 162).

(iii) *Agential evolution:* Building on this idea further, the third major theoretical consequence that is suggested to follow from adopting the agential stance lies in the reframing of adaptive population change as something that individual organisms, as *agents*, 'enact' through the control of their own development: "the agential perspective […] explicitly represents evolution as the consequence of organisms' pursuit of their goals" (Walsh & Rupik, 2023, p. 10).

      Instead of conceptualising evolutionary change as arising 'genotype-first' from the passive processes of genetic mutation, inheritance, drift, and natural selection—with the progressive adaptation of lineages emerging as a statistical consequence—the theory posits that "adaptive evolution is caused by the adaptiveness of organismal development" (Walsh, 2015, p. 158), which, in turn, is explained by the organism's purposive agency [see (ii)]. This argument is explicit in Walsh's claim that "by locating the cause of the adaptive bias in evolution in the adaptive activities of organisms, particularly in their development, the [agential stance] does not need to invoke natural selection to do the job" (Walsh, 2015, p. 158).

This view is thus congruent with the idea of 'phenotype-first' evolution (West-Eberhard, 2003), where, in this version, the evolutionary trajectory of a species arises from adaptive developmental plasticity of individuals, which enables them to actively create novel forms that are better adapted to new environments. Genetics can then supposedly catch up by selecting either pre-existing or new genetic variants that predispose to these new phenotypes.

In this chapter, we argue that none of these three implications is well supported. Our reasoning for this is two-fold. First, the core claims of the agential stance on development need to be interpreted literally in order for the conclusions drawn in (i)–(iii) to follow from them. Yet we argue that there is insufficient evidence to support a literal interpretation. Such a perspective might conceivably be supported if one focuses primarily or solely on *adaptive* developmental processes, but it dissipates given a broader sample of the developmental evidence-base which sufficiently takes into account non-adaptive and maladaptive processes and outcomes.

Importantly, some proponents of the agential stance do expressly distance themselves from the literal interpretation of these claims, explicitly stating that there is an important sense in which the organism's 'pursuit of its goals' during development is to be understood non-causally (see Fulda, 2023 for an extensive articulation of this position). As our second line of argument, we contend that if the claims of the agential stance are to be interpreted metaphorically or heuristically in this way, then the implications above simply do not follow from it—most notably (ii) and (iii).

Before turning to these arguments, let us first examine the claims of the agential stance in more detail. To do this, we break the approach down into its two constituent dimensions: a 'causal dimension' and a 'purposive dimension.'

8.2.2 The Causal Dimension: Holistic Causation

The causal dimension of the agential stance focuses on how to operationalise the idea that the organism can be a causal contributor to development *in its own right*. That means identifying the whole organism as a *locus* of causation of (at least some) developmental effects in a way that is not reducible to or derivable from the causal contributions of its component parts.

An important aspect of the agential stance's commitment to organism-level control is that it entails a so-called anti-gene-centrism. By demonstrating that the organism has control over its own development, proponents of this view explicitly challenge standard approaches in biology which are perceived to emphasise the genetic causes of phenotypic development. Thus, one of the position's main theoretical consequences is the proposed conceptual shift from "genes as the causal driver of development to agents constructing their own development" (Snell-Rood & Ehlman, 2023, p. 4). On this model, development is an activity that the organism 'does,' and not merely a passive consequence of its (genetic) inheritance.

A holistic, dynamical systems view of development is certainly appropriate. Rather than a simplistic view of a direct, isolatable relationship between specific genes and specific traits, it is well established that genes work in concert with one another, forming emergent gene regulatory networks that reciprocally constrain, influence, and regulate the activity of their constituent genes. Gene regulatory networks therefore represent a system in which, *contra* reductionism, the dynamics of the parts (i.e., genes) are in fact dependent on the dynamics of the whole. At a level up, gene regulation, including epigenetic mechanisms, is sensitive to (and hence causally influenced by) the cellular system of which it is a part. We can then zoom out further to find cellular networks that dynamically constrain and influence the activities of individual cells. This continues until, ultimately, one is forced to recognise that this reciprocal, mutual dependence between parts and wholes occurs at all scales within the developing system—with 'wholes' at one scale appearing as 'parts' at another—right up to the level of the whole organism.

From this perspective, it becomes clear that a developing embryo or foetus is a deeply integrated and interconnected whole, with higher levels of organisation influencing lower levels in an entirely natural, non-mysterious way (Oyama, 2000; Jaeger, 2022). This causal holism provides a natural mechanism through which developing organisms might causally contribute to their own development *in their own right*—as is necessary to motivate the causal dimension of an agential perspective.

However, causal holism, on its own, is not sufficient to attribute agency or 'executive control' to the developing organism. This kind of loopy, reciprocally causal dynamic is a fundamental feature of all complex systems, even non-living ones: when elements of these systems become mutually entrained and interdependent, a dynamic emerges in which the activity of any individual part is influenced by the whole it is embedded in, whilst concurrently contributing to those global dynamics itself (Juarrero, 1999; Pessoa, 2022). In development, these holistic dynamics have already been recognised and codified in the form of DST (Oyama, 2000), without being taken as evidence of organismal agency (see Nadolski & Moczek, 2023). Instead, the agential stance also requires a purposive dimension—an argument to support the additional connotation that embryos influence their own development in real time 'with purpose.'

8.2.3 The Purposive Dimension: Goal-Directedness

In conventional usage, agency is not just irreducible, holistic causation. It also entails *purposiveness*, i.e., some sense in which the agent caused the effect *for a reason* or *in the pursuit of a goal* (Nadolski & Moczek, 2023; Potter & Mitchell, 2022; Mitchell, 2023). Since goals cannot be free-floating, such a teleological framing requires the existence of a subject to which the goal or purpose is attributed in some sort of behaviour-informing capacity. In line with this, most articulations of the agential stance include a purposive dimension, which attempts to operationalise the idea that "the organism *purposefully* molds itself" in ontogeny (Baedke, 2021, p. 6; emphasis added).

While this claim applies to typical development, support for it derives from observations where developing organisms, in virtue of their tightly integrated architectures, respond to perturbations that occur during development with holistic re-adjustments to their dynamical regimes. For example, an environmental cue detected in a localised area of the embryo can trigger a seemingly co-ordinated whole-system response to that cue, while damage to a particular tissue or cell can trigger whole-system changes that activate compensatory mechanisms and pathways. These self-organising or self-regulating capacities of the organism are referred to in the agential stance literature as 'plasticity' (Walsh, 2006), capturing the notion of the responsive "mutual adjustment among variable parts in development" (Uller, 2022, p. 11). This focuses on the real-time responses of the developmental system to individual perturbations (both internal and external) with an emphasis on these not just occurring within the organism or being carried out by the organism, but being *caused by* the organism.

Sometimes these responses will involve buffering perturbations—such as genetic mutations, molecular noise, and environmental disturbances, even including lesions that split the embryo in half—in ways that maintain functional stability. On longer timescales, this manifests as developmental robustness or canalisation (Waddington, 1957; Wagner, 2013) and contributes to what biologist Ludwig von Bertalanffy (1968) called 'equifinality'—i.e., the propensity to reach the same end state, despite different starting conditions and different developmental trajectories. Other times they involve adopting alternative dynamical regimes and developmental trajectories in response to particular environmental cues, in ways that favour organismal forms that are better fitted to those conditions. This manifests as what we call here phenotypic plasticity, referring to outcomes of *processes* of developmental plasticity that result in altered phenotypes, as opposed to those which result in robust attainment of typical phenotypes. These variations in outcome can be merely quantitative (e.g., smaller growth under restrictive nutrient conditions) or, in some striking cases, can manifest as qualitatively distinct phenotypes (e.g., temperature-dependent sex determination). In all cases, plastic self-organisation plays an essential role in enabling morphogenesis to reach an adaptive outcome.

The crucial point for the agential stance is the inference that the developing organism does not plastically readjust *randomly* or *passively* in the face of perturbations. Rather, its whole-system responses are, in some sense, directed towards adaptivity. It is for this reason that developmental plasticity is taken as evidence of the developing organism exerting a *goal-directed* or *purposive* influence over its own development. The types of plastic, holistic re-adjustments the embryo tends to undergo in response to genetic and environmental variation are often precisely those which enable it to robustly attain and maintain *its* 'goal' of persisting: "[…] robust, plastic organisms produce the responses they do precisely because, under the circumstances, those responses are conducive to an organism's survival" (Walsh, 2015, p. 202).

These two dimensions therefore jointly motivate the view that developing embryos really are the agents of their own development. The apparent adaptivity of developmental robustness and phenotypic plasticity, in virtue of being 'underwritten' by the embryo's self-regulating capacities, is posited to sufficiently establish the developing organism as actively and purposefully controlling and directing its own ontogenetic trajectory towards well-adapted outcomes—and, thus, to constitute a naturalised account of the agential stance on development (see also Walsh & Sultan, this volume). In turn, this agential view of development, with the organism itself as the 'executive controller,' is argued to motivate and justify the three theoretical implications of (i) *organism-level control,* (ii) *agential adaptation*, and (iii) *agential evolution.*

Our aim in the remainder of this chapter is to contest these three implications, at least insofar as they are derived from an agential stance on development. Of particular note is the suggestion that this capacity to produce alternate phenotypic forms in a goal-directed, adaptive way is the primary source of evolutionary novelties, thereby enabling 'phenotype-first' evolution in an adaptive fashion:

> Organismal plasticity, a manifestation of goal-directed purposiveness, underwrites much of the production of evolutionary novelty. These novelties are not adaptively neutral [...]; they are adaptively biased. The goal-directed purposiveness of organisms appears to be essential to the explanation of the origin of evolutionary novelties.
>
> (Walsh & Rupik, 2023, p. 12)

8.3 A Critique of the Literal Interpretation of the Agential Stance

With the arguments underlying the approach laid out more explicitly, we can now revisit the agential stance's core claims in order to highlight an area of possible ambiguity. The claim that "development [...] is under the active control of the developing organism" (Sultan et al., 2022, p. 9)—or that embryos purposively "direct their own development" in pursuit of their goals—can be interpreted in one of two ways. First, it could be seen not only as ascribing causal power to the embryo-as-a-whole, but as ascribing some sort of forward-looking or *directive* causal power—a means by which the organism can be viewed as *trying* to attain certain outcomes, even in a deflationary sense. We will call this the *literal interpretation* because we take it that this is generally what it means for a subject to actively direct a process towards an outcome, particularly when this is characterised as the subject 'pursuing its own *goal.*' On this reading, the two dimensions of the agential stance would be intimately combined; it is not merely that the developing organism exerts an identifiable influence over development, it is that it exerts *the* influence that explains why a particular outcome occurs (rather than another) in virtue of its own goals and purposes.

Importantly, many authors of the agential stance expressly state that "[a]scribing agency to a system in no way imputes to it intentions or desires" (Sultan et al., 2022, p. 5); it "is not a claim that living systems must have cognitive representations and desires that guide their activity" (Nadolski & Moczek, 2023, p. 2). While we agree that intentionality or cognition are not implied or required, we take it that these statements are still compatible with a minimal notion of directive causal power—in the manner implied by a literal reading of the claim that developing organism are the "unit that exerts *executive control* over development" (Walsh, 2015, p. 157; emphasis added).

Alternatively, as is sometimes suggested, the core claims of the agential stance could also be interpreted metaphorically or heuristically, in a sense that emphasises the important insight that organisms-as-a-whole exert a causal influence over development that is real and often necessary for attaining adaptive end states, but without committing oneself to the implication that it literally exerts some sort of *directive* causal influence, in real-time. On this reading, the stance's two dimensions come apart. Embryos causally contribute to the processes of development (via holistic causation) and these causal contributions are necessary to enable developmental processes that are amenable to a particular (teleological) mode of explanation that posits and cites the organism's 'goal-directedness' as its *explanans*. But crucially, the latter is not intended to be a *consequence* of the former. Instead, the organism's 'goals' are taken to explain its developmental trajectory and outcome in a distinctly non-causal manner (Fulda, 2023).

For reasons we discuss in the next section, we contend that only a literal interpretation of the agential stance can justify the implications outlined in (i)–(iii). However, as we will now argue, such an interpretation does not appear to be empirically well supported.

8.3.1 The Problem of Non-Adaptive Plasticity

The agential stance leans heavily on observations of adaptive directionality in development to justify its purposive dimension. Developmental plasticity is claimed to be a manifestation of organismal purposiveness precisely *because* it is seen as representing the organism's 'ability' to reliably direct its development towards adaptive outcomes (its 'goal'), via supple, holistic, real-time responses to changing conditions. On a literal reading of these claims, this implies that these whole-system, self-organising responses are in some sense inherently directed towards adaptivity, so as to justify being described as the organism's "goal-directed pursuit of its ways of life" (Walsh & Rupik, 2023, p. 9). Yet, such a reading does not appear to fit with the empirical evidence.

Developmental plasticity is indeed a ubiquitous property of organismal development; it is an essential component of development's general robustness to noise, mutation, and environmental variance, without which species could not survive or evolve (Wagner, 2013). However, it is presumably uncontroversial to note that the capacity to robustly buffer or attune to the contingencies of development—and thus tend towards well-adapted outcomes—is not unlimited. Many environmental insults or genetic mutations do in fact cause the developing system to dynamically re-adjust in atypical and non-adaptive ways, often resulting in malformations of various tissues or organs, or the complete failure of development and death.

As the agential stance highlights, the robustness of developing systems prevents many genetic mutations from having deleterious phenotypic effects. But it is acknowledged by proponents of the view that this is only true until it is not. Many single mutations have very strong, sometimes devastating effects on organismal development. And specific combinations of these—or even just the overall burden of mutations—can also often push the developing organism towards a non-adaptive outcome.

It is less common in the agential stance literature to see it acknowledged that sometimes it is the processes of developmental plasticity *themselves* that cause the emergence of a pathological state. In many cases, the proximate effect of some perturbation may be fairly innocuous, but it may lead to a set of reactive, cascading effects that take the dynamics of the whole system out of its typical regime and into a pathological one. In some cases, this involves moving the dynamical system to an alternate, sometimes qualitatively novel, but maladaptive attractor state, one that has not been selected for but that simply arises as an unpredictable 'failure mode' of a dynamical system with complex, non-linear interdependencies. This is well studied, for example, in the case of epileptogenesis (Neuberger et al., 2019; Lignani et al., 2020), and is also posited for many of the emergent symptoms of neurodevelopmental disorders (Mitchell, 2015; Durstewitz et al., 2021; Bartsch et al., 2023).

Thus, it is not only the case that the processes of developmental plasticity often fail to produce an adaptive outcome, but they also can, under some circumstances, 'actively' produce a maladaptive outcome. This poses an important problem for the literal interpretation of the agential stance. What justifies the claim that adaptive developmental outcomes are the organism's 'goal,' and that developmental plasticity represents its active, real-time pursuit of that goal, given these cases of maladaptive plasticity and pathological outcomes? An agential approach would presumably need to account for the latter as situations in which the agent *fails* to attain its goal, yet there appears little empirical support for this. If anything, evidence of the full range of developmental processes and phenotypes seems to contradict the stance's original hypothesis that plasticity is a purposive phenomenon and hence requires an agential explanation.

Similar concerns relate to the agential interpretation of developing organisms' capacity to respond to environmental cues in adaptive ways, so as to produce alternative phenotypic outcomes that are well fitted to their respective contexts. A number of striking examples of phenotypic plasticity, across a menagerie of strange and wonderful creatures, are often presented as evidence of the organism 'actively directing' its own development. But these cases are noteworthy precisely because they are unusual. Most animals do not exhibit such plasticity with alternate, well-fitted phenes, and most environmentally sensitive phenotypic variation is not of this sort. On the contrary, in general, deviations from the 'wild-type' pheno-type of most animals are non-adaptive (Ghalambor et al., 2015). Again, this leaves the agential stance with some difficult conceptual problems, particularly if it is to be interpreted literally: why is adaptive phenotypic plasticity not a more widely observed phenomenon, if embryos in general are taken as having the power to actively and purposively pursue adaptive end states during development, including the production of 'novel' phenotypes? And what justifies positing organisms as actively pursuing adaptive outcomes, in a literal sense, when most environmentally sensitive phenotypic variation does not fit this description?

For our purposes, these questions serve an important rhetorical function. They demonstrate how taking into consideration evidence of non-adaptive and maladap-tive plasticity can effectively undermine the impression that the 'internal logic of development' (Alberch, 1989) is inherently and actively adaptive, in such a way that might demand explanation in terms of the organism's real-time, agential pur-suit of its goal to persist. When one focuses solely or primarily on cases of *adaptive* robustness and plasticity and ignores or downplays those times when development is not able to buffer perturbations or when alternate phenotypic outcomes are actually maladaptive, it can conceivably create a false impression of the scale to which the self-organising and self-regulating dynamics of developmental plasticity are *necessarily* directed towards adaptivity (Figure 8.1). And, in turn, create the impression that these processes are a manifestation, or consequence, of the "supple goal-directed, compensatory capacities" *of the organism* (Walsh, 2006, p. 773; see also Walsh & Rupik, 2023)—akin to a real-time "decision-making process" (Snell-Rood & Ehlman, 2023, p. 5). But, as we have seen, this is not a fair reflection of the full developmental evidence-base.

As such, there would appear to be no principled reason to take developmental plasticity as *evidence* of 'organismal purposiveness' or of the organism 'actively pursuing its goal' of adaptation. At least, not if these claims are to be interpreted in the literal sense demanded by the assumption that (ii) *agential adaptation* and (iii) *agential evolution* follow from adopting the agential stance. These purported implications rely on the organism's agency (i.e., its purposiveness or goal-directedness) being able to offer a novel or distinctive explanation for *why* indi-vidual cases of plasticity—and development, more generally—turn out adaptive (rather than non-adaptive or maladaptive): "Organismal purposiveness *underlies* the contribution of development to adaptive evolution" (Walsh, 2015, p. 159; emphasis added).

However, it appears as though 'organismal purposiveness' is just another way of *describing* instances where development robustly tends towards an adaptive outcome. We see no reason to think it 'underlies' or 'causes' that adaptivity, in a manner that could support a literal interpretation of the theory. And, thus, we see no distinctive role for this developmental form of agency in helping us explain adaptive population change.

We therefore suggest that a literal interpretation of the agential stance's claims is untenable, or at least unsubstantiated, and only *appears* viable under a highly selective reading of the developmental evidence-base.

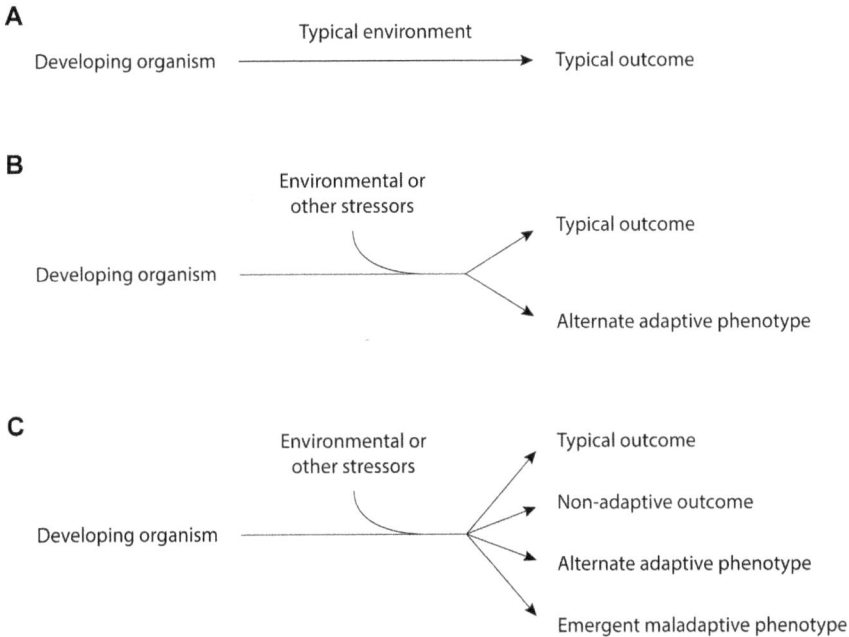

Figure 8.1 Alternate views of phenotypic variation. (A) The observation that organisms develop toward species-typical outcomes under species-typical environmental conditions is congruent with the standard view of development as a passive 'happening,' the necessary 'playing out' of interactions between inherited components, with no subject in charge. (B) The observation that developmental plasticity, under atypical environmental conditions or the presence of other stressors (including genetic variants), results in either robust attainment of a typical outcome or adaptive phenotypic plasticity implies an apparent purposiveness, which is taken to support a novel perspective in which the organism actively controls its development in a goal-directed manner. (C) A more even-handed survey of the evidence, which includes non-adaptive outcomes and 'actively' maladaptive ones, deflates this impression of purposiveness and leaves a more neutral view of developmental robustness and plasticity as evolved, holistic but passive tendencies of developmental systems that do not entail real-time executive control.

8.4 A Critique of the Heuristic Interpretation of the Agential Stance

Some proponents of the agential stance pre-empt these challenges to the literal interpretation of their claims, by articulating a more heuristic version of the theory. On this view, the value of invoking concepts of agency, purposiveness, and goals in development is said to lie, not in its identification of a literal, directive causal power exerted by the organism, but in elucidating a teleological aspect of organismal development. By appealing to the embryo's purposiveness, it is argued that we obtain a previously unexplored explanatory lens for understanding development, one which can answer *why* a particular developmental process or outcome occurred and not just *how* it occurred (Walsh, 2015; Fulda, 2023).

Importantly, on this model, the goal that is imputed to the organism—which provides the answer to 'why' a developmental event occurs—does not *cause* the event in question. Instead, goals and events stand in a non-causal relation to one another called 'hypothetical necessity,' wherein "without the action [or event] in question, the goal would not have occurred and, with it, the goal [...] occurs reliably" (Walsh, 2018, p. 173). Here, goals 'hypothetically necessitate' developmental events insofar as, if the organism has a particular goal, it necessitates (non-causally) that specific developmental events will occur. Goals are therefore taken to provide non-causal *explanations* of events using an analogous logical structure to causal explanations and hence are argued to represent a scientifically valid form of explanation (see Walsh, 2015, Chapter 9).

Accepting this alternative teleological (or agential) form of explanation is then said to equip us with the explanatory apparatus to explain *why* developmental events and processes occur in the way they do (rather than another possible way), i.e., embryos respond to their conditions in precisely the way that conduces to their goal of adaptation. A crucial supposition is that this sort of explanatory power is missing from existing evolutionary models. Hence, a (heuristic) agential stance on development is justified and necessary precisely *because* it offers a distinctive and ineliminable means of explaining individual-level adaptation [see (ii)] and population-level adaptation [see (iii)]. In essence, developmental and evolutionary processes can be seen as tending towards morphological adaptation because individual organisms are (hypothetically) 'pursuing their goals' during development, sometimes producing novel, adaptive phenotypes in the process:

> Novel phenotypes [...] occur when they do *precisely because* they contribute to the organism's goals of survival and reproduction. These are not chance occurrences. We need to invoke the capacity of organisms to pursue goals in order to explain the origin of adaptive novelties.
>
> (Walsh, 2015, p. 203; emphasis in original)

We suggest that this heuristic interpretation of the agential stance faces the same set of questions and problems discussed in sub-section 8.3.1 if it is taken to apply to development in general, i.e., what justifies characterising developmental processes as generally 'purposive,' in this sense, when it is often not evident that they are conducing to a distinctly adaptive morphological end state? To avoid this challenge, one could say that it is only *adaptive* developmental processes that are amenable to the proposed teleological or agential form of explanation, but, whenever organisms *do* develop adaptively, invoking concepts of agency and purposiveness still gives us an ineliminable insight into *why* they are responding to perturbations in the way they are.

We take it that this heuristic perspective is appropriate insofar as it goes. It does valuable epistemic work in highlighting the importance of developmental robustness in securing well-adapted morphological outcomes. But we take issue with it on two fronts. First, we suggest that deploying the language of organismal agency to formulate this perspective is misleading. These concepts actively invite confusion, with reasonable readers primed to interpret discussion of the embryo's 'pursuit of its goals' as implying a literal, *directive* causal power, guiding development towards an adaptive end. This confusion is compounded by the frequent use of active verb constructions, often explicitly tagged with the modifier 'active,' which frame the organism as the subject of development, actively 'using,' 'marshalling,' 'altering,' and 'directing' its own genes, proteins, cells, tissues, and organs. This clearly does not correspond to the picture painted by the heuristic version of the agential stance, which, instead, depicts the organism's agency and purposiveness in its development as consisting in the (non-causal) relation of hypothetical necessity that holds whenever developmental processes are (robustly) adaptive. Supplementing this with distinctly causal language is needlessly misleading, if a literal interpretation is not what is intended.

Second, we argue that this approach does not offer any sort of distinctive or ineliminable explanation for *why development tends towards adaptive outcomes,* and thus why evolution is adaptively biased, beyond what is already entailed by the Darwinian logic underwriting existing evolutionary models and explanations (cf. Walsh & Sultan, this volume). In order to do so, the embryo's agential 'pursuit of its goals' would somehow need to (literally) underlie or bring about the adaptive directionality of its development—and, consequently, of its lineage's evolution—in such a way that can be differentiated from the biasing effects of natural selection. Yet there seems little empirical support for this, and, in any case, it is not what is claimed by the heuristic interpretation of the agential stance. On the contrary, as we show in the next sub-section, the *adaptive* robustness—and, hence, the apparent 'goal-directedness'—of embryonic development, which is what makes development amenable to the relevant form of teleological explanation, can readily be explained as a selected trait. The tendency of developmental processes to robustly attain well-adapted morphological outcomes is therefore more parsimoniously explained as the passive consequence of the configuration of the system, such that it has the statistical tendency to converge onto a viable phenotype—without the need to posit additional explanatory factors, such as the organism's active, real-time pursuit of its goal to persist.

8.4.1 Robustness and Phenotypic Plasticity Are Selected Traits

In general, any complex physical dynamical system that persists far from thermodynamic equilibrium, across changeable environments, will tend to adopt configurations that confer robustness across diverse conditions (Chvykov et al., 2021; England, 2022). This includes the appearance of a sort of 'memory' that lets the system rapidly shift to latent attractor states that were stable in previously encountered environments. This is consistent with the idea that complex systems that persist necessarily come to model aspects of their environment in their own physical structures (Conant & Ashby, 1970; Still et al., 2012). In effect, these tendencies reflect the outcome of a tautological selection process that selects simply for physical persistence. Where living systems are concerned, which do not just persist but also reproduce, natural selection will do the same job by selecting embryonic configurations that can robustly generate viable individuals over a range of experienced environments (Wagner, 2013; Watson & Szathmáry, 2016; Szilagyi et al., 2020).

Indeed, there is a wealth of evidence supporting the idea that robustness is itself a selected trait of the developmental system (Wagner, 2013; Alon, 2006). First, there is the well-known and general observation that phenotypes resulting from genetic mutation are not just different from the wild-type phenotype, but more variable. This was noted by Conrad Hal Waddington already in 1957. He also observed that some genotypes are associated with greater levels of developmental robustness than others:

> Another essential point about histogenesis (and morphogenesis) is that the *degree* of canalisation is under genetic control. That is to say, individuals of some genotypes show a more powerful tendency to regulate to the normal canalised paths of development than do others.
>
> (Waddington, 1957, p. 20; emphasis in original)

It is even possible to experimentally screen for mutant genotypes that specifically increase *variance* of a quantitative phenotype without altering the mean value (Ayroles et al., 2015; Hallgrimsson et al., 2019; Mestek-Boukhibar & Barkoulas, 2016). Different genotypes are thus linked to differing levels of developmental robustness, and, consequently, different amounts of inherent developmental variability (manifest as phenotypic variability across clones of genetically identical individuals reared in the same environments) (Vogt, 2015; Mitchell, 2018).

Similarly, the capacity to produce alternate, adaptive phenes in response to varying environmental conditions or factors is a species-specific trait which is selected for and, again, appears to have a genetic basis (Lea et al., 2017; Pigliucci, 2005). Many closely related species to those with celebrated examples of phenotypic plasticity (e.g., *Daphnia*, locusts, and aphids) *do not show* such plasticity when exposed to the same environmental conditions or factors. This makes sense, in that any such capacity must rely on some specifically configured biochemical and cellular components and the network of regulatory interactions between them that actually mediate the developmental processes involved in producing the alternate phenotypes (Beldade et al., 2011). There is thus every reason to expect that observed instances of adaptive phenotypic plasticity reflect prior selection for that capacity (Pigliucci, 2005; Ghalambor et al., 2015). Hence, these capacities are more akin to a pre-configured control 'policy' *implemented at the organism-level*, than an active or novel real-time 'decision' made *by the organism*. The selected nature of these capacities therefore does not support the view that organisms can agentially produce genuinely novel, *adaptive* phenotypes that could be the substrate for a distinctive form of 'phenotype-first' evolution.

Instead, these observations provide strong evidence that the robustness of an organism's development (i.e., its apparent 'goal-directedness'), whether attained through adaptive plasticity or through canalisation, is the consequence of a selection process with a predominantly genetic basis. Natural selection favours not just molecular systems that can mediate specific developmental processes, but systems that do so robustly. This is evident in gene regulatory network and signal transduction motifs employed in multicellular development, where the specific subset of possible motifs actually observed are notably robust, in engineering terms (Alon, 2006). And it is evident in the observed distributed robustness of networks at multiple cellular and physiological levels (Félix & Barkoulas, 2015; Payne et al., 2014; Nijhout et al., 2017).

We therefore do not need to appeal to the organism's purposiveness, or its active pursuit of a goal, in any sense, to explain the observation that its developmental processes resiliently tend towards adaptive outcomes. Doing so not only faces some conceptual difficulties with regard to non-adaptive and maladaptive developmental outcomes, but it also obscures the fact that existing selectionist models of adaptation already have the conceptual toolkit for more parsimonious and better empirically supported explanations of why developmental processes, such as canalisation and plasticity, often tend towards adaptive outcomes (rather than other possible outcomes).

At best, then, the agential stance provides a new vocabulary for *describing* certain (holistic) processes that are necessary to realise or enact the adaptive tendencies of development and evolution. However, this does not translate into any sort of novel, radical, or improved understanding of *why* these processes exhibit this adaptive character (as opposed to being non-adaptive or maladaptive), as is often claimed. We therefore contend that, while a literal interpretation of the agential stance is untenable, a heuristic interpretation is simply not strong enough to support the implications of (ii) *agential adaptation* and (iii) *agential evolution*. This view appears to merely represent the well-established passive 'goal' of development (i.e., a statistical tendency to produce a well-adapted individual organism) as an active 'goal' of the developing organism (i.e., a real-time pursuit of adaptation), without sufficient justification for doing so. If one's object of explanation is the *adaptive bias* observed in both development and evolution, then we suggest that standard selection-based approaches are more parsimonious and already appear entirely consistent with the purported insights of the agential stance on development.

8.5 Organism-Level Control

What about the implication outlined in (i) *organism-level control*? If one's object of explanation is the *causes of morphological development*, then some of the vocabulary afforded by the agential stance may indeed carry some noteworthy, *prima facie* strengths. First, it explicitly foregrounds the essential role that emergent levels of organisation (including the whole organism level) play in altering the causal landscape of the developing system and, thus, in securing well-adapted morphological outcomes (when they occur). Second, it strongly emphasises the robustness—or equifinality—of adaptive development, which is often missed from more reductionist, typically gene-focused, perspectives.

We therefore do not disagree that there is value in adopting an *organism-centric* perspective on development (see Walsh, 2018). Our concern is that, for the reasons given above, framing this organism-centric perspective in active, agential language often obscures the fact that this is only one perspective (or level) at which the multi-scale causal dynamics of development can be analysed. Both implicitly and explicitly, it gives the impression of an argument *against* other perspectives (e.g., gene-centrism or genotype-first adaptationism) and an argument for endorsing 'top-level causation' *over* 'multi-scale causation.'

By contrast, we would contend that the lesson from complexity science is that there is no 'privileged' level *or* timescale of causality (at least with regard to the processes of development)—no 'executive controller' at all (genetic, organismic, or otherwise). Observations of plastic self-organisation do not only imply evidence of a lack of an 'external director,' but they also speak to the absence of an 'internal director.' The whole point is that order emerges from the dynamics of a collective, as a 'happening,' not a 'doing.'

As Susan Oyama describes in the DST perspective: "Form emerges in successive interactions. Far from being imposed by some agent, it is a function of the reactivity of matter at many hierarchical levels, and of the responsiveness of those interactions to each other" (Oyama, 2000, p. 22).

We therefore suggest that the language of agency, purpose, and goals is not well suited to the problem at hand. Instead, we should indeed recognise that development is a holistic, whole-system, non-decomposable set of complex processes. Organisms exhibit reactive dynamics of self-organisation which manifest (sometimes) as adaptive developmental robustness or phenotypic plasticity. These are system-level properties that are crucial to understanding the causes of morphological development, but that does not mean they require the exercise of active, real-time control *by a subject*. Rather, any complex physical system that persists through time, including hurricanes and candle flames, will exhibit precisely these kinds of robust dynamics to environmental variation and other perturbations (Meena et al., 2023). In living systems, these dynamics explicitly reflect the past effects of natural selection in the way the zygote (and developmental environment) is configured such that these tendencies obtain, often in an adaptively biased manner. We therefore submit that it is entirely appropriate, and more parsimonious, to continue to think of them as passive tendencies (things that tend to happen) rather than capacities that need to be actively exercised by a subject (i.e., the 'doings' of an agent).

Acknowledgements

HP was supported by a Provost's Fund grant (1481.9050961) from Trinity College Dublin. We are grateful to two anonymous reviewers for very thoughtful critiques and helpful suggestions. And we thank members of the Basal Cognition Group for very helpful feedback on an earlier draft.

Note

1 Our analysis is restricted specifically to claims about *organismal* agency, wherein developing organisms are argued to 'actively control' their own development by *directly*

influencing their internal states. We therefore do not consider the role of cellular agency (e.g., Jaeger, 2022), developmental niche construction (e.g., Schwab et al., 2017), or general behavioural embryology (Gottlieb, 1976) during development.

References

Alberch, P. (1989). The logic of monsters: Evidence for internal constraint in development and evolution. *Geobios, 22*, 21–57.

Alon, U. (2006). *An introduction to systems biology: Design principles of biological circuits* (1st ed.). Chapman & Hall/CRC.

Ayroles, J. F., Buchanan, S. M., O'Leary, C., Skutt-Kakaria, K., Grenier, J. K., Clark, A. G., Hartl, D. L., & De Bivort, B. L. (2015). Behavioral idiosyncrasy reveals genetic control of phenotypic variability. *Proceedings of the National Academy of Sciences, 112*(21), 6706–6711.

Baedke, J. (2021). What's wrong with evolutionary causation? *Acta Biotheoretica, 69*(1), 79–89.

Bartsch, J. C., Schott, B. H., & Behr, J. (2023). Hippocampal dysfunction in schizophrenia and aberrant hippocampal synaptic plasticity in rodent model psychosis: A selective review. *Pharmacopsychiatry, 56*(2), 57–63.

Beldade, P., Mateus, A. R. A., & Keller, R. A. (2011). Evolution and molecular mechanisms of adaptive developmental plasticity. *Molecular Ecology, 20*(7), 1347–1363.

Bertalanffy, L. v. (1968). *General system theory: Foundations, development, applications.* George Braziller.

Chvykov, P., Berrueta, T. A., Vardhan, A., Savoie, W., Samland, A., Murphey, T. D., Wiesenfeld, K., Goldman, D. I., & England, J. L. (2021). Low rattling: A predictive principle for self-organization in active collectives. *Science, 371*(6524), 90–95.

Conant, R. C., & Ashby, W. R. (1970). Every good regulator of a system must be a model of that system. *International Journal of Systems Science, 1*(2), 89–97.

Durstewitz, D., Huys, Q. J., & Koppe, G. (2021). Psychiatric illnesses as disorders of network dynamics. *Biological Psychiatry: Cognitive Neuroscience and Neuroimaging, 6*(9), 865–876.

England, J. L. (2022). Self-organized computation in the far-from-equilibrium cell. *Biophysics Reviews, 3*(4), 041303.

Fábregas-Tejeda, A., & Baedke, J. (2023). Teleology, organisms, and genes: A commentary on Haig. In T. E. Dickins & B. J. A. Dickins (Eds.), *Evolutionary biology: Contemporary and historical reflections upon core theory* (pp. 249–264). Springer.

Félix, M.-A., & Barkoulas, M. (2015). Pervasive robustness in biological systems. *Nature Reviews Genetics, 16*(8), 483–496.

Fulda, F. C. (2023). Agential autonomy and biological individuality. *Evolution & Development, 25*(6), 353–370.

Ghalambor, C. K., Hoke, K. L., Ruell, E. W., Fischer, E. K., Reznick, D. N., & Hughes, K. A. (2015). Non-adaptive plasticity potentiates rapid adaptive evolution of gene expression in nature. *Nature, 525*(7569), 372–375.

Gottlieb, G. (1976). Behavioral embryology. *Recherche, 7*(71), 833–841.

Hallgrimsson, B., Green, R. M., Katz, D. C., Fish, J. L., Bernier, F. P., Roseman, C. C., Young, N. M., Cheverud, J. M., & Marcucio, R. S. (2019). The developmental-genetics of canalization. *Seminars in Cell & Developmental Biology, 88*, 67–79.

Jaeger, J. (2022). Ontogenesis, organization, and organismal agency. OSF. https://doi.org/ https://doi.org/10.31219/osf.io/x3uny

Juarrero, A. (1999). *Dynamics in action.* MIT Press.

Lea, A. J., Tung, J., Archie, E. A., & Alberts, S. C. (2017). Developmental plasticity: Bridging research in evolution and human health. *Evolution, Medicine, and Public Health, 2017*(1), 162–175.

Lignani, G., Baldelli, P., & Marra, V. (2020). Homeostatic plasticity in epilepsy. *Frontiers in Cellular Neuroscience, 14,* 197. https://doi.org/10.3389/fncel.2020.00197

Meena, C., Hens, C., Acharyya, S., Haber, S., Boccaletti, S., & Barzel, B. (2023). Emergent stability in complex network dynamics. *Nature Physics, 19,* 1033–1042.

Mestek-Boukhibar, L., & Barkoulas, M. (2016). The developmental genetics of biological robustness. *Annals of Botany, 117*(5), 699–707.

Mitchell, K. J. (2015). The genetic architecture of neurodevelopmental disorders. In K. J. Mitchell (Ed.), *The genetics of neurodevelopmental disorders* (pp. 1–28). Wiley.

Mitchell, K. J. (2018). *Innate: How the wiring of our brains shapes who we are.* Princeton University Press.

Mitchell, K. J. (2023). *Free agents: How evolution gave us free will.* Princeton University Press.

Nadolski, E. M., & Moczek, A. P. (2023). Promises and limits of an agency perspective in evolutionary developmental biology. *Evolution & Development.* https://doi.org/10.1111/ ede.12432

Neuberger, E. J., Gupta, A., Subramanian, D., Korgaonkar, A. A., & Santhakumar, V. (2019). Converging early responses to brain injury pave the road to epileptogenesis. *Journal of Neuroscience Research, 97*(11), 1335–1344.

Nijhout, H. F., Sadre-Marandi, F., Best, J., & Reed, M. C. (2017). Systems biology of phenotypic robustness and plasticity. *Integrative and Comparative Biology, 57*(2), 171–184.

Oyama, S. (2000). *The ontogeny of information: Developmental systems and evolution.* Duke University Press.

Payne, J. L., Moore, J. H., & Wagner, A. (2014). Robustness, evolvability, and the logic of genetic regulation. *Artificial Life, 20*(1), 111–126.

Pessoa, L. (2022). *The entangled brain: How perception, cognition, and emotion are woven together.* MIT Press.

Pigliucci, M. (2005). Evolution of phenotypic plasticity: Where are we going now? *Trends in Ecology & Evolution, 20*(9), 481–486.

Potter, H. D., & Mitchell, K. J. (2022). Naturalizing agent causation. *Entropy, 24*(4), 472. https://doi.org/10.3390/e24040472

Russell, E. S. (1924). *The study of living things.* Methuen & Company Limited.

Schwab, D. B., Casasa, S., & Moczek, A. P. (2017). Evidence of developmental niche construction in dung beetles: Effects on growth, scaling and reproductive success. *Ecology Letters, 20*(11), 1353–1363.

Snell-Rood, E. C., & Ehlman, S. M. (2023). Developing the genotype-to-phenotype relationship in evolutionary theory: A primer of developmental features. *Evolution & Development, 25,* 393–409.

Still, S., Sivak, D. A., Bell, A. J., & Crooks, G. E. (2012). Thermodynamics of prediction. *Physical Review Letters, 109*(12), 120604.

Sultan, S. E., Moczek, A. P., & Walsh, D. (2022). Bridging the explanatory gaps: What can we learn from a biological agency perspective? *BioEssays, 44*(1), 2100185.

Szilágyi, A., Szabó, P., Santos, M., & Szathmáry, E. (2020). Phenotypes to remember: Evolutionary developmental memory capacity and robustness. *PLoS Computational Biology, 16*(11), e1008425.

Uller, T. (2022). Agency, goal-orientation and evolutionary explanations. OSF. https://doi.org/https://doi.org/10.31219/osf.io/49qrs

Vogt, G. (2015). Stochastic developmental variation, an epigenetic source of phenotypic diversity with far-reaching biological consequences. *Journal of Biosciences, 40*, 159–204.

Waddington, C. H. (1957). *The strategy of the genes*. Macmillan.

Wagner, A. (2013). *Robustness and evolvability in living systems*. Princeton University Press.

Walsh, D. M. (2006). Organisms as natural purposes: The contemporary evolutionary perspective. *Studies in History and Philosophy of Biological and Biomedical Sciences, 37*(4), 771–791.

Walsh, D. M. (2015). *Organisms, agency, and evolution*. Cambridge University Press.

Walsh, D. M. (2018). Objectcy and agency: Towards a methodological vitalism. In D. J. Nicholson & J. Dupré (Eds.), *Everything flows: Towards a processual philosophy of biology* (pp. 167–186). Oxford University Press.

Walsh, D. M., & Rupik, G. (2023). The agential perspective: Countermapping the modern synthesis. *Evolution & Development, 25*, 335–352.

Watson, R. A., & Szathmáry, E. (2016). How can evolution learn? *Trends in Ecology & Evolution, 31*(2), 147–157.

West-Eberhard, M. J. (2003). *Developmental plasticity and evolution*. Oxford University Press.

Part III

Behaviour, Scientific Practice, and Self-Individuation

9 In Defense of the Whole

Behavioral Novelty as a Reflection of Organismal Agency

Gregory M. Kohn

9.1 Introduction

In his book *The Directiveness of Organic Activities*, E. S. Russell described the behavior of a caddisfly larva. A caddisfly spends most of its life in its larval stage. The larva uses the silk glands in its mouth to cement small pebbles and sand into a tubular case. Russell (1945) recounts a series of studies where researchers selectively damaged a portion of the case and then described in detail the various ways the caddisfly repaired it. If the posterior part of the case was damaged, some larvae responded by reversing in their tube and fixing the damage directly, others extended the tube until it was the length typical of the species, while others built a new case on top of the damaged one and severed the connection with the damaged case upon opening. Overall, Russell (1945) noted that "of fifty-four cases no two were restored in exactly the same way" (p. 21).

This example—while on the surface seems trivial—contains a distinctive property of living systems: the ability to flexibly adapt to changing circumstances. This is ubiquitous across all organisms; it is seen in the taxis behavior in single-celled archaea, to the elaborate motor patterns that constitute social interactions in vertebrates. But what explains the emergence of such flexible actions? Are they simply noisy responses to changing circumstances without direct functional relevance, or are they specifically utilized to satisfying specific future functions, or goals?

The relationship between goals and behavior has its roots in the writings of Aristotle (Farquharson, 1985; see also Esposito, this volume). In his *De motu animalium*, Aristotle locates the source of animacy in animals within an inborn *pneuma*, or creative spirit, which was the source of desire. Desire represents the activation of some internal processes that compel the animal to move toward the fulfillment of goals. Aristotle states: "[...] living creatures are impelled to move and to act, and desire is the last or immediate cause of movement, and desire arises after perception or after imagination and conception" (Aristotle, as translated in Farquharson, 1985, p. 1092). In making a comparison between an automaton, a toy wagon, and the activity of an organism, Aristotle discusses how the wagon must be animated from the outside and is, therefore, an entirely passive object. In contrast to the wagon, the organism moves because it desired to move toward some future

DOI: 10.4324/9781003413318-12

state. It moved because it possessed goals of its own that *caused* the behavior. However, despite Aristotle's attempt to center the study of behavior around animating spirits and goals, the science of behavior moved in a different direction.

Sir Isaac Newton's three laws of motion proposed that no object, no matter how complex, could move without some local cause acting upon it first. All objects in the universe behaved in a series of stops and starts, where objects at rest were stirred into motion by external physical forces. The pursuit of goals seems to violate this. The goal itself could not cause the behavior, as this would suggest that ends transcend the forward movement of time to cause the behavior in the first place.

Nonetheless, the behavior of organisms seems distinctly non-Newtonian. The ability of organisms to move up gradients, to grow and develop, to create new behaviors and evolve new levels of complexity seems to work in opposition to Newtonian forces, not because of them. Due to this, there have been increasing discussions on how to naturalize agency to show that organisms are goal-directed systems while avoiding the unseen forces of the past. In this chapter, I propose a perspective wherein traces of agency can be observed, assessed, and measured in behavioral novelties.

First, I discuss the origins of goal-directed behaviors and argue that the autopoietic organization of organisms allows them to be goal-possessing systems, and that behavior is an observable and measurable outcome of this organization. Next, I discuss the structure of goal-directed behaviors and I propose that many goal-directed behaviors reflect the use of behavioral variability to construct or sustain an invariant relationship with the environment. I then move to discuss the control of goal-directed behavior, wherein the animal develops an agential niche that allows it to control how the environment influences itself.

9.2 The Origin of Goal-Directed Behavior: What Systems Possess Goals?

In his essay "Cybernetics and Purpose," Hans Jonas (1966/2001) made a distinction between two types of systems: (i) those that *act* on goals and (ii) those that *possess* goals. A thermostat and a guided missile are systems that act on a goal but do not possess them. A guided missile can sense the presence of a moving target via sensors and use this information to direct its movements through a guidance system. This monitors input from the sensory system, and when it detects the target deviating from a preferred orientation, it makes compensatory adjustments to the missile's behavior, aiming to place the target directly in the center of its flight path.

The behavior of guided missiles and thermostats is primarily driven by the stimulus itself. These objects possess negative feedback mechanisms that respond to changing patterns of input from a target stimulus. While the behaviors of these objects are directed toward specific goal states, they are also pre-specified, as experiencing the same stimuli in the same way will result in the same behaviors. The behavior is completely stimulus-driven and, despite its apparent flexibility, fixed. These objects do not seek out specific contexts or situations when no relevant stimuli are present but respond appropriately to a stimulus if placed in an appropriate context. As such, these artifacts do not possess goals in themselves but are programmed with goals by humans.

Organisms are the only systems known that can both possess goals and act according to them (Jonas, 1966/2001). I view this capacity as a cornerstone of organismal agency. Definitions of agency vary from "the capacity to perform acts" (Di Paolo et al., 2017, p. 6) to the "capacity of a system to cope with its setting to obtain a goal by responding to affordances *as* affordances" (Walsh, 2015, p. 210; emphasis in original) to the ability to produce a "variation of means to bring about an end" (Turvey, 2018, p. 305). In all these definitions, organisms use their behavior to bring about a particular state of affairs necessary to them. Be it a caddisfly larva repairing its case, a male zebra finch attempting to force nesting material into a cavity, or a hungry green anole tracking an ant as it runs across a branch, most behavior seems directed toward particular states. But what is distinct about such behaviors from those in a guided missile? I argue that one clear difference is that organisms possess and acquire goals and use these to motivate their behavior.

But what type of shared biological organization allows an organism to possess goals? The project to naturalize agency identifies the necessary and sufficient factors for possessing goals. While it is beyond the scope of this chapter to systematically review the project to naturalize agency, most work in this area starts from the fact that organisms exist in a thermodynamically far-from-equilibrium state and need to maintain this state from birth until death. To uphold this state, organisms need to be self-creating, continually exchanging energy and matter with their surrounding environment to maintain themselves against entropic forces (Maturana & Varela, 1991).

A minimal living system is characterized by its ability to create and sustain its individuality. This feature has many different names. For Jonas it was called "metabolism" (Jonas, 1966/2001), and for Maturana and Varela (1991) it was called "autopoiesis" (Weber & Varela, 2002; for discussion, see also Michelini, this volume). According to these proposals, a minimal living system is a bounded set of molecular components and reactions that are also produced by that system. Such components can either be produced endogenously or extracted from the environment through the behavior of that system. This shares similarities with autocatalytic sets, where each molecule in a network of chemical reactions is made by another interaction in that same network. Autocatalytic sets have been discovered in non-organic molecules (Miras et al., 2020), *Escherichia coli* (Sousa et al., 2015), and have been hypothesized to be an essential component in the emergence of life on Earth (Hordijk et al., 2010; Kauffman, 1993).

Perturbed autocatalytic sets initiate internally consistent signals to other components when the reaction is not able to recreate itself (Hordijk, 2019; Loutchko, 2023). These signals are cascading changes across the whole network of reactions that cease once the network regains its self-generating capabilities. Autocatalytic sets act in reference to constructing and maintaining their network of reactions in the face of perturbation. They do not just sustain themselves while at rest but move toward sustaining themselves when the context changes. Such reactions represent a unique self-referential, self-determining, and self-constructing chemistry.

While autocatalytic sets create networks capable of reinforcing and recreating themselves, they do not yet have the distinction of being autonomous entities. To become an autonomous organism, a physical boundary needs to emerge that delineates internal autocatalytic processes from external environments. This boundary—be it a cell membrane, skin, cell wall, or chitinous exoskeleton—sets the stage for self-individualization (Varela et al., 1991; Varela, 1991). The activity of the internal process comes to define and sustain itself through the reinforcement of this boundary.

As stated by Jonas, an individual is "essentially its own function, its own concern, its own continuous achievement" (Jonas, 1966/2001, p. 80). This creates a normative relationship with the world, wherein the whole organism seeks out features of the environment that sustain and recreate its individuality. At this juncture we see the emergence of agency. Organisms no longer merely respond passively to external forces but can assess the context and move toward conditions that sustain their integrity and away from conditions that do not. Individuals perform actions and decisions directed toward persisting. As such, all living organisms possess basal goals directed toward preserving the integrity of the processes that generate and sustain their individual existence.

9.3 How Do Goal-Possessing Systems Pursue Goals?

[A]n insect which has lost a leg will at once change its style of walking to make up for the loss. This may involve a complete alteration of the normal method, limbs which were advanced alternately being now advanced simultaneously. The activities […] are directed to a definite end, the forward movement of the animal—it uses whatever means are at its disposal and is not limited to particular pathways.

(Adrian, 1933, p. 468, as cited in Russell, 1945, p. 127)

The project of naturalizing agency remains an open and challenging area of investigation. While progress is being made, there is no consensus on which specific biological criteria are sufficient for agency (Barandiaran et al., 2009; Lidgard & Nyhart, 2017; Peng et al., 2022). Empirical approaches, such as identifying autocatalytic sets, pose significant challenges, and the *in situ* study of autocatalytic reactions is still in its early stages (Hordijk, 2019). My approach begins at the other end, by asking if the structure of behavior itself can be used to assess if organisms possess and act on goals. As stated by Walsh, "[a]gency is observable in the sense that we see when we observe an agent in its dynamics, the way the agent negotiates the situation" (Walsh, 2015, p. 210).

But which aspects of behavior carry the signature of agency? What do we mean when we observe the way an "agent negotiates the situation"? Like all modern sciences, the behavioral sciences inherited the mechanist foundation of Newton and hold that internal and external factors that precede the behavior are its causes (Cziko, 2000; Nicholson, 2012; Riskin, 2016). Instead of viewing organisms as capable of initiating self-movement toward goals, most biologists and psychologists take a mechanistic view, wherein organisms contain pre-established instructions for specific inputs. These inputs *cause* the behavior that then moves the organism toward the end state, which is held to be a byproduct of those initial causes.

This mechanistic approach has been tremendously fruitful in heuristic and explanatory terms. By investigating the input–output structure of their components, we have discovered more about how living systems function in comparison to agency-centered approaches (Anctil, 2022). Here, the ability to sustain a far-from-equilibrium state reflects the utilization of an evolved "toolkit" transmitted across generations. This toolkit contains a set of genetic and neural mechanisms tailored to specific features of the environment. In this view, the organism is a machine "designed" by selection and is nothing more than a series of adaptive mechanisms that process information and respond appropriately. Individuals with a less-than-optimal toolkit fail to survive and reproduce and therefore do not contribute to the next generation.

This view avoids a serious consideration of organismal agency. It is assumed that environmental selective pressures will be consistent enough so populations converge on adaptive stimulus-response mechanisms (Mayr, 1985). While organisms behave as if they have goals, their behavior is a set response to the stimulus' properties. Behaviors are a reflection of fixed, pre-established, and latent internal mechanisms designed to ensure consistency over time. Any observed variation is considered noise within this fixed structure.

Nonetheless, when investigating the details of behavior, it is difficult to identify which motor patterns are fixed and which are not. In practice, researchers often ignore motoric variability and rely on subjective assessments to decide when a behavior begins and ends (Golani, 2022). In nesting greylag geese (*Anser anser*), when an egg falls outside the nest individuals respond by extending their necks, placing the displaced egg under their bill, and rolling it back toward the nest. This action was considered fixed with static neural circuits instantiating every move (Lorenz & Tinbergen, 1970). However, subsequent research showed that even when the egg is placed in the same location, the details vary significantly (Lehrman, 1953). As observers, we often assume that behavior directed toward a fixed point has itself some fixed structure, whereas research has highlighted how the details of the behavior change each time it is performed (Golani, 2022; Latash, 2012; Pellis, 1985; Zilov et al., 2017).

Knowing that behavior continually changes does not itself suggest that variability is functional. Many scholars relegated variability to noise around an optimal behavioral response (Lehner, 1998). No individual is going to have the exact same experiences as their ancestors. As novel environmental challenges will create novel responses, behavioral flexibility may simply be noisy responses to novel stimuli interfering with the performance of a fixed behavior. Others have proposed the idea of adaptive flexibility, where discrete mechanisms produce functional variability in response to novel changes. Here "novel stimuli" can be pigeonholed into an overarching and predictable type of environmental stimuli.

Consider a very simple organism that moves toward chemical cues associated with food. With basic reinforcement learning, the organism classifies chemical cues as food (which initiates chemotaxis) or non-food (which does not). However,

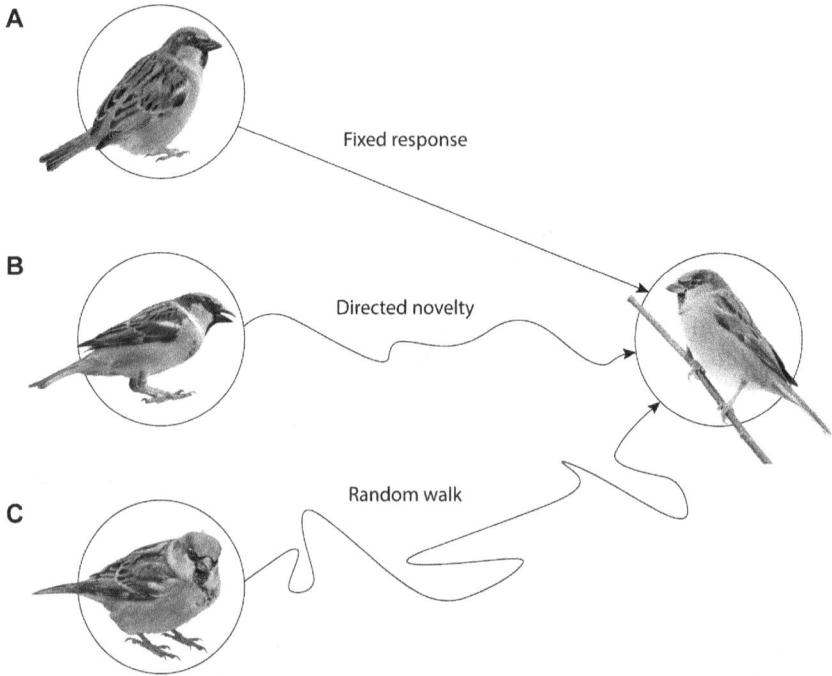

Figure 9.1 Three possible scenarios of behavioral variability toward a target object. (A) A fixed and directed path is a direct result of a fixed mechanism; the individual is not able to make compensatory adjustments. (B) A directed and novel path shows the modulation of approach–withdrawal processes to persist toward a goal when perturbed. (C) Individuals show a random walk that is neither directed nor persistent toward the targeted object.

the introduction of any novel perturbation immediately activates a program producing a random walk (Figure 9.1). This random walk increases the likelihood that the organism will encounter beneficial chemical gradients in changing conditions. What looks like an agent purposefully behaving in an immediately adaptive way is simply a stimulus-response mechanism, where the stimuli of "novel change" initiate a variable random walk response.

From this perspective, variability is the result of a pre-specified mechanism tailored to changing environments. One difficulty with this view is that there are no unifying stimulus properties that characterize a changing environment. The way that environments change is simply too diverse to develop a single mechanism to address all of them. Saying flexibility is simply a response to changing environments leaves open the question about what the organism was responding to, and exactly why it responded in the way it did.

I argue that in behavioral variability we see signatures of agency. Novel motor patterns are not mere responses to changing environments, but competent, directed actions aimed at obtaining goals in unpredictable contexts. For instance, the variations in the egg-retrieval behavior are specifically tailored to the individual's immediate experience, allowing the goose to cope with unpredictable changes in the position and movements of the eggs while drawing them back into the nest (Lehrman, 1953). Expecting the organism to come with a pre-specified mechanism for every environmental stimulus is not realistic. Nonetheless, almost all individuals navigate changing environmental conditions and thrive. How organisms do this points to systems that can both possess and act on goals.

The ability to express immediately adaptive actions in novel contexts suggests behavior is not entirely stimulus-driven. Individuals would not have the necessary experiences (individually or phylogenetically) to acquire a stimulus-response system tailored to each context. This can be seen in some of Tolman's studies where rats are trained to swim through a maze toward a goal point (Tolman, 1949). After learning to swim the route successfully, they were placed in the maze with most of the water drained. Despite shifting their behavior from swimming to wading, individuals were able to successfully navigate to the end of the maze. What was controlling the behavior was not a set response to a particular stimulus, but active accommodations made to facilitate the acquisition of a goal. The challenge of identifying signatures of agency therefore lies in disentangling the stimuli-driven aspects of behavior from the goal-driven aspects of behavior. This task involves identifying which factors control and constrain the structure a behavior takes in naturalistic contexts.

9.4 Structure of Goal-Directed Behavior: Concerted Action toward Constructing Invariants

> The various forms, variations, and motions a plant or animal expresses are not mere unions or mechanical movements of insignificant matter: because each has an inseparable relationship to the whole, each should be regarded as a developmental expression of one unifying self.
>
> (Nishida, 1992, p. 63)

To act on a goal, individuals must direct their behavior toward specific future states. Following Sommerhoff (1950), Nagel (1977) proposed that all goal-directed actions are characterized by *persistence* and *plasticity* toward these states. Persistence is the resilience shown in goal-directed actions. If perturbed when moving toward a goal, an organism can "correct" itself and proceed. A protist undergoing chemotaxis toward an area with a high-food concentration can easily readjust if an obstacle is placed within its path. Plasticity (also called "equifinality," see Bertalanffy, 1968) is the ability of a system to move from different starting points toward a singular goal state. The protist can be released in any location in a tank, but at each location, it will inevitably find its way to the high-food concentration area.

Persistence and plasticity were supposed to capture the necessary and sufficient conditions for all goal-directed behavior. But many scholars issued strong objections to this, for instance, such criteria were so overarching that they captured many systems that are not agents (Heylighen, 2023). One example is a ball spinning downwards in a cone. This system shows persistence as most perturbations to the ball's trajectory will not stop it from reaching the bottom of the cone. It also shows plasticity as the ball may start spinning from any position along the cone but will ultimately reach the bottom. Despite meeting these two criteria, few people would assign goals to the ball.

To contend with this, some proposed that goals reflect a shared factor that creates and controls persistence and plasticity. This factor, known as "concerted action," reflects coordination across all levels and components of the organism necessary to act toward a goal (Heylighen, 2022). A fly alighting on the tip of a grass blade needs its eyes, wings, metabolism, and nervous system to work in a synchronized fashion to land. As stated by Trestman (2012), an animal "is composed of many and diverse subsystems whose responsive dispositions relate to each other and to the external environment as a highly complex coherent structure of interrelated dynamical convergences" (p. 216).

The concept of concerted action shares many similarities with the concept of behavioral gradients. A behavioral gradient is a network of coordinated components and processes that result in a specific behavior. The concept of the behavior gradient was first proposed by Kuo (1976) to show that in each behavior the "whole organism is involved." Kuo (1967, pp. 92–93) stated that behavior was the "most complex and complicated biological phenomena" as they "include activities of every part and every organ of the whole animals as well as their feedbacks and interactions." Due to the interactions of so many different components, "qualitative and quantitative variations from moment to moment" emerge. In other words, the result of behavioral gradients is that behavior exhibits some level of novelty.

The existence of behavioral gradients does not mean that all parts of organisms are equally involved. For instance, one could make the case that the digestive system does not play a significant role in courtship behaviors. While behavior requires a network of coordinated interactions across the organism, different parts of the network may play a more or less substantial role, resulting in a gradient of influences across the organism. For the fly alighting on the grass blade, the role of the eyes is substantial, while the role of reproductive physiology might be less important. Nonetheless, even reproductive physiology cannot be discounted as inconsequential, considering that the fly may have landed on the grass to get close to a potential mate.

However, does the existence of behavioral gradients imply individuals move toward goals? Kuo resisted that interpretation. He maintained that viewing animals as goal-possessing agents creates a focus on the "end result of the performances of the animal subject than in the processes of reorganization of the behavioral gradient patterns" (Kuo, 1967, p. 142). He thought the "start-end" nature of goal-directed behavior caused the continual stream of actions to be "arbitrarily sectionalized or fractionalized to serve the purpose of the observer or experimenter" (Kuo, 1967, p. 14). He regarded this as finalistic and stated that behaviors do not have sharp boundaries and should be viewed as a continual flow of movements from birth to death.

As such, Kuo saw behavioral gradients in opposition to goal-directed actions. Gradients focused on the dynamics of actions; goals focused on their ends. He proposed that a complete understanding of all interacting internal and external factors involved would result in a complete explanation for the behaviors without recourse to any end state. As he maintained, "[b]etween the action of a man and the movement of a stone is merely a difference in complexity" (Kuo, 1928, p. 417). Behavioral gradients were seen as elaborate stimulus-response relationships that differed only by virtue of their complexity to the movements of non-living objects.

Despite this, other conceptualizations of goals emerged that avoided arbitrary finalism. According to them, behaviors are not compelled to some final state by an unseen force but emerge as individuals seek to construct relationships with relevant features of their environment. Here each gradient is a compensatory adjustment to create and sustain preferred invariant relationships with objects in their environment. Behavioral gradients do not just reflect where the organism has been, but also where it wants to go. Re-conceptualizing goal states as the active construction of invariants, rather than end states themselves, avoids the finalistic problems raised by Kuo.

But what do I mean by an invariant relationship? An invariant captures a spatial relationship between an organism's body and its environment that is sustained over time. Work by Golani (1981, 2022), Pellis (1985), Pellis et al. (2009, 2014), Fentress (1983), Fentress and McLeod (1986), and others (e.g., Gomez-Marin et al., 2016; Schleidt, 1974; Schleidt & Crawley, 1980) developed methods to classify invariants within the continual stream of behaviors. Invariants can be as simple as a stable angle of orientation toward a static object, to complex postures oriented toward another individual. For instance, in the pre-copulatory behavior of Tasmanian devils (*Sarcophilus harrisii*), two individuals maintain a cheek-to-cheek orientation through a diversity of behaviors such as rolling, wrestling, and revolving (Golani, 1981). In crickets (*Gryllus bimaculatus*), males maintain a parallel posture with the ground regardless of the slope of the substrate. As the slope changes, crickets modify their leg positions to maintain a parallel posture. In Mallard (*Anas platyrhynchos*) courtship, individuals perform a changing array of displays; yet despite substantial variation in the structure of the displays, each one is characterized by a fixed bill-to-water distance and this distance is sustained even across changes in the water level brought by, for example, waves (Finley et al., 1983).

The invariant is not an end-point to the behavior; it represents a goal not because the behavior stops once it is created, but because the organism has to work to maintain it despite environmental obstacles. To maintain an invariant in a changing environment, individuals must adapt to changes in sensory input and make compensatory adjustments in their behavioral gradients. The result of this process is the production of directed novelties. A directed novelty is shown in the unique motor patterns directed toward constructing an invariant.

Directed novelties only make sense in relation to the aim of creating and maintaining an invariant. In many cases, physical forces may be acting in direct opposition to the behavior of an organism. For instance, an organism that reorients its posture to climb up an incline works against the opposing force of gravity. While many physical systems show action toward specific states, the details of these actions can be predicted based on initial conditions. With a description of the exact structure of the ball (weight, material) and its environment (sitting 5 cm from a 45-degree slope), we can predict with accuracy what the ball will do if an external force is applied to it.

No such descriptions exist for describing most biological motion. Complete knowledge of the context and structure of the organism at time point t will not predict the exact motor patterns employed at time t_{+1}. This point can be seen in the work of Nikolai Bernstein. He was a neurophysiologist who was the first to use movement-tracking technology to map the exact motor patterns used in basic goal-directed tasks (Bernstein, 1967; Latash, 2020). Bernstein discovered that even stereotyped behaviors were characterized as "repetition without repetition." Motor abundance, variability, and novelty were ubiquitous in goal-directed action. Because of this, Bernstein noted that the degrees of freedom in how an animal could move (as constrained by their morphology) were often orders of magnitude greater than the constraints within a typical context.

As such, each behavior contains a near-infinite number of possible motor combinations. Let us imagine a bird flying across a room to a perch where another individual is seated. The bird could take any number of paths constrained by the shape of the room. Within each path, the bird is free to vary the number, depth, and posture associated with their wing beats. Going down further to the muscle systems governing the wing strokes, there is a near-infinite number of combinations and coordinated patterns of muscle contractions that could be employed to maintain the overall posture and motion during each wing beat.

As proposed by Bernstein, an individual's actions are not just repeated but reconstructed anew, and the purpose of research is to identify the level that regulates the reconstruction (Latash, 2020). An important task for behavioral research is to show which variables control behavioral gradients and reduce the near-infinite number of possible patterns to the observed ones. Since the exact form of behavior is novel, it is difficult to see how each pattern could be wholly instantiated in any pre-specified mechanism. Therefore, the control of behavior likely does not come solely from neural activation, muscle contractions, environmental constraints, or stimulus properties, but from transactions across them.

9.5 Control of Goal-Directed Behavior: Developing an Agential Niche

Control is a negotiation between the present and the future. To be in control is to utilize information in the current context in the service of constructing a specific future context. In the previous section, I looked across behaviors to highlight how directed novelties are produced to create an invariant. In this section, I aim to understand how the *prospectivity* of these directed novelties develops. Prospectivity reflects a characteristic of directed novelties to serve as a means to both anticipate and construct future goals.

For some scholars, prospectivity is a simple outcome of learning. The organism learns which stimuli are predictive of the goals in a future context. However, even with learned associations one must contend with the variability in the path between stimuli and goals. Learned associations generalize across contexts where individuals must cope with intervening factors in the environment. The variability shown as the animal navigates from stimuli to goal is often not part of the originally learned association. Thus, even if a stimulus is predictive of future goals, the often-winding path taken by the organism from the exposure to the stimuli to the goal is unexplained by learning alone.

Other perspectives beyond learning are needed to map the path an animal forms as it moves toward their goals. From cybernetics (Wiener, 1961) to ecological psychology (Turvey, 2018) to perceptual control theory (Powers, 1978), many different models and approaches have been proposed to capture how individuals anticipate and direct their behavior toward future states. Nonetheless, a common thread throughout all these theories is that individuals inhabit a niche of meaningful objects around them that allows them to control how the environment influences them, and how they influence the environment (Figure 9.2).

I call this niche the *agential niche*. From the perspective of ecological psychology, an agential niche comprises all affordances, or "possibilities for action," at each time point. Heft (1989) defines affordances as a perceived "relation to some

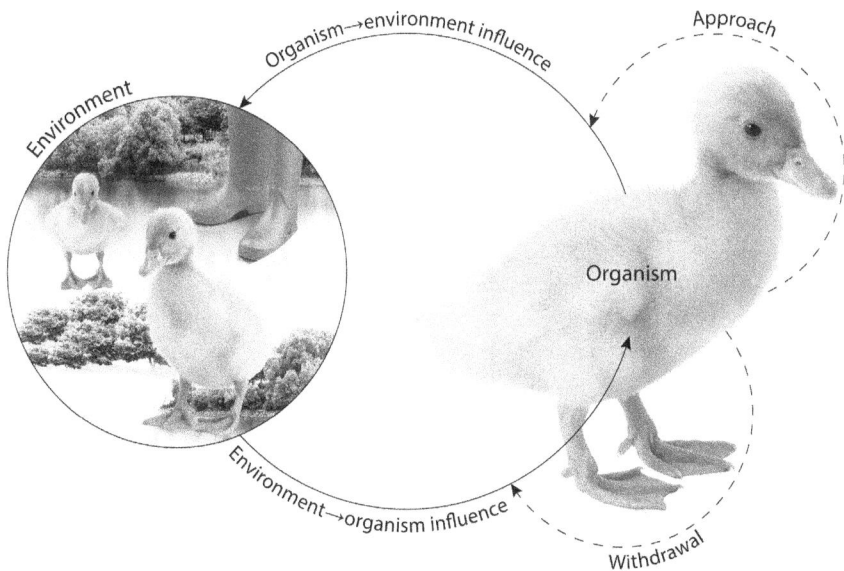

Figure 9.2 Organismal control in an agential niche. Organisms are not passive recipients of environmental influences. By selectively approaching objects, they control how they shape their environments, and by selectively withdrawing from objects, they control how the environment influences them.

intentional act, not only in relation to the body's physical dimensions" (p. 13; for discussion, see also Michelini, this volume). At any time point, there are many potential invariants that the organism could form. The invariants for a potential mate are very different from those that could be formed with a food source, even though both may be present simultaneously. The question of prospectivity thus becomes centered on: (i) how individuals decide which invariants will be prioritized and (ii) how that prioritization creates future-oriented directed novelties toward that invariant.

What does it mean to prioritize an invariant? This starts with recognizing objects in the environment and selectively attending to them. To do this, some sensory information needs to become more salient than others. Because of their autopoietic organization, even the most basal organisms must be able to sense aspects of the environment that facilitate their self-maintenance. Such sensory systems can be as simple as a receptor on the membrane of a single-celled organism, to complicated structures such as an eye. These sensory systems curate the types of information the organism picks up from the surrounding environment.

Understanding how a sensed object becomes a prioritized object requires an understanding of how goals, interests, and needs develop. During development, the cumulative influence of an individual's experiences will increase the salience of certain sensory information. An example of this can be seen in the response to maternal contact calls in fledgling wood ducklings (Gottlieb, 1997). Upon fledging, wood ducklings follow the sound of the maternal contact call to their mother and sustain a pattern of close proximity with her. Embryos require exposure to the sound of their siblings' embryonic vocalizations to develop this response to the maternal contact call. If eggs are raised alone in incubators, they do not show the ability to respond to maternal contact calls, and the calls lose their relevance to the individual. Experiences—including non-obvious ones—allow the wood duck to prioritize the maternal contact call above and beyond other types of sensory information in its agential niche.

Across generations, individuals with similar sensory systems are likely to have similar experiences resulting in the predictable prioritization of sensory information (Gottlieb, 2007). Simply having the right body, at the right time, with the right history of experiences assures that some objects will have more relevance than others. Now, once an object is prioritized, what controls how an individual directs future actions toward it? The basis of this is the modulation of an animal's approach–withdrawal system to control its experiences in the current agential niche (Figure 9.2).

The ability to approach and withdraw from specific objects is central to all models of goal-directed behavior from cybernetics to perceptual control theory (Weiner, 1961; Powers, 1978). It represents the most foundational way to exert control over your spatial relationships. These basic processes of approach–withdrawal are common to nearly all animals during early development and provide the basis for the ontogeny of more complex behavior (Schneirla, 1959, 1965). Even sessile animals such as barnacles can exhibit approach–withdrawal through the extension and retraction of their cirri. As stated by Weber and Varela (2002, p. 117), these approach–withdrawal processes allow the organism to "acquire a valence which is dual at its basis: attraction or rejection, approach or escape."

But do approach–withdrawal systems shape the development of invariant relationships? Almost all invariants contain spatial relationships to particular objects. For example, strangers, friends, potential threats, food, mates, and others are usually kept at a specific distance from the individual. If we start with a simple goal of maintaining a set distance from an object, we see how these shared approach–withdraw processes are molded into actions aimed at sustaining a specific invariant. As Golani (1981) and others (e.g., Bell, 2014; Pellis & Bell, 2020) have shown, individuals maintain a stable distance by modulating how they approach and withdraw from changes in the paired object's movement.

During early development moderate-to-gentle onsets of sensory activation produce an approach, with a sudden onset of sensory activation producing withdrawal (Schneirla, 1959). An invariant may be acquired as individuals incorporate the changes in sensory activation associated with an object into their behavioral repertoire. A stimulus that is more likely to be experienced with a sudden onset will lead individuals to maintain a larger distance from that object to buffer against the potential of strong activation of sensory information, while an object more likely to present itself gently may be held closer. For instance, aggression is usually characterized by sudden and drastic changes in sensory activation, and as a result, individuals often acquire a larger spatial invariant when interacting with an aggressive individual. In this fashion, some of the first invariant relationships with objects are developed.

The contingent flow of experiences from conception onward shapes the prioritization of specific objects and the invariant relationships with them. Then again, what about the directed novelties individuals use to navigate toward the object on a moment-to-moment basis? To move toward a goal, the organism needs to have the ability to coordinate its approach–withdrawal abilities from past niches and in anticipation of future ones (Turvey, 2018, p. 305). Thus, future invariant relationships become the goal, and novel behaviors are produced to satisfy that goal. Let us return to the example of a bird flying across a room to its partner. Imagine we pause the behavior a fraction of a second after the bird alights from its perch. In this present niche, the bird is frozen in mid-air clearly oriented toward the object of relevance, the partner at the other end of the cage. The bird's behavior and posture reflect its past agential niche but are directed toward a future one. In this past niche, the bird was pushing off the perch but had not yet gained flight. Here the bird had to coordinate its body posture and center of mass to match the sturdiness and grip of the perch against its feet; the angle of the initial push to flight also reflected the angle necessary to direct it to the partner at the end of the cage.

There is nothing inherent about the stimuli of the bird's partner that caused it to angle and lift off the perch in that particular way. No single receptor or mechanism recognizes the bird's partner and initiates flight from the perch at that specific angle, and there is likely no learned association between that specific angle and the goal of being close to the bird. No description of the physical forces acting in the cage compels a naive observer to predict how the bird would fly across the cage. The common thread is that in each moment, the organism reoriented its attention and behavior toward relationships with specific objects to satisfy its future interests, goals, and needs (Russell, 1945). Its behavior reflects withdrawal from objects unnecessary or in opposition to its goal and approaching those necessary to it at each moment in time (Figure 9.2).

9.6 Conclusion: The Emergence of Organismal Creativity

It is the nature of living systems to be radically indeterminate to continually construct their—our—own future, albeit in circumstances not of our own choosing.

(Rose, 2003, p. 7)

Walsh (2015, p. 210) proposed that organismal agency is observable in "the way the agent negotiates the situation." But what are the commonalities across organisms that make agency so apparent? What allows us to pluck out agency from behavior? In this chapter, I attempted to highlight some shared behavioral characteristics that allow us to identify agency in the behavior of living systems. The directed novelty inherent to most behaviors cannot just be relegated to noise or a pre-specified function but as a means to navigate the moment-by-moment challenges all organisms face. It is this novelty that carries the signatures of agency which can be seen, measured, and assessed.

In my discussion, I did not prioritize any specific approach to explain the specifics of goal-directed behavior. Whether individuals employ the feedback of cybernetics or perceptual control theory or instead use conscious intentions, planning, or imagination will depend on the particulars of their biology and developmental history. I aimed to find commonalities across approaches that capture the sufficient conditions necessary to move from behavioral observations to agency. The common thread that binds these perspectives together is that the prioritization of meaningful objects leads an individual to modulate approach–withdrawal processes to create a fixed relationship with that object in the future.

In the behavioral sciences, we are faced with two contrasting views of organisms. One view maintains that organisms are adaptive machines, honed collections of mechanistic answers to environmental problems. The apparent goal-directedness of behaviors is to be treated metaphorically, as they simply wait for the right mechanism to be uncovered to remove the veneer of goals. It emphasizes the fixed nature of behavioral response, wherein an organism gets from point A to point B utilizing a set series of tailored input–output responses.

The other view looks at organisms as agents. Here organisms are *loci* of meaning, significance, and action due to their ability to develop goals relevant to them. This view emphasizes the flexibility of behavior, one where the path from A to B will never be fixed, but requires creativity not seen in fixed mechanisms. Creativity is not just the blind production of variation, but the production of variation with a purpose; variation directed toward goals in the face of obstacles one can neither avoid nor predict. This creativity—shown as the continual production of directed novelties—allows individuals to exert control in the face of the unpredictability of life.

The ability to construct and maintain relationships with your surroundings is an essential component in securing a way of life. As agents, individual organisms are not simply blindly responding to environmental contingencies, but creating them, actively controlling what they can to secure their needs. Creativity is thus not a specialized cognitive feature found in some select vertebrates, but a shared characteristic necessary for many organisms to persist in the world. It allows individuals to solve problems that have not yet occurred in their developmental or phylogenetic histories. It is seen in everything from the righting behavior in the single cellular *Arcella*, to the novel behaviors in introduced Quaker parrots (*Myiopsitta monachus*) as they forge new lifestyles to thrive in a novel landscape.

Organisms are products of, and producers of, their own activities. By virtue of their self-creating autopoietic organization, all living things inhabit environments of significance with objects spanning the gamut between beneficial to harmful. As stated by Dewey (1934), the

> live animal does not have to project emotions into the objects experienced. Nature is kind and hateful, bland and morose, irritating and comforting, long before she is mathematically qualified or even a congeries of "secondary" qualities like colors and their shapes.
>
> (p. 16)

When negotiating this landscape of significance, indeterminacy and novelty are not mere noise, but the guidance systems that ensure organisms persist in maintaining their lives against the ubiquitous force of entropy that acts to unravel them.

References

Anctil, M. (2022). *Animal as machine: The quest to understand how animals work and adapt*. McGill-Queen's Press-MQUP.

Barandiaran, X. E., Di Paolo, E., & Rohde, M. (2009). Defining agency: Individuality, normativity, asymmetry, and spatio-temporality in action. *Adaptive Behavior, 17*(5), 367–386.

Bell, H. C. (2014). Behavioral variability in the service of constancy. *International Journal of Comparative Psychology, 27*(2), 196–217.

Bernstein, N. (1967). *The co-ordination and regulation of movements*. Pergamon Press.

Cziko, G. (2000). *The things we do: Using the lessons of Bernard and Darwin to understand the what, how, and why of our behavior*. MIT Press.

Dewey, J. (1934). *Art as experience*. Minton, Balch.

Di Paolo, E., Buhrmann, T., & Barandiaran, X. (2017). *Sensorimotor life: An enactive proposal*. Oxford University Press.

Farquharson, A. S. L. (1985). Movements of animals. In J. Barnes (Ed.), *The complete works of Aristotle, volume one: The revised Oxford translation* (pp. 1087–1096). Princeton University Press.

Fentress, J. C. (1983). The analysis of behavioral networks. In J.-P. Ewert, R. R. Capranica, & D. J. Ingle (Eds.), *Advances in vertebrate neuroethology* (pp. 939–968). Springer.

Fentress, J. C., & McLeod, P. J. (1986). Motor patterns in development. In E. M. Blass (Ed.), *Developmental psychobiology and developmental neurobiology* (pp. 35–97). Springer.

Finley, J., Ireton, D., Schleidt, W. M., & Thompson, T. A. (1983). A new look at the features of mallard courtship displays. *Animal Behaviour, 31*(2), 348–354.

Golani, I. (1981). The search for invariants in motor behavior. In K. Immelmann, G. W. Barlow, L. Petrinovich & M. Main (Eds.), *Behavioral development* (pp. 372–390). Cambridge University Press.

Golani, I. (2022). On the growth and form of animal behavior. *OSF Preprints.* https://doi.org/10.31219/osf.io/mqeg3

Gomez-Marin, A., Oron, E., Gakamsky, A., Valente, D., Benjamini, Y., & Golani, I. (2016). Generative rules of *Drosophila* locomotor behavior as a candidate homology across phyla. *Scientific Reports, 6*(1), 27555.

Gottlieb, G. (1997). *Synthesizing nature-nurture: Prenatal roots of instinctive behavior.* Lawrence Erlbaum Associates Publishers.

Gottlieb, G. (2007). Probabilistic epigenesis. *Developmental Science, 10*(1), 1–11.

Heft, H. (1989). Affordances and the body: An intentional analysis of Gibson's ecological approach to visual perception. *Journal for the Theory of Social Behaviour, 19*, 1–30.

Heylighen, F. (2023). The meaning and origin of goal-directedness: A dynamical systems perspective. *Biological Journal of the Linnean Society, 139*(4), 1–18.

Hordijk, W. (2019). A history of autocatalytic sets. *Biological Theory, 14*(4), 224–246.

Hordijk, W., Hein, J., & Steel, M. (2010). Autocatalytic sets and the origin of life. *Entropy, 12*(7), 1733–1742.

Jonas, H. (2001). *The phenomenon of life: Toward a philosophical biology.* Northwestern University Press (Original work published 1966).

Kauffman, S. A. (1993). *The origins of order: Self-organization and selection in evolution.* Oxford University Press.

Kuo, Z. Y. (1928). The fundamental error of the concept of purpose and the trial and error fallacy. *Psychological Review, 35*(5), 414–433.

Kuo, Z. Y. (1967). *The dynamics of behavior development.* Random House Press.

Kuo, Z. Y. (1976). *The dynamics of behavior development: An epigenetic view.* Plenum.

Latash, M. L. (2012). The bliss (not the problem) of motor abundance (not redundancy). *Experimental Brain Research, 217*(1), 1–5.

Latash, M. L. (2020). *Bernstein's construction of movements: The original text and commentaries.* Routledge.

Lehner, P. N. (1998). *Handbook of ethological methods.* Cambridge University Press.

Lehrman, D. S. (1953). A critique of Konrad Lorenz's theory of instinctive behavior. *The Quarterly Review of Biology, 28*(4), 337–363.

Lidgard, S., & Nyhart, L. K. (Eds.). (2017). *Biological individuality: Integrating scientific, philosophical, and historical perspectives.* University of Chicago Press.

Lorenz, K., & Tinbergen, N. (1970). Taxis and instinctive behaviour pattern in egg-rolling by the Greylag goose. *Studies in Animal and Human Behavior, 1*, 316–359.

Loutchko, D. (2023). An algebraic characterization of self-generating chemical reaction networks using semigroup models. *Journal of Mathematical Biology, 86*(5), 76.

Maturana, H. R., & Varela, F. J. (1991). *Autopoiesis and cognition: The realization of the living.* Springer.

Mayr, E. (1985). Teleological and teleonomic, a new analysis. In R. S. Cohen & M. W. Wartofsky (Eds.), *A portrait of twenty-five years: Boston colloquium for the philosophy of science 1960–1985* (pp. 133–159). Springer.

Miras, H. N., Mathis, C., Xuan, W., Long, D.-L., Pow, R., & Cronin, L. (2020). Spontaneous formation of autocatalytic sets with self-replicating inorganic metal oxide clusters. *Proceedings of the National Academy of Sciences, 117*(20), 10699–10705.

Nagel, E. (1977). Goal-directed processes in biology. *The Journal of Philosophy, 74*(5), 261–279.

Nicholson, D. J. (2012). The concept of mechanism in biology. *Studies in History and Philosophy of Science Part C : Studies in History and Philosophy of Biological and Biomedical Sciences, 43*(1), 152–163.

Nishida, K. (1992). *An inquiry into the good.* Yale University Press.

Pellis, S. M. (1985). What is "Fixed" in a fixed action pattern? A problem of methodology. *Bird Behavior, 6*(1), 10–15.

Pellis, S. M., & Bell, H. C. (2020). Unraveling the dynamics of dyadic interactions: Perceptual control in animal contests. In W. Mansell (Ed.), *The interdisciplinary handbook of perceptual control theory* (pp. 75–99). Academic Press.

Pellis, S. M., Gray, D., & Cade, W. H. (2009). The judder of the cricket: The variance underlying the invariance in behavior. *International Journal of Comparative Psychology, 22*(4).

Pellis, S. M., Pellis, V. C., & Iwaniuk, A. N. (2014). Pattern in behavior: The characterization, origins, and evolution of behavior patterns. In M. Naguib, L. Barrett, H. J. Brockmann, S. Healy, J. C. Mitani, T. J. Roper, & L. W. Simmons (Eds.), *Advances in the study of behavior* (pp. 127–189). Academic Press.

Peng, Z., Linderoth, J., & Baum, D. A. (2022). The hierarchical organization of autocatalytic reaction networks and its relevance to the origin of life. *PLOS Computational Biology, 18*(9), e1010498.

Powers, W. T. (1978). Quantitative analysis of purposive systems: Some spadework at the foundations of scientific psychology. *Psychological Review, 85*(5), 417.

Riskin, J. (2016). *The restless clock.* University of Chicago Press.

Rose, S. (2003). *Lifelines: Life beyond the gene.* Oxford University Press.

Russell, E. S. (1945). *The directiveness of organic activities.* Cambridge University Press.

Schleidt, W. M. (1974). How "fixed" is the fixed action pattern? *Zeitschrift für Tierpsychologie, 36*, 184–211.

Schleidt, W. M., & Crawley, J. N. (1980). Patterns in the behaviour of organisms. *Journal of Social and Biological Structures, 3*(1), 1–15.

Schneirla, T. C. (1959). An evolutionary and developmental theory of biphasic processes underlying approach and withdrawal. In *Nebraska symposium on motivation, 1959* (pp. 1–42). University of Nebraska Press.

Schneirla, T. C. (1965). Aspects of stimulation and organization in approach/withdrawal processes underlying vertebrate behavioral development. *Advances in the Study of Behavior, 1*, 1–74.

Sommerhoff, G. (1950). *Analytical biology.* Oxford University Press.

Sousa, F. L., Hordijk, W., Steel, M., & Martin, W. F. (2015). Autocatalytic sets in *E. coli* metabolism. *Journal of Systems Chemistry, 6*(1), 4.

Tolman, E. C. (1949). *Purposive behavior in animal and men.* Appleton.

Trestman, M. A. (2012). Implicit and explicit goal-directedness. *Erkenntnis, 77*(2), 207–236.

Turvey, M. T. (2018). *Lectures on perception: An ecological perspective.* Routledge.

Varela, F. J., Maturana, H. R., & Uribe, R. (1991). Autopoiesis: The organization of living systems, its characterization and a model. In G. J. Klir (Ed.), *Facets of Systems Science* (pp. 559–569). Springer.

Varela, F. J. (1991). Organism: A meshwork of selfless selves. In A. I. Tauber (Ed.), *Organism and the origins of self* (pp. 79–107). Springer.

von Bertalanffy, L. (1968). General systems theory as integrating factor in contemporary science. *Akten des XIV. Internationalen Kongresses für Philosophie, 2*, 335–340.

Walsh, D. M. (2015). *Organisms, agency, and evolution.* Cambridge University Press.

Weber, A., & Varela, F. J. (2002). Life after Kant: Natural purposes and the autopoietic foundations of biological individuality. *Phenomenology and the Cognitive Sciences, 1*(2), 97–125.

Wiener, N. (1961). *Cybernetics, or, control and communication in the animal and the machine.* MIT Press.

Zilov, V. G., Eskov, V. M., Khadartsev, A. A., & Eskov, V. V. (2017). Experimental verification of the Bernstein effect "Repetition without Repetition." *Bulletin of Experimental Biology and Medicine, 163*(1), 1–5.

10 Chimpanzees as "Resisters" or "Collaborators"

Animal Agency in Biomedical and Psychology Experiments at the Pasteur Institute in Paris and the Yale Laboratories for Primate Biology in the US (1903–1930)

Marion Thomas

10.1 Introduction

Recently, animal studies scholars have gone beyond documenting the impact of non-human animals on human history and recovering their subjective experiences to explore their capacity for agency (e.g., Hribal, 2007; Montgomery & Kalof, 2010; Woods et al., 2017, pp. 1–14; Rees, 2017a; Pearson, 2017).[1] Among the multiple ways to show that animals can be purposeful, capable agents, as well as capable of interacting with humans, I want to focus here on two of them: *resisting* and *collaborating*. The former has mainly captured the attention of animal studies scholars who have tightly linked it to organismal agency. Why has so much emphasis been put on animal resistance? Maybe a pragmatic reason is at stake. As Amanda Rees underscored: "[i]t is harder to see non-human agency when animals are behaving in accordance with human intent [...]: conflict is easier to spot than is the relationship of mutual accommodation on which cooperation depends" (Rees, 2017b, pp. 127–128). Or it could be explained by a reason rooted in political concerns. As noted by Chris Pearson, "animals are often portrayed as innocent victims of, or active resisters against, human oppression" (Pearson, 2015, p. 712). For instance, Jason Hribal argued for intentional acts of resistance on the part of captive animals, mostly in circuses or zoos, such as escaping from the miserable conditions of their confinement, refusing to perform tricks, or breaking out from their handlers. Hribal was not daunted by the prospect of comparing these rebelling animals to active agents in their own liberation against oppression (Hribal, 2010, pp. 28–29). Critical about the anthropomorphic earmark of the concept of resistance, Pearson (2015) rather uses the less problematic, and also less politically loaded, term of "blocking." Moreover, Pearson argues for rethinking non-human agency "beyond resistance" and shifting the gaze to "less dramatic forms of non-human agency and activity" to account for the full extent of non-human agency (Pearson, 2015, p. 713). Drawing from the cases of river rescue dogs and animals on the battlefield (such as horses, dogs, and pigeons), Pearson aims to explore not only how

DOI: 10.4324/9781003413318-13

"non-humans unintentionally 'resist' human objectives and activities," but also "'enable' interspecies entanglements, collaborations and partnerships" (Pearson, 2015, pp. 713–714). Also tackling the issue of animal resistance, Violette Pouillard encourages to address it from a flexible and relational perspective. Using the cases of an orangutan and a tiger held at the London Zoo in the early twentieth century, she retrieves a "broader spectrum of forms of resistance" from "intentional resistance" to "resistance from within, or silent resistance" as epitomized by stereotypic behaviors (Pouillard, 2023, p. 50).

Other animal studies scholars have challenged the concept of resistance, while emphasizing animal agents as "collaborators." In *When Species Meet*, Donna Haraway argues that the interactions of humans with non-human animals, especially domestic ones, can be characterized by mutual adaptation rather than exploitation (Haraway, 2008, p. 207). Arguing along the same lines, Benson states that it is "possible to see domestication more as a partnership, however unequal, than as a simple case of domination" (Benson, 2011, p. 5). In the context of psychological investigations, where researchers often have long relationships with the animals they study, familiarity is often a condition for the successful completion of experiments. For instance, the American psychologist and primatologist Robert Yerkes (1876–1956) reckoned that: "Confidence is also the essential basis of cooperation. Once an investigator has won the complete trust of his subject and its affection, he may count on obedience and cooperation to extraordinary lengths" (Yerkes, 1939, p. 110). This shows that scientists themselves readily acknowledged how important it is to take animal agency in advancing their research. So it is high time for historians to do the same.

In this chapter, I attempt to retrieve the agency of chimpanzees that were enrolled in biomedical or psychological experiments at the Pasteur Institutes in Paris and Kindia (Guinea), and the Yale Laboratories for Primate Biology, in the US. My aim is to engage more deeply with their lived experiences and explore the ways in which they are intertwined with human histories and the extent to which they shape them. More specifically, I want to show that they were not only passive, i.e., manipulated by experimenters who strive to derive knowledge from them but also had the capacity to impinge on the experiments either by facilitating them, slowing them down, or blocking them, eventually influencing the production of scientific knowledge.

To do so, and in the line of work that focuses on individual animals (Montgomery, 2009; Nelson, 2010; Munz, 2011; Baratay, 2017), I reconstruct the stories of specific chimpanzees by using the traces that they left on the historical records. At the Pasteur Institute in Paris, I focus on Edwige, but also Edouard, who was subjected to syphilitic experiments, and Nicole, but also Farce, who were enrolled in psychological trials. At the Yale Laboratories of Primate Biology, I shed light on the trajectory of Moos, who served in psychobiological studies. Following the journeys of those chimps, I want to offer insights into what they did as experimental animals. On the one hand, I examine the extent to which medicine and psychology shaped these animals by imposing upon them the role of victims of disease and turning them into experimental material. On the other hand, I explore the ways in which these animals shaped knowledge production, as well as the particular responses they exhibited as individuals with personalities, both to the equipment they were exposed to and the staff who administered to them. I show how these laboratory chimpanzees follow a spectrum arc, from notably resistant animals, to partly resistant and partly collaborative animals, to notably collaborative animals. Finally, I elucidate the ways in which the animal agency—whether real or imagined— could contribute to the presentation of the scientists' work and public image.[2]

10.2 Animals as "Resisters": Edouard and Edwige, Roux and Metchnikoff's Syphilitic Animals

In 1903, when Élie Metchnikoff (1845–1916) and Émile Roux (1853–1933) began their experimental syphilis-related studies on primates, particularly chimpanzees, the latter were still infrequently utilized in research due to their high cost and scarcity. For Roux, who had been working with horses in the context of his work on diphtheria, and Metchnikoff, who had mostly benched over his microscope to observe minute animals, primates were quite unfamiliar experimental animals. Even if Metchnikoff and Roux were well-versed in the new science of bacteriology, it was undoubtedly a challenge for them to fashion primates into laboratory animals and use them as a model to examine syphilis.

Among the 22 chimpanzees Metchnikoff used in his lab, two were singled out: Edwige and Edouard (*born ca. 1900*). Edouard was a gift from Émile Oustalet (1844–1905), professor of zoology, also in charge of the *Ménagerie* at the National Museum of Natural History (Roux, 1903a). As pointed out by a journalist, Edouard was very clever and endowed with special skills. He could paw like a well-trained poodle, and evince a passion for sweeping, something he did with cries of joy and vigor (Anonymous, 1903, October 1, p. 1). No doubt Edouard acquired those skills at the *Ménagerie*, where animals were tamed to entertain the public. In the hands of Metchnikoff, however, Edouard's former compliance vanished and Metchnikoff found unorthodox ways to subdue this chimp, as we will see later.

Edwige (*ca.* 1901–1903), on her end, was purchased through British traders, in Liverpool, by a representative of the Pasteur Institute. A very young chimp, not yet 2-year-old, she hailed from the French Congo forests, her mother probably having been killed when she was captured (Anonymous, 1903, July 29, p. 1). She cost an exorbitant price of 1,200 francs, which was taken out from Metchnikoff's Moscow Prize (Vikhanski, 2016, p. 186). Her stay at the Pasteur Institute was brief: inoculated with syphilis shortly after her arrival in June 1903, she died on September 27, 1903, from pneumonia (Anonymous, 1903, October 1, p. 1). Edwige's inoculation was Metchnikoff's first try, and it was a success. One month after being inoculated with fluid taken from a man's syphilitic ulcer, the symptoms characteristic of the primary period of the disease appeared: an indurated ulcer and swollen glands, especially in the right groin (Metchnikoff & Roux, 1903, pp. 814–815). Stricken by these results, Metchnikoff, a laboratory man, needed to confront the verdict of clinicians. On July 28, he and Roux stepped into the Academy of Medicine, accompanied by a porter who carried a cage containing Edwige, to meet

Alfred Fournier (1832–1914), one of the most famous figures on the question of syphilis of the time. Fournier was literally amazed when he saw all the characteristic signs of the venereal disease. Then "the study of syphilis, until then purely clinical, entered at last into the field of experimental science" (Metchnikoff, 1921, p. 191). Edwige played a pivotal role in this process: she contributed to bridging the gap between the laboratory and the clinic and to validating the mutual exchanges between these two epistemological spheres.

The success at the academy with an apparently quiet animal did not reflect the difficulties encountered in the laboratory. Metchnikoff and his assistants struggled in keeping the chimpanzees and submitting them to trials as the primates' strength often exceeded theirs. As Metchnikoff (1980a, p. 117) wrote to his wife Olga,

> Edouard turned out to be most unfriendly and difficult to handle. He's was supposed to be only 3.5 years old but he was extraordinarily strong. He broke off the chain, ran off and was caught. Getting him back into the cage proved enormously difficult.

Two days after struggling with this animal, Metchnikoff (1980b, p. 120) tried a trick: "he got him drunk with absinthe. The idea was to put him to sleep so he could inoculate the animal with syphilis." Unfortunately, Edouard "was very suspicious about the drink." Metchnikoff finally managed "to get him used to it gradually." Edouard only "agreed to eat pieces of sugar soaked with absinthe." As Metchnikoff quipped to his wife: "I'm spreading not only syphilis but also drunkenness."[3]

The abuse of animals in the laboratory for the sake of science leaked to the press and was quite irresistible to reporters. Journalists poked fun at Edouard's alcoholic behavior, his preference for wine, and his resistance to absinthe (Anonymous, 1903, September 3, p. 1). Edwige, whose display at the Academy of Medicine made a buzz in the popular press, was pictured holding a bottle. Of course, it was a fake picture, but journalists were not daunted by the prospect of conflating what took place in the laboratory with what was displayed in the pristine environment of the academy. Roux (1903b) also made light of Metchnikoff's unconventional scientific protocol and, in his usual sarcastic way, warned Metchnikoff about Edouard's possible retaliation: "absinthic Edouard could be much more furious and dangerous than sober Edouard!"

Interestingly, the resistance of the chimpanzee also provides insight into the ways in which scientists were confused about where to place chimpanzees at the nether regions of the "Great Chain of Being." For Roux (1903b), Edouard's aggressiveness was explained by his African origins, and he feared that new apes arriving directly from Africa would be uncontrollable, as they would have "only been impregnated with an African civilization." Let's note that Roux himself had experienced Africa. In August 1883, as an outbreak of cholera was raging in Egypt, Louis Pasteur (1822–1895) sent him with other pupils to study the disease on the spot and develop ways to control it. It happened that Roux was exasperated by the Egyptians' resistance to hygiene measures, and the fact that their culture (especially funeral rites) hampered the Pastorians in the anatomo-pathological examination of cadavers, a major task they were entrusted with during the epidemics. As he wrote vehemently to his mother shortly after his arrival: "The Egyptians are pigs. They deserved what happened to them" (Roux, as cited in Cressac, 1950, pp. 74–75). Roux's mean and racist words disclosed how much he viewed the Egyptians as standing in the way of the Western scientists. As noted by the historian Kathleen Pierce, "Roux's own description of his simian research subjects reveals the ties he felt between them and the indigenous subjects of the French Empire, while also underscoring the threat to the scientists constituted by both" (Pierce, 2018, p. 160), and, one could add, the resistance of both to experimental science.

Certainly, Metchnikoff was joking when he said he was spreading syphilis and alcoholism among his primates. These two ailments, however, happened to operate together in the context of the Paris *Belle Époque*, as well as they were intertwined with the general concern about prostitution, as prostitutes were potentially affected by them. Interestingly, Edwige comes in again, and it is worth exploring how her death yielded fascinating posthumous fantasies in the press, something we could describe in terms of *imagined agency*. Already during her lifetime, Edwige's love life captured the attention of the press. For Le Passant, a journalist from the conservative *Le Figaro*, she has been declared married to the chimpanzee Edouard (Le Passant, 1903, p. 1). Once dead, her marital status changed radically. Doctor P. Drack, who was given an open opinion column in the Republican journal *Le Radical*, viewed her as a "virgin" and as "another victim to record in the martyrology of vivisection." As Drack (1903) wrote:

> Edwige, a virgin, like the forests, where her childhood blossomed in the midst of lush vegetation, and from which she was ripped away, was sold to the famous institute of bacteriology [i.e., the Institute Pasteur] to be "damaged" [i.e., infected with syphilis] by means of a Pravaz syringe.
>
> (p. 2)

Not only had Edwige been experimented on by scientists, but she had also been shamelessly exposed to the gaze of physicians and journalists at the Academy of Medicine. As Dr. Drack carried on, Edwige "suffer[ed] from the indiscretions of snapshots." And he added: "[h]er photograph rivaled in price with that of [...] popular semi-socialites" (Drack, 1903, p. 2). Rather than comparing Edwige with figures of the Parisian elite, some people linked Edwige's appeal to another class of the Parisian population. For the Parisian mob, Edwige was closer to young provincial girls, who would be seduced into coming to Paris to seek fame and fortune, only to find that they had to work for themselves, and finally end up in a brothel.

I found evidence showing that Edwige operated as a stand-in for Parisian prostitutes. For instance, there was a song written in 1904 by a singer of the Butte Montmartre on the base of an imagined farewell letter from Edwige to her lover Jocko. While the song was touching, as a letter by a dying woman to her beloved would have been, it also targeted Roux and Metchnikoff as cruel and vicious scientists (Mérall & Ryp, 1904). Metchnikoff was impervious to such criticisms. For example, he was baffled by the popular emotional outpouring triggered by

Edwige's death. While he acknowledged that Edwige "had this somewhat melancholic and old-fashioned aspect of all the wild animals deprived of freedom," he also assessed that she "did not feel any serious pain" (Anonymous, 1903, August 1, p. 2). For Metchnikoff, like for Roux (1903c), chimpanzees were mostly "material for work." This consideration chimed with Metchnikoff's utilitarian argument to legitimize experiments on living animals (Metchnikoff, 1908, p. 303). It also enabled him to diminish the realities of the animals' pain and suffering in the laboratory, as well as ward off the criticisms voiced by anti-vivisectionists, especially in England and in the US. Arguing along a similar line, Roux confessed that: "It is almost a murder to experiment on such a close cousin, but there are some questions important enough to justify even the sacrifice of a charming relative like the chimpanzee" (Roux, as cited in Pierce, 2018, p. 162).

Then, in her posthumous life, the ways in which Edwige was popularized reflected how syphilis was socially laden (let's remember that prostitutes often came from underprivileged families, and this status was often a source of familial shame, if not outright social scorn), as well as fueled debates about vivisections.[4] Edwige's imagined agency encapsulated sex and death, as well as was deeply morally and politically laden.

10.3   From Resistance to Cooperation: Nicole, the Poster-Child of Guillaume and Meyerson's Chimpanzee Studies at the Pasteur Institute in Paris

Roux and Metchnikoff did not succeed in identifying the causal agent of syphilis, much less finding a cure.[5] They paved the way, however, for new forms of bacteriological research with primates as central animal models. Following in their footsteps, the bacteriologist Albert Calmette (1863–1933) ordered the construction of a new ape house at the Pasteur Institute in the mid-1920s. It was intended to provide animals to the laboratory directed by Jean Troisier (1881–1945), an authority on tuberculosis and cancer.[6] Calmette also held the conviction that ape psychology was an unexplored field, and he offered the French psychologists Paul Guillaume (1878–1962) and Ignace Meyerson (1888–1983) to work with chimpanzees, who arrived at the Pasteur Institute for the purpose of biomedical research.

Fervent advocates of Gestalt theory, Guillaume and Meyerson aimed at reproducing pioneering studies on the intelligence of chimpanzees that the German psychologist Wolfgang Köhler conducted between 1913 and 1920 in Tenerife, Canary Islands.[7] As ardent supporters of laboratory-based psychology, they also envisaged the experimental study of ape intelligence as a means to contribute to the emergence of psychology as a scientific discipline in France. At the Pasteur Institute, Guillaume and Meyerson worked with six chimpanzees: Dubreka, Daphnis (born *ca.* 1915), Griot (*ca.* 1921–1928), Track (*ca.* 1924–1928), Farce (*ca.* 1919–1929), and Nicole (*ca.* 1917–1929). Of them, Nicole stood out. Guillaume and Meyerson noticed that the medical experiments their chimps had gone through in Troisier's lab, were often a source of trauma, and impinged on the ways they behaved later in the context of their intelligence trials. For instance, the male Daphnis was "very fearful and very violent," and Guillaume and Meyerson assumed that he "certainly [had] painful memories of the surgical procedures he had gone through at Troisier's laboratory." Afraid of men, it took Daphnis a while to get used to Guillaume and Meyerson, but he never was able to work calmly with them. As for the female Dubreka, she was "of very little intelligence, with a sweet, cuddly character, but very shy and very unresponsive. What scared her were not men, but machines." As Guillaume and Meyerson noticed: "her slowness and indifference were discouraging." She gave "the fewest results" (Guillaume & Meyerson, 1930a, pp. 179–180). It is worth mentioning that in the context of blood group research, Dubreka and Daphnis resisted blood collection, and Troisier mentioned how much he required assistance from the head of the ape house to aseptically collect serum from the reluctant animals (Troisier, 1928, pp. 367, 370–371).[8] As an exception, Nicole had been kicked out from Troisier's lab for her alleged bad temper and aggressiveness. Her resistance somehow saved her from the traumatic biomedical experiments that her chimpmates went through. Interestingly also, Nicole turned out to be a smart and cooperative animal in Guillaume and Meyerson's hands.

The investigations of Guillaume and Meyerson on the intelligence of chimpanzees in problem-solving aptitude mostly drew from Köhler's "detour test." Contrary to Köhler's chimpanzees, who could move freely in an enclosure and use an appropriate detour path to reach a desired object, Guillaume and Meyerson's apes were locked up in cages and had to move the desired object toward them to get it. Guillaume and Meyerson imagined different kinds of experiments that involved catching a fruit placed either in a box or on a writing board outside the cage. These experiments involved using an instrument, such as a stick, to retrieve the fruit. Nicole proved to be a "remarkable chimpanzee" (Meyerson, 1929). She was described as "very intelligent, persevering, and displaying a constantly keen sense of curiosity towards the experimental devices" (Guillaume & Meyerson, 1930a, p. 179). In 1927, as noticed by Guillaume and Meyerson, "her physical and psychological development was peaking" (Guillaume & Meyerson, 1930b, p. 93).[9] "The intellectual gap between Nicole and her chimpmates was particularly striking" (Guillaume & Meyerson, 1930a, p. 179). For instance, subjected to identical trials, the female chimpanzee Farce (as well as Dubreka) proved "far less gifted." Although Farce was of "average intelligence" and could "learn relatively quickly," "her impatience and abruptness hampered her in succeeding in trials, which required calm and precision." In contrast with Nicole, who pondered and looked around before acting, Farce behaved in "restless, violent and disorderly ways" (Guillaume & Meyerson, 1930a, pp. 204–206). Originating from French Guinea, where she had been subjected to cancer experiments at the Pasteur Institute in Kindia, Farce was then shipped to the Pasteur Institute in Paris. Her "wild, defiant, if not violent behaviour" (Guillaume & Meyerson, 1930a, p. 179) displayed in Meyerson and Guillaume's experiments might have originated from the invasive experiments she had endured in the Kindia Laboratories.

Nicole's cooperative character and cleverness certainly explained why she was the only chimp casted in a short film, out of which image captures were used to illustrate Guillaume and Meyerson's scientific articles (see *"Recherches sur l'intelligence du chimpanzé*," 1928). Thereby Nicole contributed to the recognition of Guillaume and Meyerson's psychological work at the Pasteur Institute. Interestingly also, the footage features a smiling Meyerson, from which one can infer a gentle handling, as well as a feeling for this specific chimp. This positive interaction certainly helped to obtain good and non-biased results. As noticed by other primate scientists, working with a non-stressed, i.e., a cooperative, animal was essential to the success of an experiment. For instance, the French zoologist Louis Boutan (1859–1934), who, in the mid-1910s, worked on the cognitive abilities of a female gibbon called Pépée, reckoned that "treating an animal with tenderness, one can make experiments on them with absolute confidence" (Boutan, 1926, p. 355, as cited in Thomas, 2005, p. 443). Doing otherwise—working with a stressed animal—"would introduce a strong bias into the result of the study" (Boutan, 1914, p. 295, as cited in Thomas, 2005, p. 443).

10.4 Cooperation as an Epistemic Value in the Laboratory

Another chimp testifies to the importance of a cooperative behavior in the success of laboratory experiments. We now turn to the male chimpanzee Moos (*ca.* 1927–1937) from the Yale Laboratories for Primate Biology directed by Yerkes. Moussa (then renamed Moos) was one of 16 chimps that were presented as a gift to the Yale Laboratory for Primate Biology by the Pasteur Institute in Kindia.[10] These Guinean chimps were brought to the US in June 1930 by the American primatologist Henry Wieghorst Nissen (1901–1958), a disciple of Yerkes. Moos spent almost all his American life in New Haven where he served as an experimental subject for seven years and eventually died of dysentery shortly after his transfer to Orange Park, Florida in June 1937 (Wesleyan University, 2006a).

Nissen described Moos as "always […] unusually friendly toward people, very cooperative, and exhibit[ing] great confidence in the investigator" (Nissen & Crawford, 1936, p. 392). On many occasions, Moos indeed proved his cheerfulness and easiness in interacting with humans. One day, as he was playing with the veterinarian De Vita, he reacted by dancing in a see-saw fashion, when De Vita, in a squatting position, started to jump up and down, while clapping his hands. Sometimes, Moos' mouth was "wide open, upper teeth showing, tongue hanging out," and De Vita exclaimed: "Look at him laugh!" (Nissen, August 28, 1930). In *Chimpanzees: A Laboratory Colony* (1943), Yerkes describes an incident that also supports Moos' good nature and cooperative behavior. As Yerkes narrates:

> Moos had been ill, and we noticed that he was still refusing hard foods. Suspecting that this might be due to some unusual condition of the teeth or gums, a member of the staff [Nissen] entered the animal's cage and indicated that he wanted to make a dental inspection. Moos readily responded, but the observer failed to detect anything wrong. Satisfied with his examination he turned to leave the room, but Moos took hold of his coat, drew him back, and raising his upper lip with one hand pointed with a finger of the other hand to a spot on his upper jaw. There proved to be a slight swelling and subsequent examination revealed that a

permanent canine was in process of eruption and in all probability causing some discomfort, especially in connection with chewing. Our examiner naturally felt somewhat chagrined at having to be assisted in diagnosis by the animal himself.

(Yerkes, 1943, p. 192)

In Nissen's eyes, Moos' "emotional and attitudinal characteristics, probably combined with excellent intelligence [...] made him an ideal experimental subject" (Nissen & Crawford, 1936, p. 392). This also fitted in the characteristics that Yerkes posited for the ideal experimental chimpanzee, that is, to say "very adaptable, active, inventive, not destructive, cooperative; naturally gentle and quickly tamed, not a fighter, affectively stable [...], altruistic, frank, docile, easy to handle, good-natured and even-tempered [...]" (Yerkes & Spragg, 1937, p. 452). In similar ways, Yerkes also insisted "that many experimental procedures, even those painful to the animal, should be done with the animals' intelligent cooperation, not by constraint and compulsion" (Yerkes, as cited in Haraway, 1989, p. 77). For instance, the researchers abandoned the use of a specially constructed restraint chair, while realizing that seating the chimpanzee in a chair before an experimental device and letting it understand the requirements of the experiment made it more inclined to cooperate. This recommendation ended up as the rule of action for Yerkes' laboratory: "Expect and command voluntary and intelligent behavior" (Yerkes, 1937, pp. 256–257). As underscored by Donna Haraway, "[c]allous treatment of animals was not permitted in Yerkes' laboratories for *scientific* as well as humane reasons" (Haraway, 1989, p. 77; emphasis in original). From the animal's point of view, a cooperative behavior certainly meant a better life expectancy in the laboratory. In contrast, an uncooperative and aggressive behavior could lead to a risky, if not fatal situation. For the reason that she did not "give positive cooperation," but instead "vigor and visciousness" (Nissen, 1931a), the female chimpanzee Kindia, also from the Guinean convoy, was transferred to the laboratory of the Yale neurophysiologist John F. Fulton (1899–1960).[11] Kindia died a few weeks after the relocation from a cerebral operation experiment (Nissen, 1931b; see also Wesleyan University, 2006b).

Cooperative behavior was also socially laden. Chimpanzees were crucial partners in Yerkes' utopic plan of "re-creating man himself in the image of a generally acceptable ideal." He was indeed convinced on how much knowledge drawn from behavioral experiments conducted with chimpanzees could help in improving human nature (Yerkes, 1943, p. 10). In Yerkes' eyes, patient inquiry about ways of controlling chimpanzee behavior could help transform humans into more cooperative, unselfish, and honest beings (Yerkes, 1937, p. 270). Making cooperation an epistemic value in the laboratory aimed to demonstrate, for Yerkes, the possibility for a new society where altruism would prevail.

10.5 Conclusions

Edouard, Edwige, Nicole, Farce, and Moos are a few examples chosen from the thousands of chimps who passed through French and American laboratories during the colonial period in the first part of the twentieth century. Exploring the agencies of those chimpanzees helped us to understand how they represent important junctures in the history of biomedical and psychological primate research. These chimpanzees served as paradigmatic cases as they were subjected to both psychological and biomedical experiments. Understanding their significance sheds light on the ways in which these chimpanzees played a role in the scientists' recognition and public image. Edwige became part of Metchnikoff's fame and his battle to liberate sex from shame. Metchnikoff indeed fought against the idea that syphilis was labeled a sexual taboo and voiced out the idea of liberating sex from shame in a posthumous writing titled "*Sur la fonction sexuelle*" (Metchnikoff, 1917). In 1930, Nicole was casted in a scientific documentary to promote primate studies and challenged the idea of human uniqueness.

As we have seen, chimpanzees, thanks to their impressive muscle strength, often showed resistance and altered the experiments. In response to this, scientists needed assistance from their caretakers or resorted to unorthodox ways to subdue them into cooperative experimental animals.[12] It is no surprise that cooperation was looked for both in the psychological and the physiological laboratory. At the Pasteur Institute in Kindia, the colonial military veterinarians Maurice Delorme (1890–1943) and Robert Wilbert (1877–1945) were also convinced that cooperative behavior could help to better handle chimpanzees in the bacteriological laboratory: they trained their young chimps, even using psychological tests, to make them more tractable, i.e., more cooperative in the laboratory (Wilbert & Delorme, 1931, p. 140). In Paris, the biologist Auguste Pettit (1869–1939), who used baboons and chimpanzees to produce a poliomyelitis vaccine, gave up a "hard" way of handling them in the laboratory for a "method of softness and cuddles." This induced more cooperative behavior in the laboratory animals, as shown by "a chimpanzee called Mamadou, who came freely to the operating table and stretched out his arm for being bled" (Mathis, 1940, p. 287). It is key to observe as well that the topic of cooperation was so crucial that it was addressed as a proper scientific one at the Yale Laboratory of Primate Biology in the late 1930s and early 1940s (Crawford, 1937, 1941).

Retrieving the chimpanzee agency in the laboratory also enabled us to make visible the agency of additional actors who have been overlooked by scientific literature—the laboratory assistants and caretakers. For instance, Troisier underscored that a chimpanzee he was using for his experimental work on cancer "had been kept in a stall [and] cared for by sturdy West Africans according to the accurate plan of Professor Calmette" (Troisier & Limousin, 1928, p. 381). In the liner that took Farce to France, the Guineans Fode Bangoura and Abou Bokari took care of the chimps bound to the Pasteur Institute in Paris (Delorme, 1928; *Certificat du 8 septembre 1928*, signed by Maurice Delorme, 1928). Some of them did remain in Paris to facilitate the animals' acclimatization and to play the role of what Éric Baratay called an "emotional shield" (Baratay, 2017, pp. 39–41, 51). Albeit fundamental for the well-being of the animals (which also included their feeding), and also to the progress of the experiments, the role of the Guinean caregivers was poorly acknowledged, even disparaged (Calmette, 1930).

Finally, using animal agency as a lens to explore biomedical and psychological experiments with primates has helped to retrieve the ways in which scientific discourses and practices are often intertwined with discourses of domination either based on race or/and gender, as well as to identify the particular experiences of domination of both animals and colonized people, and the extent to which they could overlap. Edwige epitomized women prostitutes, whose health was endangered by their job, and who happened to be often abused. At the Pasteur Institute in Kindia, Western scientists tried to "educate" their chimps (and Farce could be one of them) through a "permanent" contact with "civilized man" (Donatien, 1931, p. 528), in much the same way they did with the indigenous populations.

Notes

1 Let's note that the concept of agency also connects with Bruno Latour's ANT model and the concept of actants, which aims to include "non-humans" (animals, but also plants and inanimate things) in sociological and anthropological analysis.
2 I thank Gregory Radick for making me aware of the category of "imagined agency." See also Munz (2011, p. 412).
3 I am very grateful and thankful to Luba Vikhanski for locating these letters and translating them for me.
4 We can note that in the late nineteenth century, syphilitic experiments on women, among them prostitutes, took place. These women were either inoculated with syphilis or, when affected by the disease, treated with mercury (see Chamayou, 2008, pp. 284, 289–291). These experiments were highly criticized in the popular press, a criticism which somehow resurfaced with Edwige's case.
5 Fritz Schaudinn (1871–1906) and Erich Hoffman (1868–1959) identified the causal agent of syphilis in the tissue of human patients at the Charité Hospital in Berlin, in 1905.
6 From 1926 to 1931, Troisier pursued different lines of research on non-human primates, among them, transplants of human cancers in chimpanzees, tuberculosis, typhus fever, and studies of chimpanzees' blood groups (*Jean Troisier*, 2023).
7 In 1927, Paul Guillaume translated Wolfgang Köhler's book on the mentality of apes, *Intelligenzprüfungen an Anthropoiden* (1917) under the title *L'Intelligence des singes supérieur* (see Köhler, 1927). One decade later, Guillaume wrote *La Psychologie de la forme*, which is still regarded as one of the best accounts of the principles of Gestalt theory (see Guillaume, 1937).
8 In Troisier (1928, pp. 370–371), we can find tables of blood collection showing the names of Daphnis and Dubreka.
9 Shortly after that, Nicole's health declined. She suffered from mobility impairment, which eventually affected the execution of some experiments.
10 This chimpanzee was first called Moussa from the name of Nissen's favorite field assistant in Guinea, Moussa Camara. The chimp's name was shortened into Moos, after he landed in the US in June 1930.
11 In addition, Kindia's fight with her chimpmates did not help to make her a more amiable animal. As Yerkes (1931) wrote to Nissen:

> I hope you can get rid of Kindia soon and if necessary segregate the other animals so as to minimize noise, for I fear from what you have written me that the colony may

now be a very considerable nuisance in the community, and as such may jeopardize the continuance of our work.

12 It is important to mention that the Russian biologist Ilya Ivanov (1870–1932), whose name is attached to controversial studies on human-ape hybrid experiments, also could get primates drunk, as shown in a video featuring an orangutan downing glasses of Moldavian red wine Kagor with the help of Ivanov. In the context of the zoo, the drinking of the orangutan did not intend to make it sleep but rather perform uncontrolled antics in front of a hilarious public (see Universidad Nacional de Educación a Distancia, 2014).

References

Unpublished Sources

Calmette, A. (1930, September 23). *Letter to Robert Wilbert*. Archives de l'Institut Pasteur de Kindia, Guinea.
Certificat du 8 septembre 1928, signed by Maurice Delorme. (1928). Archives de l'Institut Pasteur de Kindia, Guinea.
Delorme, M. (1928, September 9). *Letter to Albert Calmette*. Archives de l'Institut Pasteur de Kindia, Guinea.
Meyerson, I. (1929, June 17). *Letter to Robert Yerkes on behalf of Madame* (name not given). Robert Mearns Yerkes Papers (group 569, series I, box 34, folder 650), Sterling Memorial Library, Yale University, New Haven, United States of America.
Nissen, H. W. (1930, August 28). Notebook #5, pp. 70–72 (verso). Yerkes National Primate Research Center Records (box 77, EU14), Stuart A. Rose Manuscript Archives and Rare Book Library, Emory University, Atlanta, United States of America.
Nissen, H. W. (1931a, March 10). *Letter to Robert M. Yerkes*. Robert Mearns Yerkes Papers (group 659, series I, box 36, folder 690), Sterling Memorial Library, Yale University, New Haven, United States of America.
Nissen, H. W. (1931b, June 4). *Letter to Robert M. Yerkes*. Robert Mearns Yerkes Papers (group 659, series I, box 36, folder 690), Sterling Memorial Library, Yale University, New Haven, United States of America.
Roux, É. (1903a, August 24). *Letter to Élie Metchnikoff*. Archives de l'Institut Pasteur (MTC.2, folder "Correspondance Émile Roux et Élie/Olga Metchnikoff"), Paris, France.
Roux, É. (1903b, August 28). *Letter to Élie Metchnikoff*. Archives de l'Institut Pasteur (MTC.2, folder "Correspondance Émile Roux et Élie/Olga Metchnikoff"), Paris, France.
Roux, É. (1903c, September 13). *Letter to Élie and Olga Metchnikoff*. Archives de l'Institut Pasteur (MTC.2, folder "Correspondance Émile Roux et Élie/Olga Metchnikoff"), Paris, France.
Yerkes, R. M. (1931, May 23). *Letter to Henry Nissen*. Robert Mearns Yerkes Papers (group 659, series I, box 36, folder 690), Sterling Memorial Library, Yale University, New Haven, United States of America.

Published Sources

Anonymous. (1903, July 29). Est-elle transmissible? L'avarie chez les singes. Une communication sensationnelle à l'Académie de médecine. Les expériences de MM. Metchnikoff et Roux, L'opinion de M. Fournier, Présentation du malade, Une visite à "Edwige". *Le Matin*, *20*(7094), 4.

Anonymous. (1903, August 1). Edwige. Une guenon historique. A l'Institut Pasteur – Entretien avec le professeur Metchnikoff – L'état de la malade – A travers la ménagerie. *Le Matin, 20*(7097), 2.

Anonymous. (1903, September 3). Petites histoires. *Le Figaro, 49*(246), 1.

Anonymous. (1903, October 1). La mort d'Edwige. A l'Institut Pasteur – Une mauvaise nouvelle – Il y a quatre jours – L'opinion du docteur Salomon. *Le Matin, 20*(7158), 1.

Baratay, É. (2017). *Biographies animales. Des vies retrouvées.* Le Seuil.

Benson, E. (2011). Animal writes: Historiography, disciplinarity, and the animal trace. In L. Kalof & G. M. Montgomery (Eds.), *Making animal meaning* (pp. 3–16). Michigan State University Press.

Chamayou, G. (2008). *Les Corps vils. Expérimenter sur les êtres humains au XVIIIe et XIXe siècles.* LaDécouverte.

Crawford, M. P. (1937). The cooperative solving of problems by young chimpanzees. *Comparative Psychology Monographs, 14*(2), 1–88.

Crawford, M. P. (1941). The cooperative solving by chimpanzees of problems requiring serial responses to color cues. *The Journal of Social Psychology, 13*(2), 259–280.

Cressac, M. (1950). *Le Docteur Roux. Mon oncle.* L'Arche.

Donatien, A. (1931). "Pastoria", Centre de recherches biologiques et d'élevage sur les singes (Institut Pasteur de Kindia, Guinée française). *Revue vétérinaire et Journal de médicine vétérinaire et de zootechnie réunis, LXXXIII*, 528–531.

Drack, P. (1903, October 6). Chronique médicale. Vierge & Martyre. *Le Radical, 23*(279), 2.

Guillaume, P. (1937). *La Psychologie de la forme.* Flammarion.

Guillaume, P., & Meyerson, I. (1930a). Recherches sur l'usage de l'instrument chez les singes. I Le problème du détour. *Journal de psychologie normale et pathologique, XXVII*, 177–236.

Guillaume, P., & Meyerson, I. (1930b). Quelques recherches sur l'intelligence des singes (Communication préliminaire). *Journal de psychologie normale et pathologique, XXVII*, 92–97.

Haraway, D. (1989). *Primate visions: Gender, race, and nature in the world of modern science.* Routledge.

Haraway, D. (2008). *When species meet.* University of Minnesota Press.

Hribal, J. C. (2007). Animals, agency, and class: Writing the history of animals from below. *Human Ecology Forum, 14*(1), 101–112.

Hribal, J. C. (2010). *Fear of the animal planet: The hidden history of animal resistance.* CounterPunch.

Jean Troisier. (2023, December 22). In *Wikipedia.* https://en.wikipedia.org/wiki/Jean_Troisier

Köhler, W. (1927). *L'Intelligence des signes supérieurs.* Alcan.

Le Passant. (1903, September 7). Le mariage d'Edwige. *Le Figaro, 49*(250), 1.

Mathis, M. (1940, March 12). Chroniques. Les Anthropoïdes. *La Presse médicale, 24*, 287–288.

Mérall, M., & Ryp. (1904). *Les Refrains de la Butte. Derniers succès Montmartrois. Adieu Jocko! Dernière lettre d'une guenon inoculée à son fiancé.* Librairie Chez Plessis.

Metchnikoff, É. (1908). *The prolongation of life: Optimistic studies* (P. Chalmers Mitchell, Trans.). G. P. Putnam's Sons.

Metchnikoff, É. (1917). Sur la fonction sexuelle. *Mercure de France, 120*(451), 412–418.

Metchnikoff, É. (1980a). Letter to Olga Metchnikoff from August 23, 1903. In A. E. Gaissinovich & B. V. Levshin (Eds.), *Pis'ma (1900–1914)* [Correspondence (1900-1914)] (pp. 117–118). Nauka.

Metchnikoff, É. (1980b). Letter to Olga Metchnikoff from August 25, 1903. In A. E. Gaissinovich & B. V. Levshin (Eds.), *Pis'ma (1900–1914)* [Correspondence (1900–1914)] (p. 120). Nauka.

Metchnikoff, É., & Roux, É. (1903). Études expérimentales sur la syphilis. Premier mémoire. *Annales de l'Institut Pasteur, 17*(12), 809–821.

Metchnikoff, O. (1921). *Life of Elie Metchnikoff, 1845–1916*. Houghton Mifflin Company.

Montgomery, G. M. (2009). "Infinite loneliness": The life and times of Miss Congo. *Endeavour, 33*(3), 101–105.

Montgomery, G. M., & Kalof, L. (2010). History from below: Animals as historical subjects. In M. DeMello (Ed.), *Teaching the animal: Human-animal studies across the disciplines* (pp. 35–47). Lantern Books.

Munz, T. (2011). "My goose child Martina": The multiple uses of geese in the writings of Konrad Lorenz. *Historical Studies in the Natural Sciences, 41*(4), 405–446.

Nelson, A. (2010). The legacy of Laika: Celebrity, sacrifice, and the Soviet space dogs. In D. Brantz (Ed.), *Beastly natures: Animals, humans and the study of history* (pp. 204–224). University of Virginia Press.

Nissen, H. W., & Crawford, M. P. (1936). A preliminary study of food-sharing behavior in young chimpanzees. *Journal of Comparative Psychology, 22*(3), 383–419.

Pearson, C. (2015). Beyond "resistance": Rethinking non-human agency for a "more-than-human" world. *European Review of History. Revue européenne d'histoire, 22*(5), 709–725.

Pearson, C. (2017). History and animal agencies. In L. Kalof (Ed.), *The Oxford handbook of animal studies* (pp. 240–257). Oxford University Press.

Pierce, K. (2018). Scarified skin and simian symptoms: Experimental medicine and Picasso's *Les Demoiselles d'Avignon. Nineteenth-Century Art Worldwide, 17*(2), 149–178.

Pouillard, V. (2023). The silence and the fury: Addressing animal resistance and agency through the history of human-animal relationships. In J. Dugnoille & E. Vander Meer (Eds.), *Animals matter: Resistance and transformation in animal commodification* (pp. 32–55). Brill.

Recherches sur l'intelligence du Chimpanzé [Film]. (1928). Sociedad Española de Historia de la Psicología. http://sehp.org/recursos/Experimentos%20con%20uso%20de%20instrumentos%20en%20monos%20%282%29.mp4

Rees, A. (2017a). Animal agents? Historiography, theory and the history of science in the Anthropocene. *British Journal for the History of Science Themes 2*, 1–10.

Rees, A. (2017b). Wildlife agencies: Practice, intentionality and history in twentieth-century animal field studies. *The British Journal for the History of Science Themes 2*, 127–128.

Thomas, M. (2005). Are animals just noisy machines? Louis Boutan and the co-invention of animal and child psychology in the French Third Republic. *Journal of the History of Biology, 38*(3), 425–460.

Troisier, J. (1928). Le groupe sanguin II de l'homme chez le chimpanzé. *Annales de l'Institut Pasteur, 42*(4), 363–379.

Troisier, J., & Limousin, H. (1928). Essais de transmission au chimpanzé d'un cancer digestif humain. *Annales de l'Institut Pasteur, 42*(4), 380–382.

Universidad Nacional de Educación a Distancia. (2014). *Ignace Meyerson* [Film]. RTVE. www.rtve.es/play/videos/uned/uned-19122014-ignace/2918356/

Vikhanski, L. (2016). *Immunity: How Elie Metchnikoff changed the course of modern medicine*. Chicago Review Press.

Wesleyan University. (2006a). Moos. The First 100 Chimpanzees. http://first100chimps.wesleyan.edu/moos.html

Wesleyan University. (2006b). Kindia. The First 100 Chimpanzees. http://first100chimps. wesleyan.edu/kindia.html

Wilbert, R., & Delorme, M. (1931). "Pastoria". Centre de recherches biologiques et d'élevage de singes, Institut Pasteur de Kindia, Guinée française. *Bulletin de la Société de pathologie exotique et des filiales de l'Ouest africain et de Madagascar, 24*, 131–149.

Woods, A., Bresalier, M., Cassidy, A., & Mason Dentinger, R. (2017). *Animals and the shaping of modern medicine: One health and its histories.* Palgrave Macmillan.

Yerkes, R. M. (1937). Primate cooperation and intelligence. *The American Journal of Psychology, 50*(14), 254–270.

Yerkes, R. M. (1939). The life history and personality of the chimpanzee. *The American Naturalist, LXXIII(745)*, 97–112.

Yerkes, R. M. (1943). *Chimpanzees: A laboratory colony.* Yale University Press.

Yerkes, R. M., & Spragg, S. D. (1937). La mesure du comportement adapté chez les chimpanzés. *Journal de psychologie normale et pathologique, XXXIV*, 449–475.

11 Self-Individuation, Environment, and Agency

Comparing Plessner and Autopoietic Enactivism

Francesca Michelini

11.1 Introduction

The natural philosophy of Helmuth Plessner (1892–1985) is mostly unknown in the field of contemporary philosophy of biology. This is due to different factors. First of all, a relatively contingent one: originally published in 1928, his most important work, *Die Stufen des Organischen und der Mensch: Einleitung in die philosophische Anthropologie* (Levels of Organic Life and the Human: An Introduction to Philosophical Anthropology), has been translated into English only in 2019. Rather than for his philosophical biology, Plessner has been known to date mainly for his philosophical anthropology.[1] However, even a quick look at the outline of Plessner's book reveals that only the last chapter is devoted explicitly to the sphere of humans and their "eccentric form" (Plessner, 1928/2019, pp. 267–321). For the most part, his work is a genuine treatise on philosophical biology, aimed to account for the qualitative distinctiveness of different organic forms in the continuity of life and the rootedness of human beings in it.

However, there is also another factor that is possibly even more explanatory of Plessner's lack of reception. Contrary to his view, contemporary research in philosophy of biology has no longer as its goal to understand what life is. Whereas the clarification of the assumptions, concepts, and main practices in the life sciences seems to be the main goal of today's philosophy of biology, according to Plessner, bio-philosophical inquiries should not stop at the mere elucidation of the findings of the life sciences (see, e.g., Plessner 1928/2019, pp. xxx–xxxi, 86, 119). He strongly believed that the aliveness of beings cannot be assessed merely by means of empirical research. According to him, life cannot be grasped purely in terms of exact methods, verifications, and analyses.

The method he applies is simultaneously phenomenological and hermeneutical. Despite his criticism (see, e.g., Plessner, 1928/2019, p. xvii, Chapters 1–3), Plessner can be placed within the phenomenological tradition, in that he aims to discover a kind of "evidence" concerning biological phenomena that the natural sciences assume as a general framework and cannot be corroborated nor falsified by empirical knowledge. Philosophical biology, Plessner writes, "is oriented according to the pre-scientific experience of biologists, it questions the meaning of what, in the everyday world, is understood under the concepts of plant, animal and man. Which

DOI: 10.4324/9781003413318-14

means that it is phenomenological" (Plessner, 1973, p. 398). At the same time, as I will argue again later on, Plessner's investigation is hermeneutic and—with respect to Dilthey's famous distinction between "understanding," and "explanation" or demonstration in scientific terms—aims to shed light on "the sense of the living" (Sommer, 2016, p. 100). His could potentially be Maurice Merleau-Ponty's sentence—as laid out in *La structure du comportement* (The Structure of Behavior, 1942)—"vital acts *have* a meaning" (Merleau-Ponty 1942/1963, p. 159; emphasis in original).

Nevertheless, it is also true that Plessner's philosophy developed in close exchange with empirical data and the main scientific theories of his time (see Ingensiep, 2004, p. 36; Köchy, 2022, pp. 162–163). Starting from the phenomenon in the "pre-scientific experience of biologists" means, for Plessner, especially in the years of his collaboration with the Dutch anthropologist, biologist, and psychologist, Frederik J. J. Buytendijk (1887–1974), to provide a solid foundation "for a cooperation between philosophical procedures and those of the particular sciences" (Buytendijk & Plessner, 1925/1982, pp. 75–76; my translation).[2] He studied zoology and knew the scientific *milieu* of his time very well. He was well acquainted, for instance, with the theories of Jakob Johannes von Uexküll (1864–1944) and Wolfgang Köhler (1887–1967), and he was a pupil of Hans Driesch (1867–1941). However, Plessner embraced neither the latter's neo-vitalist conception, nor mechanistic or neo-Darwinian ones. As a result, some scholars have recently seen his philosophy as a *"non-reductionist naturalism,"* which "allots complementary roles to the causal and functional investigations of the life sciences and the phenomenological and hermeneutic interpretation of the phenomenon of life in its successive levels and stages" (de Mul, 2019, p. 67; see also Pagan, 2023). Although such an interpretation is backed up by Plessner's frequent reference to scientific theories corroborating his ideas (see Lessing, 2002), one should also always bear in mind that, according to him, natural sciences by their very nature cut themselves off from other dimensions of what is intuitively given. According to Plessner (1928/2019), "[i]n the world there is much more than what can be ascertained" (p. 119). More on this point will be added in the closing section of this chapter.

Within this framework, Plessner's interest and relevance even today consists in having suggested what I call a "third way" (Michelini, 2021). He provides a viable starting point for the elaboration of a theory of the organism that not only eludes any Cartesian dualism but also overcomes the limits of any simplified organic monism. In contrast to Descartes' ontological dualism, Plessner defends a sort of "perspectivistic dualism" (de Mul, 2019, p. 70), which is based on what he terms "dual aspectivity" (*Doppelaspektivität*; Plessner, 1928/2019, pp. 84–85). According to him, depending on the point of view, the in itself "neutral unity" of life (Plessner, 1928/2019, p. 271) appears to us as an external or internal world, as a physical or psychic phenomenon, as spirit or body. I will come back to this point in Section 11.2 and at the end of Section 11.4. For now, suffice it to say that "dual-aspectivity" is seen as the distinctive feature of living beings, as different from non-living ones, and it is based on the lived body and the way it relates to its surroundings.

Already from these general considerations, it is clear why, today more than ever, it is important to engage with Plessner's philosophical biology (see also Bernstein, as cited in Plessner, 1928/2019, p. lxv). The current organism revival in biology has great similarities to what happened in Plessner's time (Michelini, 2019, 2021). A need is indisputably felt for a more "integrative" version of biology, which goes beyond restrictive or reductionist approaches, such as, for instance, the Modern Evolutionary Synthesis in genetics (Walsh, 2015; Moss, 2003), without, however, falling back on anti-scientific standpoints. From this point of view, I believe that Plessner's philosophy of the organism can find fruitful resonances in all those approaches that not only work on a more comprehensive and "relational" view of the organism (Ingold, 1990) but also still aim to understand what life is. This is the case, for instance, of autopoietic enactivism, which I will refer to in this chapter. As is well known, this is a research program in the cognitive sciences that was launched in Francisco J. Varela, Evan Thompson and Eleanor Rosch's 1991 volume *The Embodied Mind* (Varela et al., 1991), and developed since the 1980s by Varela in several essays (e.g., Varela, 1991, 1997). Over the last two decades, the program has experienced great increase and application.[3]

Plessner's philosophy and autopoietic enactivism indeed share several elements. Both approaches grant a key role to the lived experience of the body in its relation to its surroundings, and both are at variance with prevailing mechanistic and computational–representational paradigms. Similarly, they both argue that life is not reducible to a sum of properties, but it is based on a "distinctive criterion." Finally, both approaches maintain that organisms are teleological and intentional "agents" that cannot be understood independently from their systematic relationship with the environment.

This chapter focuses in particular on the latter point. From a general point of view, Plessner agrees with enactivism that organismal agency is not the exclusive prerogative of humans, but it is rather pertaining to all living beings. This is indeed a key point in enactive approaches and, more generally, in later reformulations of autopoiesis (e.g., Weber & Varela, 2002; Di Paolo, 2005; Thompson, 2007; Barandiaran et al., 2009; Moreno & Mossio, 2015; Di Paolo et al., 2017). Furthermore, besides this general convergence, Plessner also identifies at least two preconditions allowing us to say that an organism is an agent, which closely match enactivist analyses: (a) the fact that it is an identity enclosed within its limits and asymmetrical with respect to the environment (Moreno & Barandiaran, 2004; Barandiaran et al., 2009), which we can call the "autopoietic criterion"; and, equally important, (b) the fact that it is able to respond to the conditions of its environment and its internal constitution in a way that promotes the achievement and maintenance of its purposes (Walsh, 2015, p. 210), which we can call the "ecological criterion."

Sections 11.2 and 11.3 deal with the "autopoietic criterion" of agency, in particular in reference to Plessner's idea of the realisation of the boundary (*Verwirklichung der Grenze*; Plessner, 1928/2019, pp. 93–94) and the autopoietic and enactive concept of organism's self-individuation. Some open problems will be spelled out in relation to the early formulation of this second concept. The "ecological criterion" will make the object of Section 11.4, within an account on what Plessner, together with Buytendijk, call "environmental intentionality" (*Umweltintentionalität*) and apply to animal behaviour in its relationship with the environment. Echoing Jakob von Uexküll's *Umwelt* theory, the notion of "environmental intentionality" seems also to anticipate in some way J. J. Gibson's concept of "affordances." To date, the notion of *Umweltintentionalität* has not attracted sufficient scholarly interest, despite the fact that it had an impact, for instance, on Merleau-Ponty's philosophy (Sommer, 2016).[4] Plessner and Buytendijk came up with this terminology in a 1925 essay—*The Interpretation of the Mimic Expression*—mainly to point out that it is impossible to understand the meaning of living beings if one prescinds from their intimate interaction with the environment and their "orientation" to it. In this respect, analogies can be discovered with the enactive account of "sense making."

Despite the interesting convergence of the two approaches, I will argue that some significant differences still persist which prevent a complete assimilation between them. However, I also argue that based on these differences, one cannot only identify some open problems in the enactive perspective but also clarify whether and how a philosophical biology *à la* Plessner can still be supported today.

11.2 Self-Individuation and Realisation of the Boundary

The starting point of autopoiesis theory shares some striking similarities with Plessner's opening question: how to overcome all theories according to which life would be definable through a list of physical and functional properties (Thompson, 2011, p. 114). In Humberto Maturana and Francisco Varela's early inquiries, life is not understood as a sum of characteristics. On the contrary, their core idea is that an autopoietic system is organised as a network of processes that is self-producing, and that distinguishes itself from its environment as a topological unit (Maturana & Varela, 1980, pp. 78–79).

In this basic idea, Varela saw the foundations of the autonomy of living beings. According to him, living beings can be autonomous only in virtue of their self-generated identity as distinct entities (Varela, 1997, p. 73). The self-individuation process is the key definitory criterion. This amounts to saying that living systems make themselves distinct from their immediate surroundings, thereby enabling an observer to distinguish them as an identifiable entity in the environment (Di Paolo & Thompson, 2014, p. 68).

This definition of autonomy closely matches what Plessner says in *Levels of Organic Life*: "It [the organism] remains autonomous, as nothing gets close to it and nothing gains influence on or in it that it does not subject to the law of the bounded/bordered system" (Plessner, 1928/2019, p. 179). Similar to autopoiesis, the "distinctive criterion of life" rests on the relation entertained by organic bodies with their boundary (*Grenze*) and has nothing to do with vitalistic or metaphysical assumptions. According to Plessner, it is rather linked to "the intuitional structure of the so-called 'things of perception'" (p. 31). More precisely, he maintains that a spatial object which appears in its own "dual aspect" of inside and outside must at the same time display a boundary between the two. In the case of merely physical bodies, or inanimate things, the boundary is however simply identical to the border or outline of the physical body. It belongs neither to the simple body, nor to the surroundings—in Plessner's words: the "medium"—or perhaps, in a certain sense, it belongs to both of them. The border is actually a simple, virtual "in-between" between the body and the medium which comes from the reciprocal self-limiting of the body and its surroundings. On the other hand, in the case of living things, the boundary does not just mark where they stop and the adjoining medium begins; rather, the boundary and the overstepping of that boundary both belong to the body itself (Plessner, 1928/2019, pp. 93–94). Living things appear as things that embody boundaries and cross those boundaries at the same time. They have a "divergent" relation with both sides of their boundary. Plessner (1928/2019) writes: "Physical objects of intuition for which a fundamentally divergent relation-ship between outer and inner objectively figures as part of their being are called *living*" (p. 84; emphasis in original). Plessner's technical term for this relation-ship with the constituting boundary is "positionality" (pp. 118–119). Positionality means that a living being does not simply fill a space, like inanimate things that are simply located somewhere, but rather takes *its* place or claims its position in the environment. This idea has two major implications.

The first one is that positionality is not a static concept, unlike what the word might suggest, but rather a radically dynamic one. It is an activity of the lived body. As Marjorie Grene aptly puts it, it is "the whole way in which an organism takes its place in an environment, arises in it, is dependent on it, yet opposes itself to it" (Grene, 1966, p. 254). Living beings are embodied. That means, on the one hand, that they constantly have to "realize" themselves, building and maintaining their own boundaries against the environment. Boundary realisation is "a life-long task"; it lasts until the living being dies (de Mul, 2019, p. 72). It also means, on the other hand, that a living being is always already "placed within its boundaries" (Plessner, 1928/2019, p. 271). This amounts to saying that, although it occupies a space like all physical bodies, its "center is nevertheless not a spatial center, it is a core which transcends spatiality and at the same time controls the spatiality of the body whose core it is" (Grene, 1966, p. 263). In other words, it is a self (Plessner, 1928/2019, p. 148), an identity that is neither spatially measurable nor empirically quantifiable.[5]

The second implication is closely connected to the first. Thanks to positionality, "the medium" develops for the lived body into *its own* environment (*Umwelt*). Organisms are structurally embedded in their environment and are equiprimordial with it. Plessner takes up and reshapes in a not-constructivist key Uexküll's claim according to which living organisms are the center of their own *Umwelt*. Uexküll understands the notion of *Umwelt* as the world of experiences and relations an animal generates, while being its sensorial and operational fulcrum. The idea of a general environment which is the same for everybody is, on the whole, also for Plessner, just an abstraction. The environment is neither "pre-given" nor a mental construc-tion. It is rather—as recently maintained by ecological psychologists mainly in reference to Uexküll—"enacted during the individual's history of development and learning" (Baggs & Chemero, 2018, p. S 2186). However, unlike Uexküll, who would assume some sort of "encompassing harmony" between an organism and its environment, Plessner's view is way more dialectical. "The organism has to fit into the medium and at the same time have enough leeway [*Spielraum*] in it to not only exist within the fixed forms of harmony," he writes, "but also to survive dangers with them" (Plessner, 1928/2019, p. 190).

Plessner identifies in living bodies a tension, or better a "radical conflict," "between the compulsion to close itself off as a physical body and the compulsion to open up as an organism" (Plessner, 1928/2019, p. 202). This tension between openness and closeness finds specific articulations in several different modes of relation between the living and the environment. While, for instance, plants are directly placed in their environment, and therefore are said to be "open" forms, animals only have a mediated relation to the environment and are therefore said to be "closed" forms (Plessner, 1928/2019, p. 209). Nevertheless, the original tension can never be "solved" nor can it ever entirely disappear. It finds instead even more specific articulations in different degrees of complexity within internal and external conflicts, which are inherent in the living condition and whose most extreme expression can be found in human "eccentricity." Plessner (1928/2019) writes that "the organism remains endangered, regardless of how secure it is [...] in peace and at war, in life and in death. That is why life means to be in danger, why existence means risk" (p. 192).

To sum up: from what has been laid out above, it emerges that self-closure and self-individuation are, also for Plessner, key requirements for the autonomy of living systems, and therefore, as I will argue, also for their agency. However, despite this key similarity, Plessner's starting point is, as I said, a phenomenological and hermeneutic one (see pp. 24–25, 68). And this prevents us from simply equating Plessner's theory and autopoiesis, as much as one might be tempted to (see Mugerauer, 2014; de Mul, 2019; Moss, 2020).

According to Maturana and Varela, in fact, the boundary of living beings can be empirically identified. It is almost a physical property of the cell's membrane that can be measured, with a contour that can be outlined.[6] Differently, for Plessner, the boundary is primarily an object of intuition. Unlike outlines and contours, "it can only be understood, not drawn" (Plessner, 1928/2019, p. xxxii). What is at stake in the realisation of *Grenze* is the relationship of the bounded living body to its boundaries, and this cannot be measured. This does not mean that Plessner rules out that his theory may have empirical correlates. On the contrary, in the 1960s, for instance, he points out that his theory is corroborated by the back then recent findings on membranes' semipermeability, including a direct reference to J. B. S. Haldane (Plessner, 1928/2019, p. 332).[7]

In what follows, the differences between the two views will be further elucidated, within the framework of a general discussion of some open problems in the early formulation of autopoiesis. These have already been abundantly discussed both from the viewpoint of later formulations of an autonomy theory and from a more specifically enactive viewpoint. The key issue concerns the interaction of an organism with its environment as a key element for the understanding of its "agency."

11.3 Autonomy, Agency, and Environment

In its first formulation by Maturana and Varela, autopoiesis was presented as a necessary and sufficient condition to understand the self-organisation of living beings. However, subsequent analyses, particularly from the side of enactivism and autonomy theory, have shown that it is indeed a necessary condition, but not a sufficient one. In short, some weaknesses have been identified in Maturana and Varela's account concerning the organism's interaction with the environment. According to their model, these interactions are "structural couplings" (Maturana & Varela, 1980, p. 75), that is, recurring interactions that lead to structural congruence between one or more systems and which "result from" the specific (internal) identity of each autopoietic system. In this early formulation, interactions with the environment neither determine nor establish the organisation. They "do not enter" into the definition-constitution of autonomous systems; rather they follow on or simply derive from the specific internal identity of each autopoietic system (Moreno & Mossio, 2015, p. xxvii). This amounts to saying that the relations of the system to the environment are simply "necessary extensions of this primary internal organisation" and are understood as actions that the system performs "on its own behalf" (cf. Ruiz-Mirazo & Moreno, 2004, p. 236). Now, besides the fact that this approach still leaves unresolved the question of how such interactions are regulated by the organism (Di Paolo, 2005; more on this further on), it has been famously criticised for being an internalist or a constructivist approach, with solipsistic outcomes (e.g., Swenson, 1992).

Subsequent inquiries have thereby aimed to clarify in what respects the interaction with the environment should be included as integral to the definition of autonomy, and not handled as a simply derivative consequence of it. Two equally important and closely connected dimensions of autonomy have been thematised: the "constitutive" one, which determines the identity of the living system, and the "interactive" one, also called "agency," which "far from being a mere side effect of the constitutive dimension, deals with the inherent functional interactions that the organisms must maintain with the environment" (Moreno & Mossio, 2015, p. viii). A strong connection between the two dimensions does not exclude, though, a certain "asymmetry" between an organism and the environment. This is actually a key point when it comes to understanding agency as pertaining also to the basic levels of organisms:

> the (self) generation of an inside is *ontologically prior* in the dichotomy in-out. It is the inside that generates the asymmetry and it is in relation to this inside that an outside can be established. The interactive processes/relations are secondary for the maintenance of the system: they presuppose it (the system) since it is the internal organization of the system that controls the interactive relations.
>
> (Moreno & Barandiaran, 2004, p. 17; emphasis added)

From an enactive viewpoint, also Ezequiel di Paolo has further insisted on this asymmetry, connecting it to the notion of regulation. He writes:

Behavior defined not as physical coupling, but as its regulation is always asymmetrical, has an intentional structure, and can be said to either succeed or fail. It is only at this stage, when the organism behaves, that we may speak of an agent.

(Di Paolo 2005, p. 13)

Next to self-individuation and asymmetry, a third element needs to be taken into account in order to have a full picture of "genuine" agency, namely the fact that organisms self-regulate their interactions with the environment based on self-produced *norms* (Barandiaran et al., 2009, pp. 367–368).

The normative character of living beings together with their teleology has been emphasised also by Evan Thompson (2007, pp. 128–129), who also tries to expand earlier formulations with references to philosophical theories such as Maurice Merleau-Ponty's phenomenology, Hans Jonas' theory of metabolism, and Gilbert Simondon's idea of self-individuation (Thompson 2007, 2011). It is then interesting to remark that in enactive accounts no direct reference can be found to Plessner and his theory of positionality. However, as previously anticipated, there are strong reasons to believe that Plessner's contributions, in particular concerning the organism's interactions with the environment, might still have some valuable input to offer.

Plessner's theory of positionality outlines in fact a relation to the environment, which is neither simply derivative nor a structural coupling. His idea of self-individuation has little to do with the constitution of a topological unit, which, once "constituted," would establish an interaction with its surroundings (see Mitscherlich, 2007, p. 134: also Fischer, 2018).[8] Plessner's focus falls instead primarily on the activity of "positioning" of the lived body in relationship with the environment, which becomes its environment precisely through this positioning. In this original positioning activity, an "asymmetry" can certainly be identified in relation to the environment, which is instead missing in the case of the non-living. Precisely thanks to this asymmetry, a living being establishes a system with the environment (Plessner, 1928/2019, p. 144). The reference to a "system" to talk about non-living things would not make much sense. Without the environment, Plessner writes, the organism is "only half its life" (p. 180). The environment and the organism are therefore equiprimordial. The relation between an organism and its *Umwelt* is reciprocally constituting and reciprocally effecting. Each distinctive feature of living beings—what Plessner calls "organic modals" (p. 115)—such as organisation, regulation, system, but also characters like intentionality and teleology, can be seen as a configuration of positionality, understood as the living's original way to set itself in a specific relation with its environment. This latter is not a projection of meaning or a form of radical construction of the organism but its "opposing field" (*Gegenfeld*). The organism is "with and against it" (Plessner, 1928/2019, p. 186). From this viewpoint, one could argue that the autopoietic approach and the ecological approach, within the notion of positionality, are not mutually exclusive but rather reciprocally integrated.[9]

All in all, the main pitfall of the early formulation of autopoiesis—as shown by later enactive efforts to integrate it with widely understood phenomenological approaches—is to have missed embodied corporeality (*Leib*) as a starting point. Only a *Leib* can, in fact, react with a genuine action to the stimuli of the environment. This latter is not to be understood as a generic abstract surrounding, but rather as the set of opportunities and possibilities it affords, in other words, as the set of "affordances" peculiar to it: "an action is a response initiated by an agent, to a set of affordances, in pursuit of her goal" (Walsh, 2015, p. 216). Furthermore, also the viewpoint of the observer to the action needs to be taken into account. The meaning of the behaviour of a living being can be understood only by embodied beings. A merely physiological perspective is of little use here. In *Levels* (Plessner, 1928/2019, p. 19), Plessner echoes Dilthey's famous statement: "Life can be only known by life," which Jonas would later make his own (Michelini, 2019).

The notion of environmental intentionality (*Umweltintentionalität*), as developed by Plessner, together with Buytendijk, brings together these two aspects. As we shall see in the next section, this notion not only provides an interesting contribution to the topic of agency, but it also resembles the enactive notion of "sense-making."

11.4 Environmental Intentionality and Sense-Making

While aiming to build a phenomenology of expression in opposition to behaviourism, together with Buytendijk, Plessner focused on the relationship between sense and animal behaviour. Especially relevant for our topic, in the first part, the two authors deal in general with animal behaviour, and how its meaning can only be grasped in close connection with the environment. Those who overlook such a relation, as behaviourists do, tend to see the organism not as a "lived body" but as a *belebtes Körper*, "a merely animate body." This leads behaviourist physiologists to interpret behaviour in purely mechanistic terms. As a result, animal agency is reduced to measurable locomotor processes (for discussion, see also Kohn, this volume). As soon as we—as a thought experiment—cut the organism out of its relation to the environment, we are left with nothing but a living body in motion, whose behaviour objectively displays a truly incomprehensible chain of changes of place. Of course—the authors say—such a merely animate body receives or exerts effects but can never really act. The animal "no longer grasps [...] seeks, threatens, flees" but only "shows tactical reactions" (Buytendijk & Plessner, 1925/1982, p. 80; my translation).

Seeking, seeing, and grasping are, for Buytendijk and Plessner, "environmental-intentional" concepts, connected to our direct perception. We perceive "the way in which" humans and animals behave, and we do not see them as mere operators of movement. As embodied observers, what is given to us directly is the whole movement of an animal, a *Gestalt* of movement (*Bewegungsgestalt*), which is unitary also when extended over a given period of time.[10] In the contact with the *Umwelt*, animal behaviour shows, Buytendijk and Plessner (1925/1982) contend, a "direction" that can be grasped only in the light of the structure of the lived body–environment relation (*Leibumweltrelation*):

What the animal sees, smells, and whether it can at all do so is established via experiments [...], but [...] that the surrounding is then present to it [to the animal] in the way of hearing, seeing, etc., is clear to me based on the intuitive presentation [*Vergegenwärtigung*] of the lived body–environment relation.

(Buytendijk & Plessner, 1925/1982, p. 81; my translation)

On the one hand, the environment is present to the animal directly in the range of things that it can do with it. With reference to the lexicon introduced by J. J. Gibson several decades later (1979), one may say that, for animals, the environment is made of the set of its "affordances," that is to say its opportunities and possibilities of actions, for instance, "hearing, grasping, seeking, fleeing." The environment is never present to the animal as an objective neutral space, to which living beings eventually add meaning. On the other hand, since every observer is also a *Leib*, the relation with the environment can be grasped or vividly presentified.

Buytendijk and Plessner come up with a "technical" denomination: "environmental intentionality of the lived body" (*Umweltintentionalität des Leibes*). This is described as a "vividly [*bildhaft*] appearing relationship" between body and environment (Buytendijk & Plessner, 1925/1982, p. 79; my translation), which can be directly grasped in animal behaviour, in the unitary form of its movement. No anthropomorphism is here at work. Buytendijk and Plessner are harsh critics of what they see as anthropomorphic "cryptopsychology" (see also Plessner, 1928/2019, p. 58) based on empathy or self-identification with other living beings. What they suggest does not imply any projection of typically human experiential modes. They rather simply assess the "modal character" of animal behaviour, "which is intuitively certain [*anschaulich gewiß*] to me" (Buytendijk & Plessner, 1925/1982, p. 81; my translation).

While describing the structure of animal behaviour in its *Umwelt*, Buytendijk and Plessner also make reference to Uexküll's famous image of the *Bewegungsmelodie* or "motoric melody" (Uexküll 1921, p. 19; my translation). They state that far from being an unrelated temporal mosaic, animal behaviour has a "melody," a given dynamic form [*Gestalt*], which follows a unitary rhythm to the environment. As Salvatore Tedesco correctly remarks:

Not animate body, but *Leib*, the organism is a rhythmic whole with a gearing towards a world [...]. There is a co-belonging between the rhythm of the living and the establishing of a "direction" [*Richtung*] in the intentional relation with the environment; this amounts to saying that said rhythm can be grasped only based on the relation [...]. In the environmental interaction, in full reciprocity with the environment, the organism imposes and receives at the same time a rhythm, a direction.

(Tedesco, 2008, p. 94; my translation)

This melody metaphor is also widespread within enactivism, in reference in particular to what enactivists call "sense-making." As is well known, according to enactive theory: "living is sense-making in precarious conditions" (Thompson,

2011, p. 114). Sense-making is placed at the very origin of life and taken as the basic form of cognition. Cognition is seen "neither as retrieval nor projection," but "as embodied activity" (Varela et al., 1991, p. 172). At the most basic level, this activity means "establishing relevance": "Basic cognition, on this view, is not a matter of representing states of affairs but rather of establishing relevance through the need to maintain an identity that is constantly facing the possibility of disintegration" (Di Paolo & Thompson, 2014, p. 36).

The melody image can be found, for instance, in Thompson's comments on one of enactivism's most frequently mentioned example for cognition, the example of the bacterial chemotaxis, that is to say, the influence of substance concentration gradients, such as sucrose, on the direction of movement of living organisms:

> Bacterial chemotaxis provides [...] a case of living as sense-making in precarious conditions. Sucrose and aspartate, for example, have valence as attractants and significance as food, but only in the milieu or niche that emerges through bacterial living. Put another way, the status of these molecules as nutrients is not intrinsic to their molecular structure [...]. Rather, it belongs to the context of the cell as an individual, that is, as a self-individuating process that *behaves* as a *unity in dynamic concert* with its immediate environment.
>
> (Thompson, 2011, p. 119; emphasis added)

Sense, or relevance, is established through a relation of reciprocal determination or "equiprimordiality" between living beings and their environment (Varela et al., 1991, p. 173). It is through the activity of living beings, already present in bacteria, that an *Umwelt* emerges, where, for instance, sucrose and aspartate have food valence. What counts is, however, not the simple "perception" of sucrose by bacteria, but rather the dynamic concert between living beings and environment, which is never a simple perception but "a perceiving-directed activity" (Varela et al., 1991, p. 173).

Although bacteria are not among the organisms Buytendijk and Plessner take into account, at a more basic level, the "dynamic concert" Thompson refers to closely resembles the idea of a *Bewegungsmelodie* used to convey environmental intentionality.[11] One could even claim that Buytendijk and Plessner's ideas had at least an indirect impact on enactivism. The melody-related refrains in enactivism stem, in all likelihood, from reading Merleau-Ponty and his *The Structure of Behavior*. In this text, Merleau-Ponty (1942/1963) picks up the music metaphor, more notably from Buytendijk and not directly from Uexküll, as one may think (Sommer, 2016, pp. 92, 108). Furthermore, in *Phénoménologie de la perception* (Phenomenology of Perception, 1945), Merleau-Ponty refers to the notion of *Umweltintentionalität* (Merleau-Ponty, 1945/2008, p. 270), which shows that he read and assimilated the 1925 essay on mimic expression.

However, as previously in relation to autopoiesis' early formulation, also concerning sense-making, the differences between Plessner's (and Buytendijk's) theories and enactivism are as interesting as are their similarities. Enactivism aims to offer, in comparison to the first formulation of autopoiesis, a more balanced account of the dynamic co-emergence and mutual entrainment of living processes and their environments, establishing a middle path between "the Scylla of cognition as the finding of a given external world (realism) and the Charybdis of cognition as the projection of a given internal world (idealism)" (Varela et al., 1991, p. 172). This notwithstanding, it seems that its outcomes are not entirely free from the constructivism of the early formulation (e.g., Fultot et al., 2016, p. 303).

To stay within the scope of the bacteria and sucrose, for example, one should read, for instance, how Thompson—as Varela before him (Varela, 1997, p. 79)—explains sense-making, stating that "the reactions of the bacteria in their milieu—their tumbling and directed swimming—deposit a surplus of significance on the surfaces of molecules [of sucrose]" (Thompson, 2011, p. 119).[12] The terminology is rather ambiguous. Despite theoretical assumptions to the contrary, the relationship between organism and environment seems to be still understood as the act of the living that "deposits" (produces, projects, or builds?) a "significance" on its surroundings. Enactivism seems not to be completely out of the "Charybdis" of cognition as a projection of a given internal world, since it still tends to focus more on the subject side of the relation.

Unlike enactive theory, through the concept of environmental intentionality, Buytendijk and Plessner aim to define a "neutral" category with respect to the subject/object dichotomy. In behaviour, they identify a "layer" which is "psychophysically neutral," that is to say, prior to the differentiation of physical and psychic facts. We belong to this layer as *Leibwesen*, lived-body beings, which are not reducible to merely physiological factors nor to merely mental ones. Life is neither based on "subjective or internal experience," nor the exclusive pertinence of exact methods. As Buytendijk puts it in 1928: "Once the organic has been recognized, one easily discovers that every action, every perception, but also the wholeness of animal and environment shows the essence of the organic. The organic is thereby psychophysically neutral" (Buytendijk, 1958, p. 3; my translation). Furthermore, in behaviour also another indifference comes to the fore, as previously argued, between *Anschaulichkeit* (intuitivity) and *Verstehbarkeit* (intelligibility), that is to say, the immediate visualisation of a behaviour and its making sense. This double indifference, according to Buytendijk and Plessner, can only be found in organisms and in music (Buytendijk & Plessner, 1925/1982, p. 84), which explains why Uexküll's living melody plays such a key role.[13]

This last point leads us to a more general difference between Plessner and enactivism, which will be briefly addressed in the following conclusive section.

11.5 Conclusion

The aim of this essay was to explore whether it makes sense today to refer to Plessner's philosophical biology in order to better understand organismal agency. A positive answer has been formulated by pointing to some key notions in Plessner's work, which best match current debates and can provide valuable input to them.

According to Plessner, as we have seen, one should start from positionality as key criteria for the identification of living beings, and therefore for their agency. This brings him close to autopoietic positions. A key difference nevertheless remains. Whereas autopoiesis is considered, at least by its proponents, to be a genuinely empirical notion, according to Plessner, positionality can certainly have empirical correlates, but it is mainly based on our ordinary and pre-scientific intuition of the living: "the boundary relation in distinction to the boundedness relation cannot be demonstrated [represented], but only intuited [viewed]" (Plessner, 1928/2019, p. 120).

The same applies to the notion of *Umweltintentionalität*, which has been introduced as the "ecological" aspect of agency. This notion stands for the fact that animal behaviour always has a "direction" or "orientation," and that it can only be grasped by taking into account the relation between the lived body and its environment. Similarities can be found here with the enactive notion of sense-making; *Umweltintentionalität* is, however, free from subjectivist connotations, in that it is based upon the "neutral" character of life itself. This latter cannot be determined in merely physical nor psychic terms; it is neither subjective nor objective. Life is rather at the origin of both spheres. Animal behaviour shows precisely that, at an "original" vital level, both spheres are indifferent. Buytendijk and Plessner (1925/1982) write that "in the layer of behaviour we grasp the archetype of the forms and modes of constitution of all being: life" (p. 89; my translation).

In the 1990s, in order to reduce "the distance between subjective and objective," Varela understandably expressed the wish for a cognitive science with "a true circulation between lived experience and the biological mechanisms in *a seamless and mutually illuminating manner*" (Varela, 1995, p. 93; emphasis added; see also Varela et al., 1991, pp. 10–14, speaking about "circulation" between everyday experience and scientific experience). On this point, one might argue, Plessner would agree. However, one core issue emerges on how to understand said "seamless and mutually illuminating" process connecting the empirical level and the broader phenomenological and hermeneutic one. In this respect, some ambiguities can be still identified in enactive accounts.

It is true, in fact, that, on the one hand, enactivists have an absolutely empirical intent. Their proposal is to follow a "step-by-step procedure to answer empirically the question of whether a system is autonomous." And "the enactive concept of autonomy is entirely operational, and therefore naturalistic, though not reductionist" (Di Paolo & Thompson, 2014, p. 72). On the other hand, however, in reference to autonomy, they tend to introduce notions such as that of "genuine interiority" (Thompson, 2011), which are not biological concepts and cannot be understood at an empirical level.[14]

Plessner's position is in this respect more radical, and thereby possibly less contradictory. Although he also focuses on natural phenomena, he believes that the questions of life and its autonomy cannot be reduced to the empirical level. Despite their participation in empirical knowledge, the contents of intuition are, according to Plessner, indeterminable even by science: "All content that can only be acquired by intuition is fated to enter into experience without becoming determinable as experience progresses" (Plessner, 1928/2019, p. 111). One might nevertheless argue that also Plessner's approach is not free from ambiguity. If the question of life is not an empirical question, why—one might ask—did Plessner keep looking for empirical confirmation to his philosophical theories?[15] In 1965, he had his *Levels of the Organic* reissued without alterations but with an additional appendix meant to illustrate some recent scientific discoveries which he believed corroborated his theory (Plessner, 1928/2019, pp. 323–336).

The underlying reason for these ambiguities is possibly that notions such as agency, but also teleology, normativity, and affordances,[16] bring up the fact that between what is "empirically observable" and what is more properly phenomenologically and hermeneutically "intelligible" neither a seamless assimilation is possible, nor there is a clear-cut separation. A tension exists between the two, which is not sterile, but rather dialectical. The distinctive features of living beings are rooted in this tension, and based on it we can attempt to understand them. In this regard, then, the task of future "philosophical biology" could be to further develop and illustrate the meaning of the "mutually illuminating" relation among the two sides, without thereby giving up on either.

Acknowledgements

I am very grateful to the two anonymous reviewers and the volume's editors for their exceptionally helpful comments. Matteo Pagan and Flavio Artese also offered some useful comments on earlier drafts. Thanks to Tessa Marzotto for language support.

Notes

1 The reception of Plessner's philosophy of the organic in the English-speaking world has been long mediated mostly by Marjorie Grene's short introduction (Grene, 1966). More recently, see also de Mul (2014); Honenberger (2015); and Moss (2020).

2 In 1925, Helmuth Plessner founded the journal *Philosophischer Anzeiger* dedicated explicitly to the cooperation between philosophy and the individual sciences. It was an important publication venue for early phenomenologists and it ran until 1930 (4 volumes).

3 In this chapter, I refer specifically to its original formulation and its developments in authors such as Evan Thompson and Ezequiel Di Paolo. According to them, the autonomy of the living is what in particular marks out the enactive approach from other forms of embodied cognition (Di Paolo & Thompson, 2014, p. 69). The term "autopoietic enactivism" is taken from Hutto and Myin (2013).

4 Some exceptions include Köchy (2022, pp. 222–223); and Sommer (2016, pp. 91–107).

5 Self is here understood by Plessner (1928/2019) in mere structural terms and not in psychological terms: "A self is not yet a subject of consciousness: to have is not yet to know or to feel" (p. 148).

6 I say "almost" because things are actually more complex. According to Varela, for instance, the "boundary" is not necessarily topological, like a cell's membrane. In multicellular life forms, for instance, it can be virtual, i.e., behavioural, which means linked to stable patterns of sensor-motor couplings, which, strictly speaking, are not spatial—and thus cannot be "delineated." See also Thompson (2007, p. 44), where he claims that "a system can be autonomous without having this sort of material boundary." An ant colony is mentioned there as an example.

7 More empirical evidence could be found in the theory of protoplasm and of the immune system (on this point, see Grene, 1966, p. 255; and Mugerauer, 2014).

8 Joachim Fischer argues that, according to Plessner, living things are not autopoietic in the sense of self-generating, self-constituting things, but

> they find themselves "positioned" in their physical being in a kind of dependent independence, in an environmental boundary whose positionality they have to maintain. [...] The guiding concept for the organic is therefore not "position"—that is, taking or having a space-time position—but "positionality": being "set" in a space-time position in order to hold it.
>
> (Fischer 2018, p. 172; my translation)

9 Although enactivism and ecological psychology share significant theoretical analogies, the two views have rarely established a dialogue (Feiten et al., 2020, p. 3). It is worth remarking, though, that more recently the two views have started to cross more and more. Enactivists have shown greater interest for the ecological structure of the environment, and ecological psychologists have started to acknowledge the subjectivity and individual viewpoint of organisms. See, for instance, Saborido and Heras-Escribano (2023); more on this in Artese and Michelini (2023).

10 Clear references can be found here to Wolfgang Köhler's (1887–1967) Gestalt theory.

11 In this regard, an interesting suggestion has been by Orth (1990–91, p. 264), according to whom the notion of *Umweltintentionalität des Leibes* would anticipate that of positionality. This amounts to extending to all organisms the type of relationship to the environment that Plessner, in 1925, saw only in animals.

12 Thompson points out that this phrasing is actually Merleau-Ponty's. On this topic, see also Varela (1997).

13 Similarly, in his *The Structure of Behavior*, Merleau-Ponty defines behaviour as a "neutral" notion, going past the opposition between mechanism and vitalism as well as past the antinomy physiological vs. psychic (Sommer, 2016, p. 102).

14 This could be one of the reasons why enactivism is seen by many as a philosophy of nature, more than as a research programme in cognitive science. See Gallangher (2017); Käufer & Chemero (2021); Meyer & Brancazio (2022); Heras-Escribano (2023).

15 See also Plessner's lectures on metaphysics, and how contemporary empirical research is referenced throughout (Lessing, 2002).

16 In this regard, I agree with Denis Walsh on the point that, in order to formulate a theory of organisms as agents, "we need a battery of theoretical concepts and methods [...]: *goal, means, affordance, repertoire, salience, reciprocal* constitution, normative requirement, hypothetical necessity, teleology" (Walsh, 2018, p. 74; emphasis in original). However, I do not believe that these concepts are empirical, although they refer to natural phenomena.

References

Artese, G. F., & Michelini, F. (2023). La ricezione di Jakob von Uexküll nella psicologia ecologica e nell'enattivismo. *Discipline Filosofiche, XXXIII*(1), 9–44.

Baggs, E., & Chemero, A. (2018). Radical embodiment in two directions. *Synthese, 198,* S2175–S2190.

Barandiaran, X., Di Paolo, E., & Rohde, M. (2009). Defining agency: Individuality, normativity, asymmetry and spatio-temporality in action. *Adaptive Behavior, 17,* 367–86.

Buytendijk, F. J. J. (1958). Anschauliche Kennzeichen des Organischen. In F. J. J. Buytendijk (Ed.), *Das Menschliche* (pp. 1–13). K. F. Koehler. (Original work published 1928).

Buytendijk, H., & Plessner, H. (1982). Die Deutung des mimischen Ausdrucks: Ein Beitrag zur Lehre vom Bewußtsein des anderen Ichs. In H. Plessner (Ed.), *Gesammelte Schriften*, Vol. VII (pp. 67–129). Suhrkamp. (Original work published 1925).

de Mul, J. (Ed.). (2014). *Plessner's philosophical anthropology: Perspectives and prospects.* Amsterdam University Press.

de Mul, J. (2019). The emergence of practical self-understanding: Human agency and downward causation in Plessner's philosophical anthropology. *Human Studies, 42,* 65–82.

Di Paolo, E. (2005). Autopoiesis, adaptivity, teleology, agency. *Phenomenology and the Cognitive Sciences*, *4*(4), 429–452.

Di Paolo, E., & Thompson, E. (2014). The enactive approach. In L. Shapiro (Ed.), *The Routledge handbook of embodied cognition* (pp. 68–79). Routledge.

Di Paolo, E., Buhrmann T., & Barandiaran X. E. (2017). *Sensorimotor life: An enactive proposal.* Oxford University Press.

Feiten, T. E., Holland K., & Chemero, A. (2020). Worlds apart? Reassessing von Uexküll's Umwelt in embodied cognition with Canguilhem, Merleau-Ponty, and Deleuze. *Journal of French and Francophone Philosophy, 28*(1), 1–26.

Fischer, J. (2018). Plessners vital turn: Ekstatik der «exzentrischen Positionalität». In H. Delitz, F., Nungesser, & R. Seyfert (Eds.), *Soziologien des Lebens* (pp. 167–198). Transcript.

Fultot, M., Nie, L., & Carello, C. (2016). Perception-action mutuality obviates mental construction. *Constructivist Foundations, 11*(2), 298–307.

Gallagher, S. (2017). *Enactivist interventions: Rethinking the mind.* Oxford University Press.

Gibson, J. J. (1979). *The ecological approach to visual perception.* Houghton Mifflin.

Grene, M. (1966). Positionality in the philosophy of Helmuth Plessner. *Review of Metaphysics, 20*, 250–277.

Heras-Escribano, M. (2023) Ecological psychology, enaction, and the quest for an embodied and situated cognitive science. In O. Conte Casper & G. F. Artese (Eds.), *Situated cognition research: Methodological foundations* (pp. 83–102). Springer.

Honenberger, P. (Ed.). (2015). *Naturalism and philosophical anthropology: Nature, life, and the human between transcendental and empirical perspectives.* Palgrave Macmillan.

Hutto, D. D., & Myin, E. (2013). *Radicalizing enactivism: Basic minds without content.* MIT Press.

Ingensiep, H. W. (2004). Lebens-Grenzen und Lebensstufen in Plessners Biophilosophie: Perspektiven moderner Biotheorie. In U. Bröckling, B. Bühler, M. Hahn, M. Schöning, & M. Weinberg (Eds.), *Disziplinen des Lebens: Zwischen Anthropologie, Literatur und Politik* (pp. 35–46). Narr.

Ingold, T. (1990). An anthropologist looks at biology. *Man, 25*(2), 208–229.

Käufer, S., & Chemero, A. (2021). *Phenomenology: An introduction* (2nd Ed.). John Wiley & Sons.

Köchy, K. (2022). *Beseelte Tiere. Umwelten und Netzwerke der Tierpsychologie.* J. B. Metzler.

Lessing, H-U. (2002). *Helmuth Plessner, Elemente der Metaphysik: Eine Vorlesung aus dem Wintersemester 1931/32.* Akademie Verlag.

Maturana, H., & Varela, F. (1980). *Autopoiesis and cognition: The realization of the living.* Reidel Publishing.

Merleau-Ponty, M. (1963). *The structure of behavior.* Beacon Press. (Original work published 1942).

Merleau-Ponty, M. (2008). *Phenomenology of perception.* Routledge. (Original work published 1945).

Meyer, R., & Brancazio, N. (2022). Putting down the revolt: Enactivism as a philosophy of nature. *Frontiers in Psychology, 13.* https://doi.org/10.3389/fpsyg.2022.948733

Michelini, F. (2019). Anthropology versus ontology: Plessner and Jonas's readings of Heidegger's philosophy. *Revue Philosophique de Louvain, 117*(2), 311–339.

Michelini, F. (2021). Neither vitalist nor mechanist, neither dualist nor idealist: Plessner's Third Way. *History and Philosophy of Life Sciences, 43,* 71. https://doi.org/10.1007/s40 656-021-00421-7

Mitscherlich, O. (2007). *Natur und Geschichte. Helmuth Plessners in sich gebrochene Lebensphilosophie.* Akademie Verlag.

Moreno, A. & Barandiaran, X. (2004). A naturalized account of the inside-outside dichotomy. *Philosophica, 73,* 11–26.

Moreno, A. & Mossio, M. (2015). *Biological autonomy: A philosophical and theoretical enquiry.* Springer.

Moss, L. (2003). *What genes can't do.* MIT Press.

Moss, L. (2020). Levels of organic life and the human: An introduction to philosophical anthropology (Review). *Notre Dame Philosophical Reviews.* https://ndpr.nd.edu/reviews/ levels-of-organic-life-and-the-human-an-introduction-to-philosophical-anthropology/

Mugerauer, R. (2014). Bi-directional boundaries: Eccentric life and its environments. In J. de Mull (Ed.), *Plessner's philosophical anthropology: Perspectives and prospects* (pp. 211–228). Amsterdam University Press.

Orth, E. W. (1990–91). Philosophische Anthropologie als Erste Philosophie: Ein Vergleich zwischen Ernst Cassirer und Helmuth Plessner. In F. Rodi (Ed.), *Dilthey-Jahrbuch für Philosophie und Geistwissenschaften, 7,* 250–274.

Pagan, M. (2023). Les degrés de l'organique «à rebours»: de l'anthropologie philosophique à la philosophie de la nature de Helmuth Plessner. In A. Charpentier, M. Dal Pozzolo, & M. Pagan (Eds.), *Repenser la nature: Dewey, Canguilhem, Plessner* (pp. 111–124). Éditions de la Rue d'Ulm.

Plessner, H. (1973). *Der Aussagewert einer philosophischen Anthropologie.* In H. Plessner (Ed.), *Gesammelte Schriften,* Vol. IX (Conditio humana) (pp. 380–399). Suhrkamp.

Plessner, H. (2019). *Levels of organic life and the human: An introduction to philosophical anthropology.* Fordham University Press. (Original work published 1928).

Ruiz-Mirazo, K., & Moreno, A. (2004). Basic autonomy as a fundamental step in the synthesis of life. *Artificial Life, 10*(3), 235–259.

Saborido, C., & Heras-Escribano, M. (2023). Affordances and organizational functions. *Biology & Philosophy, 38*(1), 1–16.

Sommer, C. (2016). Unweltintentionalität: De Merleau-Ponty à Plessner via Buytendijk. In C. Sommer & F. Burgat (Eds.), *Le phénomène du vivant* (pp. 91–107). Metis Presses.

Swenson, R. (1992). Autocatakinetics, yes–autopoiesis, no: Steps toward a unified theory of evolutionary ordering. *International Journal of General Systems, 21*(2), 207–228.

Tedesco, S. (2008). *Forme viventi: Antropologia ed estetica dell'espressione*. Mimesis.

Thompson, E. (2007). *Mind in life: Biology, phenomenology, and the sciences of mind.* Belknap Press of Harvard University Press.

Thompson, E. (2011). Living ways of sense-making. *Philosophy Today, 55*(Supplement), 114–123.

Uexküll, J. von. (1921). *Umwelt und Innenwelt der Tiere*. Springer.

Varela, F. J. (1991). Organism: A meshwork of selfless selves. In A. I. Tauber (Ed.), *Organism and the origin of the self* (pp. 79–107). Kluwer.

Varela, F. J. (1995). Resonant cell assemblies: A new approach to cognitive functions and neuronal synchrony. *Biological Research, 28*, 81–95.

Varela, F. J. (1997). Patterns of life: Intertwining identity and cognition. *Brain and Cognition, 34*, 72–87.

Varela F. J., Thompson, E., & Rosch, E. (1991). *The embodied mind: Cognitive science and human experience*. MIT Press.

Walsh, D. M. (2015). *Organisms, agency, and evolution*. Cambridge University Press.

Walsh, D. M. (2018). Objectcy and agency: Towards a methodological vitalism. In D. J. Nicholson & J. A. Dupre (Eds.), *Everything flows*. Oxford University Press.

Weber, A., & Varela, F. J. (2002). Life after Kant: Natural purposes and the autopoietic foundations of biological individuality. *Phenomenology and the Cognitive Sciences, 1*(2), 97–125.

Part IV

Theoretical and Metaphysical Frameworks for Organismal Agency

12 Agency as Internal Control

Gunnar Babcock and Daniel W. McShea

12.1 Introduction

Agency is often seen as a property of living systems, one which stems from a particular sort of functional integration that provides certain biological entities a special kind of causal power over themselves. Such self-control allows living systems or organisms to act freely and independently of their external environments (cf. Barandiaran et al., 2009; Moreno & Mossio, 2015). It allows them to take initiative, to change themselves, and to move themselves. In this chapter, we outline a view of agency that focuses on hierarchical structures, control, and how these relate to goal directedness. This view of agency is derived from our theory of goal-directed systems, what we call *field theory*. As will be seen, our view both deviates from and aligns with some common thinking about agency. In particular, we make some critical distinctions between pairs of words that are sometimes used interchangeably, especially control and determinism, and goal directedness and self-directedness, to get at our view of agency.

At the most general level, our position is that agential systems are a subset of the goal-directed ones. To show this, we begin Section 12.2 with an explanation of how goal directedness works under field theory. According to field theory, goal-directed systems have a hierarchical structure, consisting of a small entity that moves within and is directed by a larger field in which it is immersed. A paradigmatic case is sunflowers turning to track the sun throughout the day, guided by the light field emanating from the sun. Sunflowers may well be agents, but goal directedness is found in certain things that are clearly not agents. Something like a rock falling in a well is minimally goal directed on our view (discussed later), but it is not agential. In Section 12.3, we offer our positive view of agency. There we address the relationship between agency and goal directedness. We argue that systems with the right hierarchical architecture inside them have the capacity for agency. With internal hierarchical structure, agential control often becomes possible.

It has not escaped our notice that a seeming contradiction lies at the heart of our views of goal directedness and agency, given that elsewhere we have argued that goal directedness arises from fields that are external to the goal-directed entity. We address this seeming contradiction by highlighting the difference between goal directedness and self-directedness, and by contrasting our view with others in the

DOI: 10.4324/9781003413318-16

literature. In Section 12.4 we turn to the concept of control to explain why determinism would not undermine agency. This leads to some of our final points, about the graded nature of agency, in Section 12.5 where we argue that degree of agency has to do with the number and depth of hierarchical structuring—that is, of fields—inside the agent.

12.2 Field Theory

This section provides a somewhat distilled version of field theory, providing just enough groundwork to give sufficient context for the following sections on agency.[1] Field theory makes sense of goal directedness by arguing that spatial fields, external to the entities they direct, are the sources of goal-directed behaviors. At the center of the view is hierarchy theory, as it has developed mainly in biology and philosophy of science.[2] At the most general level, what all of this work shares is the idea that physical systems are composed of multilevel hierarchies, where the notion of "hierarchy" roughly means that different entities show up at different levels of nestedness. Hierarchical organization is important in field theory because a core principle of the theory is that goal directedness takes place when entities at higher levels of organization influence those at lower levels of organization nested within them. Consider the goal-directed behavior that a dung beetle exhibits when it acquires an orb of dung. It moves away from the dung pile to escape other beetles who might steal it. To achieve this goal, the beetle needs to escape the pile without circling back to it accidently. Researchers who study dung beetles find its goal-directed tenacity impressive: "a beetle's drive to adhere to its set course is so strong that it sticks to its path regardless of obstacles; over stones, through bushes and grass, across the hand of an experimenter or in an experimental arena" (Dacke & Jundi, 2018, p. R993). It needs to move in a goal-directed way (where the goal is "away from the dung pile"). From the field theoretic perspective, when a dung beetle exhibits such goal directedness, it is because a higher-level structure, in particular the polarized light in the overhead sky, directs it. (The kinds of solar light that direct beetles often depend on the species of dung beetle in question.) In this hierarchical view, we call structures at higher levels "fields." So when a beetle engages in some goal-directed activity, such as navigating away from a pile of dung, according to field theory, it is able to do so because the larger field—the light in the sky—provides the beetle with direction.

There are several points to make using the salient case of the dung beetle. The first is that there is no need to attribute intentions, desires, or the like to the light from the sky. Others, like Dennett (1984), have made this point as well. To argue, as field theory does, that polarized light directs a beetle does not require the kind of metaphysical scaffolding found in traditional teleological externalism, which places some kind of intentional deity at the outer most rung of the hierarchy and uses hierarchical organization to will its own desired ends. Under field theory, goal directedness requires no intentionality.

What is goal directedness then? To answer this, field theory follows Nagel (1979), noting that all goal-directed phenomena, no matter how disparate, share two features: *persistence* and *plasticity*. In our terms, persistence is the tendency for an entity following a particular trajectory to return to that trajectory following perturbations that might have knocked it off course (McShea, 2012). And plasticity is the tendency for an entity to find that trajectory from a variety of different starting points. Recent work on insect navigation offers many biological examples of persistence and plasticity at work. A dung beetle leaves a pile of dung in a goal-directed way, to escape competition with other beetles in the same pile. It moves persistently in that it ultimately finds a path away from the pile despite obstacles that require it to deviate briefly back toward it. It is plastic insofar as it will move away from the pile regardless of where it starts.

Given these signature behaviors, persistence and plasticity, the question arises of how goal-directed entities are able to perform those feats. To answer this, field theory takes an engineering perspective, asking: what components are needed to make such goal-directed system work? On the one hand, certain mechanisms within the beetle are undeniably necessary. Without its visual systems, motor controls, neural pathways, etc., it would not have the capacity to move away from a dung pile. The relation between these internal mechanisms and the external fields with which we are concerned is addressed in greater detail elsewhere (Babcock & McShea, 2023b). In general, field theory is compatible with mechanistic explanations, with mechanism understood in an extended sense that posits physical upper-level fields alongside mechanisms. A beetle detects polarized light with the optic lobe of its brain, which connects to the brain's central complex. And the neuroarchitecture of the central complex is composed of "layers and slices." Some of these slices are interconnected with protocerebral bridges that transmit specific signals along neural pathways that direct behaviors. Researchers have even pinpointed the activity of certain neurons, which correlate with what they call "celestial snapshots" (Dacke & Junde, 2018). These neurons direct the beetle's motor systems when it follows a straight-line trajectory away from a dung pile. Notice the number of hierarchical structures in this description and notice how the arrow of directional influence goes from higher-level fields to lower-level mechanisms. The optic lobe sends signals to the central complex, which directs slices within it, and the slices in turn direct the protocerebral bridges, and so on. So the hierarchical architecture required by field theory seems to be in place. But something is missing: guidance. None of these mechanisms can provide any new information about where the dung pile is and which way to go to walk away from it. Only the external field of polarized light carries that information. To predict where a beetle will go, knowing facts about its neural anatomy will not help. One needs to know where the dung pile is and what is going on in the sky above it. Of course, knowing the properties of a field alone is insufficient to be able to determine whether a given entity will exhibit goal directedness when it enters the field. Answering that question requires understanding its internal mechanisms, in accord with modern mechanistic explanations. Therefore, field theory is an explanation of goal directedness as it is observed, but it offers no predictions regarding whether a given entity will exhibit goal directedness.

Now it is possible to imagine that dung beetles walking away from dung piles could direct themselves entirely from "within," using only information that is built somehow into their neurons, and some who work in insect navigation argue for this view (see, e.g., Cheeseman et al., 2014). However, from a conceptual standpoint, if this was how navigation, insect or otherwise, worked it would be devastatingly fragile and unwieldy. Imagine if dung beetles had internal maps that were somehow hard-wired into their tiny brains and they had to use those maps to generate a route away from the dung pile. In such a case, any mismatch between the internal map and the actual layout of the environment could send the beetle off in the wrong direction. Another solution might be to program into the beetle's brain the entire series of muscle contractions and footfalls needed to take it in a straight line away from the dung pile. But this system is also fragile, in that any encounter that knocks the beetle off course could irreversibly disrupt its trajectory away from the dung pile. In sum, it is hard to imagine how any internal mechanism could, by itself, produce the goal directedness we find in nature. And, Dacke et al. (2013) confirm what a field theoretic view anticipates. Dung beetles do not use internal mapping for guidance. They use polarized light in the sky, and when they are unable to detect it, they are completely lost. This has been demonstrated by placing headshields on the beetles, which obscured their overhead views. Beetles with the headshield are unable to navigate away from the dung pile. From an engineering perspective this comes as no surprise. Light from the sky is a field, and it is this field that gives the beetles the capacity to recover from errors and missteps, providing guidance no matter where they wander. It means that mistakes do not result in failure. It is just the kind of robust system that one would imagine selective processes would yield.

This approach also plays out in a different context, among and within the cells of a developing multicellular organism. In an organism's ontogeny, its cells behave teleologically, persistently and plastically moving in space, changing from one cell type to another, and changing their patterns of gene activation in ways that move the organism's development forward. These behaviors are all mediated by processes within the cells that involve many different molecules, including the genes, but also myriad proteins, lipids, and other substances, all interacting mechanistically, and the proper function of these mechanisms is critical to the goal-directed performance of the cell as a whole. Significant deviations in them can send a cell on the wrong trajectory. So mechanisms are important. But they do not guide and cannot make informed "decisions" that are relevant to the embryo as whole. For the most part that guidance comes from outside the cell, from what are called "morphogenetic fields," larger-scale gradients of biochemical substances, secreted in bulk by genes inside the cells, but present on a scale far greater than any single cell, enveloping many cells at once and directing their behavior (see Levin, 2012). When a given cell deviates by chance, it is the morphogenetic field that guides it, sending signals to the cell's internal mechanism that nudge it back to the proper mechanical and biochemical trajectory. In all of this, genes are important, both in their role as components of cellular mechanisms and in their role as factories for bulk production of the molecules that constitute the fields. But they do not guide. Indeed, thinking about this from an engineering perspective, they could not guide.

Buried far down inside cells, along with the rest of the cell's mechanisms, they simply have no information at the proper scale, no information about the needs of the larger organism, no information about where the cell is supposed to go and how it is supposed to transform in order to contribute properly to the larger whole. The information must necessarily reside at a higher level.

12.3 Agency

So far we have not touched on agency. In this section we offer our positive account of agency, contrasting it with non-agential teleology. We then consider the views of agency presented in Sultan et al. (2022) and in Moreno and Mossio (2015) in an effort to illustrate some key points of agreement and divergence with our position.

The field theoretic view of agency comes from fully appreciating the scope of hierarchical structures. The most deeply nested systems, containing the most levels and the most complexity, are often organisms. Scientific practice indicates as much. It would be a challenge to enumerate all the subfields of biology that are dedicated solely to understanding levels that are contained "under the skin" of organisms, including the standard molecule, cell, tissue, organ, and organ system levels of course, but many other intermediate levels as well.

Taking this view entails an unconventional outlook on how agency and teleology are situated in relation to each other. As stated above, for sunflowers and for the falling rock, teleological guidance is external, with upper-level fields directing lower-level entities. But in agential systems, both the field and the contained directed entity lie inside the organism. It is fitting to call this kind of guidance agential, we argue, because both guiding field and guided entity are *parts* of the organism. When field-and-entity act, the causal arrow still points down, from field to entity, but since both lie within the organism, the resulting behavior is the organism's behavior. It is the organism directing itself.[3]

To see the logic of this view, contrast it with what we consider to be *non-agential goal directedness*. First, imagine a swimmer who moves away from a shore because they are caught in a rip current, an undertow. In this case, the external current is strong and carries them, directs their movement, away from shore even if the person swims against it. In such a case, like Hume's shackled prisoner, the person has lost some degree of their agency. They are, of course, an agent insofar as they might be able swim this way or that, at a very local scale when fighting the current, but if the current is strong enough, it will be to no avail. But, as field theory is a theory of goal-directed systems, its task is to determine whether the *particular* goal-directed behavior like "moving away from the shore" counts as an agential act. In the case of the swimmer, it is clear that their goal-directed trajectory away from the shore is out of their control. As an entity caught up in a field, they are goal directed, but in a non-agential way.

Now consider another person, on a beach with no rip current, who also moves away from shore but swims deliberately away in order to escape the noise from a group of kids playing in the surf near shore. Like the first case, what directs the person away from the shore is an external field, in this case the sound field emanating from the kids. However, in this case, there are also internal fields that govern movement, deeply nested hierarchical systems within the brain, perhaps a large-scale neural field corresponding to wanting peace and quiet, with a motor control center nested within it, and efferent neurons leading to muscle groups associated with swimming nested within that. When that person moves away from the shore to avoid noise, the parts responsible for their movement are their own parts. Their wants control their goal-directed movement away from the noise. Now notice that under field theory, both cases are teleological. And in both, the directing field is external, in a sense. In the undertow case, the rip current that does the directing is external to the swimmer as a whole. In the second case, the fields doing the directing—the brain states corresponding to wanting to avoid the noisy kids— are part of the person. This difference between the main sources of guidance is the key to agency. To the extent the fields that guide a person are mainly external, the person lacks agency. To the extent the fields that guide are mainly internal, they act as an agent.

The above distinction between agential and non-agential teleological systems does not deliver our full account of agency. But it clarifies something that has sometimes been a source of confusion in understanding the commitments of field theory. "External" directedness can take place *inside* an organism, even entirely inside it, provided that the guided entities lie within the guiding field. This might be the case, for example, when the guided entities are motor control centers and the guiding fields are neural activation fields (i.e., wants, desires, motivations, etc.) that are larger than and surround them. So while terminologically the notion of "external" direction might seem at odds with the idea that it could take place "inside" an organism, there is nothing contradictory here. Organisms and some other systems are deeply nested, with many levels of entities within fields, all of them entirely under the skin. This clarification is critical, in that in conventional discourse on agency—and following Kant (see Gambarotto & Nahas, 2022)—the difference between internal and external has been central to agency (e.g., Walsh, 2015; see also Nahas, this volume). Our view does not fit neatly into this discourse. In conventional terms, our view counts as internalist in that agency arises from processes occurring inside the organism (under the skin), but externalist in that the causal arrow always runs downward from a field that is external to the guided entity, that is, in the case of complex behaviors, downward from a larger neural field to a smaller motor control center. We hope this obviates some of the critiques directed at our view (e.g., Vane-Wright, 2023; Deacon & García-Valdecasas, 2023).

So field theory is at least in principle compatible with certain internalist views of agency. Still, it differs in the way it sees the relationship between agency and goal directedness. Consider the relationship between *goal directedness* and *self-directedness*.[4] In the contemporary literature, there seems to be no general agreement on how these two align. On our view, goal directedness and self-directedness are both teleological, but not all teleological systems are self-directed. For example, as McShea (2023) has recently argued, certain kinds of evolutionary trends are teleo-logical when they are directed by larger-scale ecological fields. The iterative evo-lution of flightless rails on the Aldabra atoll is a goal-directed evolutionary trend, a lineage that is persistently and plastically directed toward flightlessness. However, such trends have no agency as they are not self-directed. The direction comes from the island ecology, from a field outside the rail lineage.

Consider the simpler case of the rock falling in a well (discussed in Babcock, 2023). Contrary to conventional thinking, we consider this a teleological process, albeit only minimally teleological. Mayr (1988) agrees citing a rock falling in a well as a particular species of teleology, calling the process "teleomatic." Teleomatic processes are goal-directed processes that are driven by simple laws of nature, so simple that it is hard to see them as instances of genuine teleology. But under field theory, they are teleological for the simple reason that, clearly, they are directed toward an end. The rock is directed, by an external gravitational *field*, toward the bottom of the well. Leaving such cases out of the scope of teleological phenomena results in an impoverished understanding of teleology and goal directedness. But surely something is lacking in the falling rock, something that distinguishes it from the more impressively complex and internally motivated goal-directed behaviors of complex organisms. Field theory agrees. The rock, unlike the complex organism, has little agency. The rock is not *self*-directed. It is like the person caught in the rip current. Agency is self-direction, which in turn requires a set of internal nested entities-within-fields. Teleology is not enough.

Now, contrast the rock with an *Escherichia coli* bacterium climbing a food gradient. *E. coli* are used in several discussions of minimal agency and as examples of teleological systems (Barandiaran et al., 2009; McShea, 2012; Moreno & Mossio, 2015; Lee & McShea, 2020). Even though an *E. coli* bacterium is just a single cell, it has a nested hierarchy inside it. It has a multi-layered cell wall that surrounds the cytoplasm and in the cytoplasm is a genome. It even has a flagellum to propel it. On the field theoretic view, this hierarchical structure is teleological, consisting of upper-level fields directing lower-level mechanisms within it. The bacterium controls its flagellum. In contrast, the rock controls no part of itself. The falling rock is somewhat teleological. The bacterium is both somewhat teleological and at least somewhat agential.

The bacterium example points to a difference between field theory and what Moreno and Mossio (2015) call an "organizational account of agency" (see also Virenque, this volume). Both take a fairly expansive view of agency in which a bacterium climbing a food gradient counts as minimally agential. Also, both locate the source of agency inside the bacterium. The organizational view identifies closure and self-maintenance as key properties that confer agency. The field theoretic view also takes these internal processes seriously, and indeed, in principle, there is nothing inherent in the organizational account's notion of closure that conflicts with how field theory treats them. Of course, field theory describes them differently, pointing to the nested hierarchy of fields and contained entities, and the mechanisms that detect the food gradient and drive the flagellum.

So what distinguishes the two views? One critical difference is that Moreno and Mossio relegate the environment to the backdrop, to the role of triggering internal mechanisms. For them, "a 'taxis' is a movement of an organism triggered by a given feature of the environment, whose presence has some relevance for its self-maintenance" (Moreno & Mossio, 2015, p. 97). In contrast, field theory sees the environment as a source of guidance, of direction, for the bacterium, the source that the bacterium consults in order to know which way to go to climb the food gradient.

The significance of this difference becomes evident when we consider how each view would treat a bacterium that is *not* in a food gradient, one lacking any sort of external guidance. Inside the gradient, the bacterium undergoes a series of random tumbles followed by straight runs, taking longer straight runs when the food concentration is higher, moving it on average toward that higher concentration. Absent a gradient, the pattern of tumbles and straight runs continues but follows no particular overall pattern. In this circumstance, Moreno and Mossio would likely decline to think of its behavior as agential. The mechanisms that produce agency are in place, active, and powering the organism along its path, but they are not doing their job, which is to move the organism up any food concentration gradients that might be present. But when such a gradient is absent, those mechanisms are not, in that moment, contributing to the organism's self-maintenance.

In contrast, the field theoretic view sees the bacterium's behavior outside a food gradient as *more* agential. Unconstrained by a directing gradient it gets to do whatever it "wants." It goes wherever its internal fields and underlying mechanisms drive it. The hierarchical structures inside the bacterium, being less constrained by the outside, exercise greater control over the direction the bacterium takes. In our terminology, it has more agency than a bacterium in a gradient (see McShea, 2016; Babcock & McShea, 2023a). A bacterium directed by a gradient is still an agent, but once it is contained within the food field, the field gains some degree of directional control, constraining the behavior of the bacterium, *reducing* (or channeling) its agency. In other words, chemotaxis is a goal-directed process, but one that is governed externally and reduces the agency of a bacterium. Put another way, a person who lives outside the bounds of a political order and a set of laws has greater agency than someone who lives within a political regime with laws that direct the range of their behaviors. Being subject to external laws reduces one's agency. Which is not to say an increase in agency is always a good thing. The bacterium that remains outside a food gradient is fully agential, but it ultimately starves.

Some will notice a seeming contradiction here, arising from our earlier claim that agential systems are a subset of teleological ones, and that goal directedness is a precondition for and gives rise in some cases to agency. Because here we are giving a case where teleology reduces agency. The contradiction disappears when we see that it is only the goal directedness arising from hierarchical structures inside the organism that gives it agency. Everything changes at the organism boundary. There the logic reverses. Any external field that controls, and therefore restricts, the organism's behavior reduces its agency.

Another feature of the field theoretic view of agency that is entailed by the hierarchical structures and causal relations within a system is that nothing requires that an agential system be living, or part of a living system. Are there non-living agential systems? Our answer is emphatically yes, and one example suffices to make the point. Hurricanes form as warm wet air rises from the surface of the tropical ocean. Perturbations give rise to rotary motion, and that motion is stabilized and amplified by the bulk upward flow of air. Later in the process, the rotary motion produces a pressure differential that gives rise to an eye and eye wall with a region of relative calm in the middle. Taking the hurricane itself as our focal system, field theory tells us that it is directed from above by the huge external field of warm air rising off the ocean. That in itself confers no agency, or at least no more agency that the rock falling down a well. But when the eye and eye wall form, they are directed from above by the much larger field of rotary motion. Thus we have some degree of agency here. The hurricane as a whole directs the parts, the eye and eye wall, within itself.

It is valuable here to compare field theory with another view of agency that is present in the literature, that of Sultan et al. (2022), where "agency is a dynamical property of a system" (p. 5; see also Walsh & Sultan, this volume). We agree, arguing that in essence the dynamics between fields and mechanisms that compose a hierarchical structure are what give rise to agency. Further, both views place special importance on the environment, as in the example of the bacterium presented above. We also agree with Sultan et al. in disconnecting agency from any notion of intentionality or desire (cf. Aaby, this volume). But there are important differences. In particular, we see no reason why agency cannot extend to non-living systems. Sultan et al. (2022) state that an agent is able "to transduce, configure, and respond to the conditions it encounters" (p. 5). Hurricanes meet these criteria. They transduce warm air, using it for energy to form and maintain themselves. They can also be said to configure and respond to the conditions within themselves, using those conditions to generate an eye and an eye wall. Under field theory, in contrast to Sultan et al., agency is not a property unique to living systems.

In sum, our conclusions regarding agency align with parts of certain other views. Like some of those views, ours is fairly permissive, acknowledging agency not just in human intentionality but in simpler entities such as bacteria. However, our view differs in that we do not see goal directedness and agency as necessarily aligned. We also believe it is possible to locate some degree of agency in non-living entities. More generally, we see agency as arising from a particular relation between external and internal structures, as a function of how "control" is divided up between them. We turn to the issue of control in the next section.

12.4 Determinism, Control, and Agency

One key to the field theoretic view of agency can be found in an observation by Dennett (1984). For some, agency is bound up with the problem of determinism versus indeterminism, with some seeing determinism as a threat to agency. How can a fully determined entity have any agency? Dennett (1984) argues that the issue is not determinism but control. There are several important ways in which we do not agree with Dennett's overarching theoretical commitments, particularly when it comes to adaptationism, but on this point we are in general agreement. His insight about control means that determinism, along with our ability to construct deterministic scientific models, is no threat to the notion of self-guidance that intuitively is central to any notion of agency.

Dennett's point is that determined processes are not equivalent to controlled ones. Consider rolling dice. A person who throws a pair of dice determines how they will land, meaning that the person's physical movements in rolling the dice—including the neural signals that activate the muscles, the force and twist imparted by the muscles, and the angle of the hand—would in a deterministic world decide how the dice will land. But, of course, the person has no control over how they land. Or, in our terms, none of the field-entity combinations inside the person can exert any control over what numbers will be showing when the dice come to rest. So even though the thrower sits at the start of the deterministic causal chain directing the dice to their final position, the thrower is not an agent with respect to the outcome. On the other hand, if that person *could* somehow control how the dice landed, their thinking and decisions about the dice might be fully deterministic, but the person would nevertheless have agency with respect to the dice. Control, not determinacy, is what imparts agency, even though controlled processes are determined ones by any normal accounting of a casual chain of events (see Dennett 1984, Chapter 1; Babcock & McShea, 2023a). Our claim is that, other things being equal, the goal-directed entities contained within an organism are largely under its control, that is, controlled by higher-level fields that are also within it. Notice that here "control" is understood in a broader sense than mere intentionality (see Dennett, 1984). "Control" in this broader sense is something that Dennett attributes to B. F. Skinner: "In the Skinnerian sense of 'control' we say that *A* controls *B* if and only if changes in *A* are reliably reflected or registered in changes in *B*" (Dennett, 1984, p. 59; emphasis in original). This is what we mean by "control" as well.[5]

Now we can see why a falling rock is negligibly agential, a bacterium has a degree of agency, and a bacterium in a chemical gradient has reduced agency. Because a rock has no significant internal hierarchical structure, it has no control. External fields completely control it. A bacterium has internal structure, i.e., hierarchical levels composed of upper levels fields and mechanisms within them, and so it possesses some control over itself. When a bacterium has no upper direction— i.e., when it is *not* in a gradient—the structures inside the bacterium have complete control, even though its behavior is aimless. Once it is in a gradient, the upper-level field now exerts some control over it. However, insofar as much of the total hierarchical structure remains within the bacterium, it retains some agency. Fields and the parts within them—fields and parts that in a real sense *are* it—are doing much of the controlling.[6]

We are now in a position to respond to the challenge posed at the beginning of this chapter: how is agency possible under the field theoretic account of goal directedness? Returning to dung beetles, if beetles are externally directed by polarized light would not this entail that they are not agents? Does not it mean that they mechanistically and passively respond to light emitted from the sky and therefore lack the kind of self-determination typically believed to be a critical component of agency? From this vantage point, it might have appeared as though field theory resembles genetic or biological determinism. All it does is switch the source of the determinism from the genes to external fields.[7] But this is not the case. Solar light directs the optic lobe, and that lobe directs the central complex, which directs neural slices, and at each level there is a causal interaction that is taking place. Some *A* inside the organism is changing some *B* inside the *A*. What is key is that most of these causes are happening inside the beetle, caused by fields and parts within the beetle. Most of the causation is controlled self-causation. Solar light determines the beetle's overall path, but the beetle is in control of the details of its movement. However, returning to where we differ with other views, a beetle that is not being directed by solar light has more control because in that case, all other things being equal, all of the overall causation takes place "under the skin" of the beetle. If its eyes are covered by the experimenter, the net result will be that the beetle's behavior becomes more aimless, but it then has greater agency as it gains more control over its movement. Control, not determination, is what confers agency.

Some will consider this a step too far. Are we really arguing that an aimless organism has greater agency than one that is actively and successfully pursuing survival and reproduction by taking advantage of the affordances of the environment? If so, we would seem to be draining agency of some of its normative content. Seemingly, under field theory, greater agency is not necessarily a good thing. This is indeed part of our view, and not as a bullet we bite but as a virtue we embrace. Consistent with our thoroughly biological take on the world, agency is—like many other biological properties—most advantageous not in its most extreme form, but in some middle ground between extremes. Organisms thrive not in achieving complete self-determination, perfect agency, but in the release of some agency, in partial surrenders to the environment, in the interest of navigating it and drawing on its resources.

It is worth acknowledging that any scientific investigation of agency must confront what one might call the *deterministic threat*. If agency can be modeled deterministically, as field theory proposes, that deprives it of the freedom that intuitively lies at its core, raising the possibility that true agency is illusory. We have addressed the issue of freedom elsewhere (Babcock & McShea, 2023a). Our reply is, as argued above, that determinism is not relevant here, that what we really want out of agency is control, that field theory preserves control, and that successful modeling of agency simply bolsters our understanding of how it works.

12.5 Graded Agency

An upshot of this view of agency is that it shows that agency comes in degrees. The degree of agency roughly corresponds with the degree of nestedness and upper directedness within the agent. Generally, entities with fewer levels of internal nestedness have less agency than those with more, as more nestedness provides more places where causation can take place, and therefore greater control. So as internal hierarchies get more complex, other things being equal, the degrees of agency increase too. A dung beetle not only has deeply nested hierarchical systems inside it, but it also has other parallel hierarchical systems. So not only are the tools that it has hierarchically deep, but it also has more of them. It has some internal hierarchically organized system for guidance by polarized light, as discussed, but it also undoubtedly has another for steering around obstacles in its path. And others for detecting and responding to the odors of dung, mates, and presumably other features of the environment. Plus, it has the usual array of internal animal systems that direct growth, development, and repair, all of these hierarchically organized and goal directed, as well as a large repertoire of homeostatic mechanisms for controlling its internal physiology, each to some degree independent of the others (Figure 12.1). Our point is that each somewhat independent, internal, goal-directed system within the beetle adds another degree of control that the beetle has over itself. Each adds some increment to its agency.

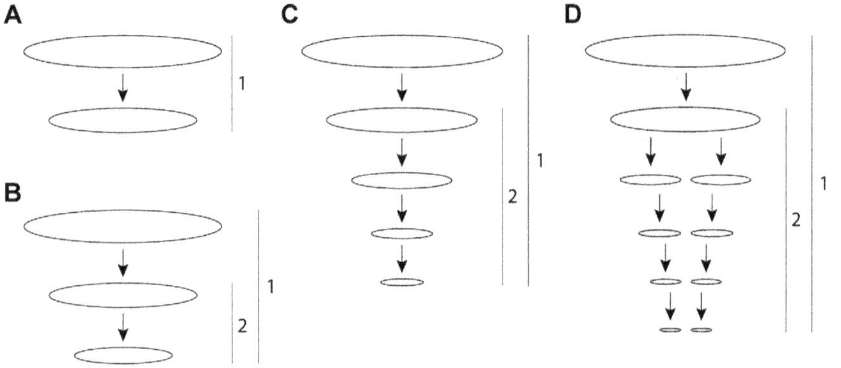

Figure 12.1 Degrees of agency. Ovals represent hierarchical levels in a system (bracket 1), and vertical arrows are causal, running downward from larger upper-level fields to contained lower-level entities. Importantly, for purposes of illustration, levels are shown as separate from each other, smaller underneath larger, but they should be understood to be contained, smaller within larger. In all cases, the top, largest oval represents a field in the environment, external to the goal-directed entity. (A) A system consisting of an upper-level external field (top oval) that directs (vertical arrow) a lower-level entity with no internal hierarchical structure of its own (bottom oval), like a rock falling in a gravitational field. The rock's trajectory is goal directed but hardly agential. (B) A case where the guiding external field (top oval) directs an entity (bracket 2) with shallow internal nesting (middle and bottom ovals). The goal direction delivered by the middle oval to the bottom oval occurs entirely within the entity, which makes the whole entity (bracket 2) somewhat agential. An example might be a bacterium following a food gradient. (C) A case in which the entity (bracket 2) is much more deeply nested internally, giving its various fields more total causal interactions, greater causal efficacy, more control over itself, and therefore greater agency. An example might be the control system inside a beetle that directs it away from a dung pile. (D) A system showing how hierarchical structures within an entity (bracket 2) can run in parallel producing yet more internal control, and therefore greater agency. Again, this might represent a dung beetle, but now taking into account more goal-directed processes than those involved in moving away from a dung pile.

The more mediation that goes on, the greater the number of internal goal-directed subsystems through which a behavior has been screened, or the more causal work they do, the more the system's agency can direct its behavioral output. And so, given that a beetle does more internal controlling, has more goal-directed systems, its agency is fairly robust compared to that of a bacterium. But, when compared to the nuanced and expansive capacity of something like human decision-making, the dung beetle's agency pales. Many human behaviors rely on countless, complex internal hierarchies found within the brain that—if Hume is right—culminate in a large and sometimes noisy chorus of motivations, passions, intentions, or "wants." The complexity of the goal-directed systems guided by these wants leads to the capacities most typically associated with the heights of agency. The beetle can move forwards, backwards, and side to side. It can even regulate its internal physiology, feed, pursue mates, and so forth. But it does not possess the internal structures needed to turn a doorknob, thread a needle, or write an essay. Because humans have so many deeply nested internal hierarchies, the range of behavioral capacities available, and the specificity of those behaviors, is greater than what are available to a dung beetle. These systems afford humans greater control over themselves because more of the field-mediated direction that guides them takes place inside them.

An important caveat: under field theory, upper direction is closely connected with control, as discussed. But as Dennett (1984) notes, control is never complete, never an all-or-none phenomenon. A beetle controls its legs to some degree, a bacterium controls its flagella to some degree, and a driver controls a car to some degree. But sometimes a leg does not do what one wants it to do, just as a minivan does not give a driver the kind of control that a Porsche gives them, even though a driver might want it to. So we underscore the heuristic nature of the correlation we are invoking between the complexity of internal hierarchy and increased degrees of agency. Simply possessing many complex, internally nested systems makes high levels of agency possible, even likely, but does not guarantee it.

A final issue to address here: in this view, as we have said, everything changes at the boundary of the individual. But as the literature on biological individuality shows, it can be difficult to locate these boundaries. If, for example, certain insect colonies are to some degree organisms then looking for certain kinds of agency in a single insect might turn out to be a lost cause. For many behaviors, the colony may be the agent, and a given ant may be mainly a part in that colony's agency. More generally, in many cases, it is not obvious where to draw the line between internal and external, and where this is true agency becomes fuzzy and uncertain. But this is not a problem for field theory. A strength of the position on agency we have adopted is that it comes in degrees, accommodating the continuous variation we see among systems in degree of individuality.

12.6 Summary

We have argued for four main points. First, the field theoretic view makes an internal–external distinction, but that is not the same as the distinction between an organism and its environment. Hierarchical structures, consisting of entities and the fields external to them, exist across many levels, with many entity-field pairs buried down deep inside the organism. Second, teleological systems can be, but need not be, agential. The falling rock is marginally goal directed, but in lacking any significant internal hierarchy, it is negligibly agential. Third, in our positive account of agency we showed that it has to do with control. Agency is control of the self by the self, regardless of how deterministic the causes are. And fourth, consistent with the continuous variation we see in virtually everything, across both biological and non-biological systems, agency comes in degrees. Agency can be greater or lesser, with greater hierarchical structuring generally associated with greater agency.

Notes

1 For other work on the theory, see McShea (2012, 2016, 2023); Lee and McShea (2020); Babcock and McShea (2021, 2023a, 2023b); and Babcock (2023).
2 See especially the work of Feibleman (1954); Campbell (1958); Simon (1962); Salthe (1985); and Wimsatt (1994, 2007).
3 Downward-pointing causal arrows suggest what is conventionally called "downward causation," a phenomenon we embrace. But the controversies surrounding this notion are beside the point here, and to avoid engaging with them, we instead adopt the phrase "upper direction" (McShea, 2012, p. 665). Thus, a goal-directed entity is said to be upper directed, that is, guided by the larger field in which it is enveloped.
4 Our notion of self-directedness should not be confused with other positions that also see a connection between teleology and hierarchy, e.g., Gontier's "self causation" (see Gontier, 2023).
5 Raginksy (2023) has recently argued that a more liberal conception of control, where "control" is more akin to interconnectedness, which is found in a new behavioral approach, helps make sense of, and model, biological autonomy. There may be fruitful work to be done in exploring this new concept of control. Also see Bich and Bechtel (2022) for another position on how "control" might be understood in organisms.
6 Levin (2012) uses the term "control" when discussing the non-local influences found in the morphogenetic fields referenced earlier.
7 See Dennett (1984) for the classic case of "sphexishness," i.e., repeated behavior exhibited by *Sphex ichneumoneus* used as an example of genetic determinism.

References

Babcock, G. (2023). Teleology and function in non-living nature. *Synthese, 201*(4). https://doi.org/10.1007/s11229-023-04099-1
Babcock, G., & McShea, D. W. (2021). An externalist teleology. *Synthese, 199*(3–4), 8755–8780.
Babcock, G., & McShea, D. W. (2023a). Resolving teleology's false dilemma. *Biological Journal of the Linnean Society, 139*(4), 415–432.
Babcock, G., & McShea, D. W. (2023b). Goal directedness and the field concept. *Philosophy of Science,* 1–10. https://doi.org/10.1017/psa.2023.121

Barandiaran, X. E., Di Paolo, E. A., & Rohde, M. (2009). Defining agency: Individuality, normativity, asymmetry, and spatio-temporality in action. *Adaptive Behavior, 17*(5), 367–386.

Bich, L., & Bechtel, W. (2022). Organization needs organization: Understanding integrated control in living organisms. *Studies in History and Philosophy of Science, 93*, 96–106.

Campbell, D. T. (1958). Common fate, similarity, and other indices of the status of aggregates of persons as social entities. *Behavioral Sciences, 3,* 14–25.

Cheeseman, J. F., Millar, C. D., Greggers, U., Lehmann, K., Pawley, M. D. M., Gallistel, C. R., Warman, G. R., & Menzel, R. (2014). Way-finding in displaced clock-shifted bees proves bees use a cognitive map. *Proceedings of the National Academy of Sciences, 111*(24), 8949–8954.

Dacke, M., Baird, E., Byrne, M. J., Scholtz, C. H., & Warrant, E. J. (2013). Dung beetles use the milky way for orientation. *Current Biology, 23*(4), 298–300.

Dacke, M., & Jundi, B. E. (2018). The dung beetle compass. *Current Biology, 28*(17), R993–R997.

Deacon, T. W., & García-Valdecasas, M. (2023). A thermodynamic basis for teleological causality. *Philosophical Transactions of the Royal Society A, 381*(2252). https://doi.org/10.1098/rsta.2022.0282

Dennett, D. (1984). *Elbow Room: The varieties of free will worth wanting.* MIT Press.

Feibleman, J. K. (1954). Theory of integrative levels. *The British Journal for the Philosophy of Science, 5*(17), 59–66.

Gambarotto, A., & Nahas, A. (2022). Teleology and the organism: Kant's controversial legacy for contemporary biology. *Studies in History and Philosophy of Science, 93*, 47–56.

Gontier, N. (2023). Teleonomy as a problem of self-causation. *Biological Journal of the Linnean Society, 139*(4), 388–414.

Lee, J. G., & McShea, D. W. (2020). Operationalizing goal directedness: An empirical route to advancing a philosophical discussion. *Philosophy, Theory and Practice in Biology, 12,* 005. https://doi.org/10.3998/ptpbio.16039257.0012.005

Levin, M. (2012). Morphogenetic fields in embryogenesis, regeneration, and cancer: Non-local control of complex patterning. *BioSystems, 109*(3), 243–261.

Mayr, E. (1988). The multiple meanings of teleological. In Mayr, E. (Ed.), *Towards a new philosophy of biology* (pp. 38–66). Harvard University Press.

McShea, D. W. (2012). Upper-directed systems: a new approach to teleology in biology. *Biology and Philosophy, 27*(5), 663–684.

McShea, D. W. (2016). Freedom and purpose in biology. *Studies in History and Philosophy of Science Part C: Studies in History and Philosophy of Biological and Biomedical Sciences, 58*, 64–72.

McShea, D. W. (2023). Evolutionary trends and goal directedness. *Synthese, 201*(5). https://doi.org/10.1007/s11229-023-04164-9

Moreno, A., & Mossio, M. (2015). *Biological autonomy: A philosophical and theoretical enquiry.* Springer.

Nagel, E. (1979). *Teleology revisited and other essays in the philosophy and history of science.* Columbia University Press.

Raginsky, M. (2023). Biological autonomy. *Biological Theory, 18*, 303–308.

Salthe, S. N. (1985). *Evolving hierarchical systems.* Columbia University Press.

Simon, H. A. (1962). The architecture of complexity. *Proceedings of the American Philosophical Society, 106*, 467–482.

Sultan, S. E., Moczek, A. P., & Walsh, D. M. (2022). Bridging the explanatory gaps: What can we learn from a biological agency perspective? *BioEssays*, *44*(1), e2100185. https://doi.org/10.1002/bies.202100185

Vane-Wright, R. I. (2023). Turning biology to life: some reflections. *Biological Journal of the Linnean Society*, *139*(4), 570–587.

Walsh, D. M., (2015). *Organisms, agency, and evolution*. Cambridge University Press.

Wimsatt, W. C. (1994). The ontology of complex systems: levels of organization, perspectives, and causal thickets. *Canadian Journal of Philosophy. Supplementary Volume, 20*, 207–274.

Wimsatt, W. C. (2007). *Re-engineering philosophy for limited beings: Piecewise approximations to reality*. Harvard University Press.

13 How Autonomy Theory Naturalizes Agency, and Why It Matters

Louis Virenque

13.1 Introduction

Agency is gradually spreading beyond the boundaries of the philosophy of mind and even beyond philosophy itself into biology (Laland et al., 2019; Sultan et al., 2022). One challenge for biologists is to fill the explanatory gaps related to the capacities of organisms to impose selection biases that are not accounted for by mainstream evolutionary theory (i.e., the modern synthesis [MS]). The most well-known of these abilities is niche construction, which is the capacity of organisms to build their environment by constructing nests, burrows, dams, etc. (Laland et al., 2015, 2019). More broadly, biologists and philosophers suggest that these explanatory gaps stem from a 'gene-centered' approach that fails to fully explain, for example, phenotypic variation, trait transmission, and the emergence of phenotypic novelty (Sultan, 2015; Sultan et al., 2022; Walsh, 2015).

Agency characterizes the intrinsic ability of organisms to act in a goal-directed way in their environment in order to survive (Laland et al., 2019; Fulda, 2016; Walsh, 2015). Some studies like those of Walsh (2015, 2018) shed light on the effects of agents within their environments and their resulting evolutionary consequences (see, e.g., Walsh & Sultan, this volume). However, they do not address the prior questions: What are the origins of a physical entity called an 'agent'? What are the distinctive features that distinguish that entity from other physical entities in the world? Not addressing these two questions implies that the Ecological Approach (EA) still lacks a theoretical framework to fully comprehend the notion of agency.[1] In particular, I think that using 'agency' in biology without a theoretical foundation leads to a two-fold problem. First, the utilization of the term 'agency' inevitably invokes the conceptual framework of theories of mind (even if unintended), given the absence of an alternative theoretical construct. Therefore, without a proper foundation, we cannot address a crucial question: How can we circumscribe the application of agency if we intend to describe more than human behavior?

Three theories compete in their understanding of the concept of agency in this context. The Standard Theory of Action (STA) restricts the concept to humans, thus neglecting a whole range of other organisms (e.g., Davidson, 1963, 1982). Neo-mechanism (NM) explains organisms like Autonomy Theory (henceforth AT) by referring to multilevel or circular causality (Craver, 2007; Bechtel, 2017; Bich

DOI: 10.4324/9781003413318-17

& Bechtel, 2021) but *cannot use* the very concept of agency because of what it implies (i.e., purposiveness). Finally, AT claims that STA's concept of agency is too restrictive and that mechanistic principles cannot account for biological organization or agency by omitting intrinsic purposiveness (Di Paolo, 2005; Barandiaran et al., 2009; Moreno & Mossio, 2015; Mossio & Bich, 2017).

In fact, on the one hand, the methodology advocated by NM to explain a phenomenon is decomposition and linear causation (Bechtel & Richardson, 1993; Machamer et al., 2000). But, in contrast to classical mechanism and by avoiding any ontological commitment regarding the nature of organisms (Machamer et al., 2000; Craver, 2007; Nicholson, 2012), NM changed its own claims as time passed. Recent developments show that NM could explain phenomena by referring to downward causation, circular causation, or even emergence (Bechtel, 2017; Lee, 2023). However, since it rejects teleology, NM still seems unable to consider organisms as agents. On the other hand, STA claims that agency is the capacity to perform an intentional action and that an action is intentional if the right mental states and events cause that action (Davidson, 1963; Schlosser, 2019). This statement excludes all systems that lack representational capacity: in other words, non-human animals are not considered genuine agents (Davidson, 1982).

AT, by contrast, examines the concept of agent in light of the organization of living systems. More specifically, living systems exhibit a particular type of circular organization, distinct from machines and less complex than cognitive systems capable of representation and intentionality. In this closure-related kind of organization, the parts do not simply cause the whole and the whole does not simply control the parts. Rather, the mutual and dependent relationship between the whole and the parts produces an organized whole, which is *self-determining* and enables agency (Barandiaran et al., 2009; Arnellos & Moreno, 2015; Moreno & Mossio, 2015). We can trace this conception of a circular organization from Kant (1790/1987) and his characterization of intrinsic teleology to the concept of autopoiesis framed, more recently, by Varela (1979) (for discussion, see Moreno & Mossio 2015). The 'biogenic' approach (Lyon, 2006) of AT allows the study of agency to start not from the restrictive assumption that representation is the only means to characterize agency, but rather from the organization of autonomous systems, avoiding both decomposition and linearity. Thus,

> Agency refers to the capacity of autonomous living beings (roughly speaking: organisms) to purposively and functionally control the interactions with the environment, and to adaptively modulate their own self-determining organization and behavior so as to maintain their own existence, construed as their intrinsic telos.
>
> (Virenque & Mossio 2024, p. 11)

In the following sections, I will argue that any theory that bases agency on representation, such as STA, is doomed to a theoretical dead end. I will then suggest that NM is a continuation of classical mechanism and that in this respect it is incapable of understanding agency, even if it promotes a form of explanatory circularity.

Subsequently, I will argue why AT is the only theory that provides a truly natural basis for agency.

13.2 The Dead End of Representation to Naturalize Agency

13.2.1 Agency as Intentional Capacity

The concept of agency can be traced back to the main themes of classical philosophy that were devoted to human action through the study of causality, i.e., 'agent' and 'patient' (Hume, 1739–1740/1978), and the body-mind relation (Descartes, 1647/1985). Theories of action that were developed during the twentieth century by philosophy of mind inherited this classical history and questioned the causal nature of human actions through mental states (desires, beliefs, etc.) and the concept of freedom. In STA, actions are full-fledged phenomena and must therefore be preceded by causes to comply with naturalism (Schlosser, 2019). STA aims to explain human actions without appealing to uncaused causes or final causes, as both are in contradiction with naturalism. The challenge, then, is to explain why an agent acts as she does, by appealing to her intentions, without at the same time mentioning final or uncaused causes, and without reducing the agent's behavior to the mechanical movements of the body. That is why, according to Davidson (1963), (a) naming propositional attitudes (i.e., wanting *P*, desiring *P*, etc.) and (b) believing that one's action is of this kind (i.e., realizing *P*) provides the reasons for an action. These reasons yield a causal explanation of why the agent acted as she did. So, if an individual acts as she does, it is because the right events and mental states—the reasons—led her to behave as she did. If Frodo drops Sauron's ring in Mount Doom, it is because (a) he wants to destroy the ring and (b) he believes that dropping it there will destroy it. Both (a) and (b) provide the primary reason that allows us to rationalize Frodo's action. Thus, the right events and mental states—(a) wanting to destroy the ring and (b) believing that dropping it into Mount Doom will destroy it—causally explain why Frodo acted as he did, at least in an event-causation theory like STA.

In fact, characterizing agency by referring only to mental states exposes STA to the criticism that the agent disappears from a causal explanation of action (Nagel, 1986; Velleman, 1992). Indeed, if the agent seems to disappear, it is primarily because of the clear distinction between intentions and actions: intentions cause actions, but they are not the actions themselves. Then, an agent appears passive rather than active when she comes to act, because actions are caused by intentions, not directly by the agent herself. This claim leads to deviant examples where the cause of an action appears to deactivate as it progresses, resulting in the agent's disappearance. This is the case of deviant causal chains (Chisholm, 1966). For example, a nephew intends to murder his uncle in order to inherit his fortune. To perform this homicide, he drives to his uncle's house. But, on the way, he hits and kills a pedestrian, who turns out to be his uncle. Here, the nephew is not the agent of the whole action, since the cause (intention) only initiated the nephew's drive but not the accidental killing. Against this odd absence of an agent, opponents of

STA have proposed a model of action in which the causal role falls to the agent, not to the events happening within her. Accordingly, the agent–causation approach argues for considering the agent as a substance that directly causes her actions (Taylor, 1966). When one acts, it is not one's intentions that cause the action but the agent herself. But, as Juarrero (1999) pointed out, agent causation may fall into the pitfall of dualism or uncaused causes by involving a substance that is different from its neurological processes or its mental events. Thus, we face two theories: STA (or event-causation) and agent causation theory. The former makes us wonder about where the agent is, while the latter makes us wonder what the nature of the agent is. According to Juarrero (1999), these two dead ends are the consequences of a modern conception of causality mixed with the Aristotelian imperative that no self-motion exists. In fact, STA as agent–causation always explains agency by reference to efficient causation, by separating the agent from her behavior or her intentions. Causes (intentions) are thus necessarily external to their effects (actions), following linear causation and conducting to the impossibility of self-motion. Juarrero thus understands self-motion as interlevel causality between the parts and the whole, i.e., with circularity rather than linearity. This kind of definition of self-movement is one of the first steps toward a new conception of agency but necessitates new theoretical foundations that naturalize the agent and circular causation.

Finally, major issues result from the mainstream conceptions of agency in the philosophy of action. First, if one considers STA to be consistent with naturalism, the agent disappears. This means that agents are superfluous since the causes of actions are intentions. Hence, extreme cases like deviant causal chains (e.g., the murderous nephew) where an intentional action becomes a non-action because of the cause that is deactivated during its realization. Subsequently, if one adopts agent causation theory, the agent re-appears, but one must give an account of what an agent is and provide a definition of an agent that complies with naturalism. However, agent causation theory seems unable to offer this (Juarerro, 1999). These problems stem not only from the fact that causality is understood as linear but also from the simple fact that the starting point for analysis, in action theories, is the human being, as if it was the only being capable of action (cf. Aaby, this volume). This not only gives rise to skepticism about the relevance of causality involved in action theories but also raises the question of whether representation as a basis for agential questions should be abandoned or not.

13.2.2 *Why Representationalism Is a Brake to Naturalism?*

Lyon (2006) examined two perspectives that approach cognition in two different ways to understand what it is and what it does. Lyon called these approaches 'anthropogenic' and 'biogenic.' On the one hand, the anthropogenic strategy investigates cognition from humans' hallmark capacities, such as mental representation or language, to understand "what sort of biological or evolutionary story might account for them" (Lyon, 2006, p. 12). On the other hand, the biogenic strategy starts from biological organization and assumes that cognition is foremost a biological function that must have been linked to reproductive and survival problems: "Biogenic approaches thus ask psychological questions as if they were biological ones" (Lyon, 2006, p. 12). I specifically mention cognition instead of agency because there is confusion regarding the relationship between the two concepts. The anthropogenic approach identifies agency with cognition, but this is still a matter of debate within AT (Bourgine & Stewart, 2004; Di Paolo et al., 2017; Virenque & Mossio, 2024), which is a biogenic approach. Identifying agency with cognition, as anthropogenic approaches do by adopting the human comparative model, leads to a clear conclusion: those who do not have human-like cognitive capacities are neither agents nor cognitive (or barely cognitive) entities. Moreover, not only does this anthropocentric position seem to be unable to explain and understand forms of cognition and action other than human ones, but, as we have just seen, it also seems to struggle to provide an account of what human cognition is, especially when it comes to action. Thus, identifying agency with cognition when the latter remains profoundly unknown (Lyon, 2006) raises a dispensable obstacle. I contend that identifying one with the other, particularly when the cognitive abilities involve high faculties such as representation or language, raises a significant problem. This becomes much more patent when we realize that representation itself already presents various issues, such as naturalizing the ability to have mental images which cause actions.

Unfortunately, this association between cognition and agency affects not only the cognitive sciences but also certain branches of biology where bacteria are described as cognitive agents (Fulda, 2017). Indeed, in his article, Fulda suggests that this confusion between agency and cognition stems from a Cartesian conception of agency, which considers agency to presuppose cognition. For Fulda (2017, pp. 76–77), Descartes separated the world into two substances and argued that organic phenomena in humans and animals could be explained by simple laws of motion, but the flexibility of human cognition required purposiveness. The division is easily made, with humans as agents and animals as automata. Specifically, according to Fulda, the idea that purposiveness is the feature that characterizes cognition results in two opposing views: *intellectualism* and *mechanism*, which represent the two extremes of the Cartesian dilemma. For intellectualists and mechanists, purposiveness is the hallmark of cognition, but while the former argues that cognition can be observed in living organisms when they adapt to new conditions in their environment, the latter contends that only human beings are agents. In fact, mechanists claim that organisms' flexible behaviors can be accounted for without referring to cognition, which requires complex capacities like language (Davidson, 1982). In short, if one intends to account for flexibility in organisms, one may fall into this dilemma because of the identification of cognition with purposive capacities, which seems the only way to naturalize purposiveness in both views. We thus need theoretical tools to account for the flexibility and plasticity of living systems (i.e., purposiveness) without being forced to conclude that prokaryotic cells are cognitive agents in the same sense as humans (but with fewer capacities), or that prokaryotic cells are mere robots to avoid vitalism.

13.2.3 Beyond the Dilemma, a Lack of Theorization

The conflation of cognition and agency is not just a dilemma, as Fulda (2017) suggests: it originates from a dearth of theoretical frameworks beyond those provided by mind theories. Lyon (2006) highlighted that extensive data on the functioning of cognition has been amassed to address the question of 'what cognition is' by examining 'how cognition works.' But no consensus has been reached on what cognition really *is*. Beyond the confusion between cognition and agency, I consider that this is in part a similar problem to the one we face with agency. In fact, in the introduction, I indicated some authors who appeal to this concept when explaining the differences agents would make in evolutionary processes (Walsh, 2015). However, the theoretical background of the EA of authors like Walsh remains incomplete (see endnote 1). In contrast, as we shall see, for AT it is essential to detail what distinguishes agents from other physical systems since this difference determines how they act.[2] Without adequate theoretical tools to account for the purposive capacity of organisms, and because of the confusion between cognition and agency, I believe that the concept of agency itself may unfortunately suffer the same fate as cognition. Although only a few biologists and philosophers of biology refer to the word 'agency' to qualify organisms' capacities, many papers consider the activities of organisms as factors (if not causes) that are irreducible to the explanations given by the MS (Levins & Lewontin, 1985; West-Eberhard, 2003; Sultan, 2015; Walsh, 2015; Laland et al., 2017). Thus, in order to circumscribe the concept of agency without falling into the same pitfalls as 'anthropogenic' perspectives and vitalism, it seems urgent to find agency on a natural basis that deals, at the same time, with agency and the structure that supports this kind of purposive behavior.

In summary, cognitive capacities have been posed to be the only epistemological support for the natural explanation of teleological behavior. However, there appears to be an impasse in the explanation of the behavior of non-human agents, for whom representation is a too complex capacity. We now have two options: either we align with MS, which focuses on molecular structures, favors bottom-up explanations, and thus rejects organisms' agency, or we turn to a theory that naturalizes the notion and considers it as a genuine explanatory tool. I first argue why neo-mechanistic methodology in MS fails to account for purposive behavior.

13.3 The Impasse of Neo-Mechanism

13.3.1 What Are the Relations of Early Neo-Mechanism to Agency?

The classical mechanism from which NM is derived emerged in the seventeenth century through the work of Galileo, Descartes, and Boyle (Nicholson, 2012). This philosophy discarded explanations through final and formal causes to prefer efficient (linear) causation (Juarrero, 1999). The methodology consisted of the reduction of a whole phenomenon to the interactions of its component parts, explainable by laws of movement. Explanation by appealing to laws was conserved and formalized into the deductive-nomological model of Hempel–Oppenheim. This model of explanation consisted of subsuming phenomena under general laws that could predict the *explanandum*. However, this model encouraged the reduction of phenomena to physical laws and was extrapolated to biology, a field in which the use of formal laws underdetermines empirical biological phenomena (Bechtel & Richardson, 1993; Bechtel & Abrahamsem, 2005). Thus, NM emerged to propose an explanatory model specific to biology that does not involve laws, but rather models of system decomposition and methods determined by the practice of biologists themselves (Bechtel & Richardson, 1993; Nicholson, 2012). Decomposition consists of analyzing a system into different parts and activities depending on the biologists' focus of study. The *explanandum* is the phenomenon on which biologists choose to focus. The different parts that result from decomposition are the basis for explaining the *explanandum* phenomenon (Craver, 2007). The *explanans* is the mechanism resulting from the interaction of parts and/or operations. In other words, biologists recompose the phenomenon on the basis of the initial and final conditions that they set beforehand (Bechtel & Richardson, 1993; Machamer et al., 2000; Bechtel & Abrahamsem, 2005). Then, it is easy to realize that this type of explanation is in complete opposition to the intellectualist descriptions of bacteria that Fulda (2017) presented. If, for example, the *explanandum* phenomenon is bacterial motility, the system will not be described as a cognitive system pursuing a goal but decomposed into the different parts and operations that produce the phenomenon in question. In other words, the bacteria as an agent goes completely unnoticed. The focus is only on the activity of the parts that produce the movement. Causality follows a unidirectional pattern from parts to whole and does not imply a goal. Instead, it is the intrinsic activity of the parts that enable the production of this phenomenon.

13.3.2 Evolution of Causal Processes Thinking within Neo-Mechanism

In 2018, Winning and Bechtel argued, however, that these mechanistic descriptions were insufficient to account for causality, the dissipative structure of living beings, and the control that organisms have over themselves to find their energy (Winning & Bechtel, 2018). The concept of 'constraint' in classical mechanics provides an answer to these three problems. A constraint can be many things, for instance, an atomic bond, the different structures of proteins, a skeleton for organisms, and an inclined plane for a ball, etcetera; in short, a constraint interacts with other entities by limiting or channeling their activities. In particular, constraints can allow a dissipative system like an organism to channel and reuse the energy that it exchanges with its environment to perform work. Winning and Bechtel (2018) then mention that these particularities went unnoticed by the early neo-mechanists, such as Machamer et al. (2000), who omitted to take into account the source of energy (that is necessary for the entities that ultimately produce the mechanism in question). The flow of matter and energy comes from the uninterrupted interaction of the system with its environment: this flow is channeled and reused by the system to perform work that essentially aims to regain energy, and so on (Kauffman, 2000; Moreno & Mossio, 2015). This is how the concept of 'constraint' provides an answer to the various problems outlined by Winning and Bechtel. It becomes the theoretical entity for explaining the origin of the causal activity of components of the system. The intrinsic activity of these components is defined as work performed through constraints that channel the energy the system exchanges with its environment. Thus, the dissipative nature of the organism and its self-maintaining activity are also explained by this concept of constraint. Furthermore, Winning and Bechtel (2018) emphasize that there is a fundamental difference between living organisms and machines: a machine's energy supply ultimately comes from an external agent, and, without this agent, the machine does nothing. In contrast, unlike machines, organisms

> have to be responsive to changing conditions: for example, routing free energy in different ways on different occasions and seeking a different form of free energy when the current one is no longer available. This requires exercising control over time-dependent, flexible constraints so that their component mechanisms perform the work required to maintain the whole set of mechanisms that constitutes the organism in a far-from-equilibrium condition.
>
> (Winning & Bechtel, 2018, p. 298)

In fact, Bechtel and Richardson (1993, p. 39) wrote that before providing a mechanistic explanation of a system, it is necessary to identify the *locus of control*, i.e., to identify what is responsible for the phenomenon. It is the role of the scientist to determine this locus during decomposition. More precisely, control is now understood as a form of constraint in biological systems, since it modifies the functioning of numerous mechanisms that maintain the system, such as the circadian cycle in cyanobacteria or the glycolysis pathway in cells, and even ensures the system's relationship with its environment (Winning & Bechtel, 2018; Bich & Bechtel, 2021). The concept of control seems to encompass a large number of mechanisms, including the adaptive capacity of organisms (Di Paolo, 2005; Bich & Bechtel, 2021). Thus, control mechanisms ensure the flexibility and the plasticity of organisms without identifying them with cognition. However, what NM conceives as a control mechanism, AT considers as an agential capacity. Thus, the same phenomenon is either conceived as a control mechanism or an agential capacity. Where does the difference lie?

13.3.3 The Missing Agency of Organisms

In their recent work, Winning and Bechtel (2018) and Bich and Bechtel (2021) use concepts coming from AT, such as 'autonomy' and 'constraint,' in an attempt to somehow blur the boundary between the two perspectives (NM and AT). Yet the concept of agency is never mentioned, even though many of the characteristics of an agent are described, such as an organization through the closure of constraints, adaptive behavior, and functional control (see the definition of 'agency' in the introduction). This endeavor conceals a fundamental disagreement between the two perspectives, which leads NM theory to describe an organism in the same terms as AT, without being able to acknowledge the fact that organisms are agents.

Despite their apparent proximity to AT, Winning and Bechtel (2018) describe the circularity of all control mechanisms through the concept of a 'feedback loop.' This concept was theorized by the founders of cybernetics, Rosenblueth, Wiener, and Bigelow, and, according to them, refers to the activity of a teleological system (for discussion, see also Nahas, this volume). To illustrate, let us imagine a system S whose objective is to reach a point P. S moves toward P and P sends signals (inputs) to S, which modify S's behavior (outputs) accordingly, and so on, until S reaches P. The mechanism is circular, since S regulates its activity (e.g., its trajectory) thanks to the information given by P. Feedback loops allow NM to understand the circular regulatory phenomena that occur in the organism. If the teleological model of cybernetics is used by NM without having to worry about the problems associated with it (e.g., uncaused cause, free will), it is for a very specific reason. According to Mossio (2020), the concept of purposiveness used by cybernetics is stripped of its causal connotation. In fact, Rosenblueth et al. (1943) defined a goal as a "final condition in which the behaving object reaches a definite correlation in time or in space with respect to another object or event." This definition reflects a concept of teleology in which the "goal is not a final cause, supposed to explain the existence of the system, but simply a stable state toward which the latter tends" (Mossio, 2020, p. 71; my translation). For instance, the system S tends toward P, but P does not explain why S exists, since no reciprocal dependency relationship exists between S and P. The definition of teleology by cybernetics is compatible with mechanism, but it loses its original significance (i.e., it loses its circular determination, where the whole determines the parts, and the parts the whole). In fact, following Mossio (2020), let us take the example of a thermostat whose sensor measures the temperature of a room, but this information is not used by the thermostat. In this case, the activity of the whole (which is the *explanandum*), such as regulating the temperature of a room, does not determine the functioning or existence of the parts of the system (the *explanans*). "The determination remains *unidirectional*" (Mossio, 2020, p. 71; emphasis added; my translation), i.e., the parts determine the activity of the whole, but not the other way around. The thermostat never fixes its own temperature sensor. The problem is that Winning and Bechtel (2018) are trying to show how, among other things, the interaction of an organism works by using feedback loops.

Thus, NM, by adopting several concepts of AT, suggests that the two perspectives may be compatible. NM even uses the concept of a feedback loop to capture the teleological circular causality at the core of AT. Yet, despite these efforts, NM still avoids the term agency, even though the systems described by NM exhibit all the characteristics of an agent according to AT's description. NM thus substitutes the concept of agency with control (but misses its scope) and the concept of teleology with feedback loops (but misses its meaning).[3] Yet these key concepts are required to develop a naturalized conception of agency that can be applied to all living beings. In this sense, I propose that the concept of agency provides a more detailed explanation of an organism's teleological and functional relationship with its environment for survival. NM cannot access this concept because agency is grounded on a teleological organization such that the component activities are determined by the whole and the whole is determined by its parts. Such a definition of intrinsic teleology is inherently inconsistent with NM and thus hinders NM from discussing an agent.

13.4 Agency as a Natural Phenomenon

13.4.1 Teleological Organization

Advocates of AT argue that living beings are characterized by teleological organization. 'Organization' does not refer to 'arrangement' but to the production of roles performed by the different parts of an organized system. In fact, while the concept of organization is taken from a tradition that defines the living by its structure and composition, it is also central to another tradition that makes it the defining foundation of purposiveness. This tradition includes Kant (1790/1987), Varela and Maturana (1980), and also the contemporary theory of biological autonomy (Barandiaran et al., 2009; Moreno & Mossio, 2015). For Kant, living beings exhibit natural purposiveness, whereby the parts determine the whole and the whole determines the parts. For Maturana and Varela, organization constitutes the unity of a system. To characterize this specific unity of living organisms, they coined the concept of 'autopoiesis.' What makes autopoiesis unique is its focus on the production relations within the system rather than the structure of its components (Maturana & Varela, 1980, pp. 78–79).

The definition remains largely abstract but emphasizes the unique structure of an autopoietic machine compared to a non-autopoietic one, such as an artifact. The parts of an autopoietic system produce each other; they are therefore interdependent and constitute the system as an organized unit. Despite its obvious proximity to Kant, Varela did not recognize and understand autopoiesis as a teleological organization until very late (Varela & Weber, 2002). Ultimately, autopoiesis appeared as a response to the boundaries imposed by Kant on human understanding of living phenomena. In fact, the relations of production and dependence between the parts and with the whole enable the autopoietic system to realize an intrinsic finality, i.e., a finality that is internal to the system. Varela's 2002 article also makes explicit that autopoietic systems should be considered as agents. However, as pointed out by Di Paolo (2005), autopoiesis was not sufficient to adequately explore the concept of agency, which implies not only interaction with the environment but also a form of adaptivity in a fluctuating environment (Barandiaran et al., 2009). I will come back to this later.

To characterize the dependence between parts within an autopoietic system, AT also refers to 'organizational closure' (Varela, 1979).[4] However, due perhaps to its abstract nature, there is still no consensus on this concept, although closure is crucial for the biological and natural grounding of key concepts including those of 'norm,' 'function,' and 'agency' (Moreno & Mossio, 2015; Montévil & Mossio, 2015). Yet, if closure generally refers to a dependency relationship between parts, it might possibly be applied to any physical system realizing a circularity, such as the geo-hydrological cycle. To face this challenge, Moreno, Mossio, and Montévil developed a more specific formulation of closure, enabling the concept to have an explanatory rather than a heuristic role in biology.

13.4.2 *Beyond Autopoiesis: Teleological Organization through Closure of Constraints*

Although the intuitive notion of closure refers to a perfectly closed system, for AT it denotes a mutual production relationship between the parts of a system that interacts with its environment. Varela (1979) situated this production interplay between the processes of the autopoietic system, as noted in the canonical definition of the concept. More recent theorists have located closure at the level of entities that have the status of constraints (Montévil & Mossio, 2015; Moreno & Mossio, 2015). Two types of theoretical entities must then be distinguished: constraints and processes. On the one hand, constraints limit and harness processes without being modified during the action (i.e., a constraint acts on a process but does not create it; the process could take place without the constraint but on a different time scale). On the other hand, processes refer to the whole set of physicochemical changes occurring within the system.[5] 'Acting on' means performing work in this context, and 'work' refers to a constrained release of energy (Kauffman, 2000). In fact, a living being is a dissipative system traversed by a flow of matter and energy that it channels and reuses essentially for its own survival. In concrete terms, constraints enable the system to generate work from this flow and to exploit it. Constraints can create a closure by performing different types of work, which depend on other constraints for its realization. This interdependence of constraints allows for the overall self-maintenance of the system and is referred to by the authors as a 'closure of constraints.'

Closure of constraints grounds natural purposiveness. The parts of the closure have effects that are directed toward the whole organization and the whole organization has effects that are directed toward its parts. Without this interconnectedness, the parts would be incapable of functioning. In contrast to Varela's conception of closure, closure of constraints consists of a causal relationship that is more specific to biological systems. The mutual dependency between the parts is specific: this is functional work. Each constraint (or part) performs a function of its own. The concomitant work effect of constraints enables the system as a whole to pursue its intrinsic goal, namely, its own existence. Furthermore, the whole's activity accounts for the activity of its constituent parts, and vice versa. Finally, once the organized system can perform this kind of work in relation to its environment in order to sustain itself, it is said to be autonomous.

13.4.3 *Naturalized Agency: Beyond Representationalism and Neo-Mechanism*

Naturalizing intrinsic purposiveness through the closure of constraints allows AT to naturalize the concepts of norm, function, and finally agency without having to resort to mental representation or dismiss these concepts by focusing on decomposition. In fact, an autonomous system is by definition a system that realizes a 'constitutive closure' and agency (Moreno & Mossio, 2015, p. 89). By 'constitutive closure,' AT refers to the metabolic 'side' of the organization, which refers to all the constraints (i.e., functions) in the closure and their dependent and enabling relations. Agency then refers to the interactive 'side' of the organization. Constitutive and interactive dimensions are indissociable since the autonomous system needs to interact with its surroundings to survive. These interactive and con-stitutive dimensions are both realized through the work of constraints. I referred to constraints as functions, which means that constitutive and interactive dimensions are sustained by functions. These interactive functions are "a set of constraints subject to closure, whose specificity lies in the fact that their effects are exerted on the boundary conditions of the whole system" (Moreno & Mossio, 2015, p. 90).

Thus, agency is understood as the set of functions that have causal effects on the system's environment to ensure the system's survival while being enabled by it (Moreno & Mossio, 2015, p. 92). As a set of functions, agency must respond, as any other function, to the intrinsic goal of the system which is survival. The most basic behaviors exhibited by agents pertain to motility, such as phototropism in plants, chemotaxis in bacteria, or foraging behavior of mammals, but are not limited to it.[6] These actions are quite basic in a stable environment but, according to Di Paolo (2005), an autonomous agent is also an adaptive agent. This means that the agent is capable of adapting to a changing environment or changing the environment itself. An organism's adaptivity is crucial as it allows it to regulate its behavior in response to its environment. For instance, an organism can move toward more profitable conditions or modify its organizational regime. The ability of organisms to 'make sense' of their environment is one of the critical implications of adaptive agency (Varela & Weber, 2002). Indeed, to survive in a dynamic envir-onment, an agent does not need mental representation, but must have the ability to detect relevant features of its surroundings and act accordingly (for discussion, see also Moreno, 2018). The concept of adaptivity facilitates progression toward more complex forms of agency and thus toward cognition (and even toward a pos-sible evolution of agency). For instance, understanding the environment is vital when discussing niche construction phenomena that affect the dynamics of an organism's evolution (Sultan, 2015; Laland et al., 2017). To bring together these different dimensions of agency within a single definition, I refer to Virenque and Mossio (2024):

> Agency refers to the capacity of autonomous living beings (roughly speaking: organisms) to purposively and functionally control the interactions with the environment, and to adaptively modulate their own self-determining organ-ization and behavior so as to maintain their own existence, construed as their intrinsic telos.
>
> (p. 11)

Agency is then defined in AT as the ability of organisms to control the interactive dimension of their organization. The adaptive capacity of systems is explicitly noted, opening the way to more complex, adaptive behavior. As it is stated, the goal of the system remains self-maintenance, but this does not exclude behaviors that are not directed toward this goal. This is a crucial matter for AT, particularly if it seeks to establish itself as a viable contender against action theory altogether. Indeed, if STA accounts for all kinds of human behavior with intentionality, then AT must at least have the ambition to account for all living behaviors, including complex human behaviors.[7] The issue is that AT formulated in this way only explains behaviors that are conducive to the system's self-maintenance. However, there are many actions that do not seem to fall into this category (e.g., reproduction for minimal agency and altruism, playing, or committing suicide for more complex agents) even though they remain profoundly teleological.[8] I leave the examination of this theoretical knot for future work.

For now, one may highlight that this definition of agency has a major consequence: agency is a form of control, but unlike how NM construes it, it cannot be understood as a control mechanism like any other, nor it can be characterized by simple feedback loops. Indeed, as a set of functions that control the interactions with the external environment, agency within AT is facilitated by the whole and, in return, enables the whole to operate correctly to achieve self-maintenance. This circular relationship between the whole and its parts generates individuality, normativity, and agency. This type of control, which AT calls agency, is, in contrast to NM and STA, what fully characterizes living organisms. Therefore, AT cannot limit the notion of agency to humans since it is not possible to speak of living beings without using it, and since it distinguishes living from non-living beings.

13.5 Conclusions

In conclusion, I have argued that AT is currently the only theory capable of providing a reliable natural basis for the concept of agency. AT endows the concept of agency with a precise role in the dynamics of living organisms, namely, enabling them to control and regulate their interactions with their environment in order to ensure their own survival. The 'biogenic' interpretation of agency put forth by AT stands in sharp contrast with any viewpoint that aims to restrict the notion solely to humans, broaden it to non-biological entities (such as nanorobots, oil droplets, etc.; for further discussions, see Moreno & Mossio, 2015, Chapter 4), or even to obliterate it entirely, as NM does. Furthermore, understanding agency from the viewpoint of AT helps to avoid conflating agency with cognition while setting the stage for investigating the connection between them. However, even if AT proposes a naturalized definition of agency, the link with evolutionary theories remains difficult to build. Indeed, as AT is a theory still in progress, there are many dimensions that need to be explored, such as the relationship between the different agents located at the cellular level and at the level of the organism as a whole. On the one hand, the interaction between cells refers to the phenomenon of development. This raises the question of purposiveness in a new way: when are the cells of a multicellular system subsumed under the purposive organization of the whole? On the other hand, the interaction of organisms with their environment brings us back to the notion of information, which is currently not defined by AT, even though it is the cornerstone of adaptive agency. In fact, the notion of information underlies all forms of interaction between adaptive agents with their respective environments. These challenges are as complex as they are stimulating.

Notes

1   Walsh (2015, 2018) has studied the consequences of agency in evolutionary theory with great rigor and Fulda (2016, 2017) has exposed with clarity two different ways of naturalizing agency in biology: the EA and Autonomy Theory. These authors consider that Autonomy Theory gives a good account of agency "by locating it in the causal structure of the physical world" (Fulda, 2017, p. 81) but fails to explain how an agent makes a difference in it. Fulda draws an analogy between agency and viscosity. The viscosity of a liquid is made causally possible by its internal structure, i.e., by the various causal micro-interactions between the different molecules that make it up. Yet the behavior of a liquid is not explained by its micro-causal structure, but by other concepts on a dynamic scale, such as its density or surface tension. For agency it would be the same thing: agency would be causally realized by interactions between parts of the system, but, as an emergent phenomenon, it would be explained by other concepts, such as those of 'repertoire,' 'purpose,' and 'affordance' proposed by EA. However, this rests on a misunderstanding of how agency is defined within Autonomy Theory. In this theory, behavior is not decoupled from its structural basis and, thus, this is a not reductive stance but the only way to naturalize intrinsic teleology. Nevertheless, a more detailed take on this problem would deserve a paper on its own.
2   In fact, their circular organization is understood as their intrinsic purpose (see Mossio & Bich, 2017).
3   A truly circular or teleological organization eliminates the distinction between *explanans* and *explanandum* by virtue of the relationship becoming circular. In fact, organization becomes an explanatory principle in AT.
4   Autopoiesis, which characterizes the relationship of production of its components, is a paradigmatic instance of organizational closure (Varela, 1979, p. 57). In fact, other relationships within organizational closure are possible without necessarily corresponding to the production relationship of autopoiesis (Varela, 1979, p. 55).
5   For instance, oxygen transported by the vascular system is an example of a constrained process that would be statistically possible without the constraint, but it would be much slower to achieve (Montévil & Mossio, 2015).
6   For example, pumping sodium–potassium ions into their environments, as cells do, is a form of agency (Barandiaran & Moreno, 2008).
7   It is crucial to note that limiting agency to cognition or representation is problematic, as I have argued, but this does not mean that AT aims to explain all behaviors in the same way. Hence, when complex behaviors are involved, advocates of AT know that they cannot limit the explanation to minimal agency (for discussion, see Mossio & Moreno, 2015, Chapter 4; Moreno, 2018).
8   Actually, the question of reproduction is already a key issue in AT. This theory suggests the idea of an inter-generational organization that maintains self-maintenance as an intrinsic goal of the system across generations. Behavior remains functional and therefore teleological (Saborido et al., 2011). Altruism is a complex problem, but it may have an 'answer' similar to that of reproduction. We can conceive of a form of organized 'social system' in which altruism is a behavior functional to the intrinsic purpose of the social system in question (e.g., a colony).

References

Arnellos, A., & Moreno, A. (2015). Multicellular agency: An organizational view. *Biology & Philosophy*, *30*(3), 333–357.

Barandiaran, X. E., Di Paolo, E., & Rohde, M. (2009). Defining agency: Individuality, normativity, asymmetry, and spatio-temporality in action. *Adaptive Behavior, 17*(5), 367–386.

Barandiaran, X., & Moreno, A. (2008). Adaptivity: From metabolism to behavior. *Adaptive Behavior*, *16*(5), 325–344.

Bechtel, W. (2017). Explicating top-down causation using networks and dynamics. *Philosophy of Science*, *84*(2), 253–274.

Bechtel, W., & Abrahamsen, A. (2005). Explanation: A mechanist alternative. *Studies in History and Philosophy of Science Part C: Studies in History and Philosophy of Biological and Biomedical Sciences*, *36*(2), 421–441.

Bechtel, W., & Richardson, R. C. (1993). *Discovering complexity: Decomposition and localization as strategies in scientific research*. MIT Press.

Bich, L., & Bechtel, W. (2021). Mechanism, autonomy and biological explanation. *Biology & Philosophy*, *36*(6), 53. https://doi.org/10.1007/s10539-021-09829-8

Bourgine, P., & Stewart, J. (2004). Autopoiesis and cognition. *Artificial life*, *10*(3), 327–345.

Chisholm, R. (1966). Freedom and action. In K. Lehrer (Ed.), *Freedom and determinism* (pp. 11–44). Random House.

Craver, C. F. (2007). *Explaining the brain: Mechanisms and the mosaic unity of neuroscience*. Clarendon Press.

Davidson, D. (1963). Actions, reasons, and causes. *The Journal of Philosophy*, *60*(23), 685–700.

Davidson, D. (1982). Rational animals. *Dialectica*, *36*(4), 317–327.

Descartes, R. (1985). Principia philosophiae. In J. Cottingham, J. T. Stoothoff, & D. Murdoch (Eds.), *The philosophical writings of Descartes*. Cambridge University Press. (Original work published 1647)

Di Paolo, E. A. (2005). Autopoiesis, adaptivity, teleology, agency. *Phenomenology and the Cognitive Sciences*, *4*(4), 429–452.

Di Paolo, E., Buhrmann, T., & Barandiaran, X. (2017). *Sensorimotor life: An enactive proposal*. Oxford University Press.

Fulda, F. C. (2016). *Natural agency: An ecological approach* [Doctoral Dissertation, Institute for the History and Philosophy of Science and Technology, University of Toronto].

Fulda, F. C. (2017). Natural agency: The case of bacterial cognition. *Journal of the American Philosophical Association*, *3*(1), 69–90.

Hume, D. (1978). *A treatise of human nature*. Oxford University Press. (Original work published 1739–1740)

Juarrero, A. (1999). *Dynamics in action: Intentional behavior as a complex system*. MIT Press.

Kant, I. (1987). *Critique of judgment*. Hackett Publishing. (Original work published 1790)

Kauffman, S. (2000). *Investigations*. Oxford University Press.

Laland, K. N., Uller, T., Feldman, M. W., Sterelny, K., Müller, G. B., Moczek, A., Jablonka, E., & Odling-Smee, J. (2015). The extended evolutionary synthesis: Its structure, assumptions and predictions. *Proceedings of the Royal Society B: Biological Sciences*, *282*(1813), 20151019. https://doi.org/10.1098/rspb.2015.1019

Laland, K., Odling-Smee, J., & Endler, J. (2017). Niche construction, sources of selection and trait coevolution. *Interface Focus*, *7*(5), 20160147. https://doi.org/10.1098/rsfs.2016.0147

Laland. K. N., Odling-Smee. J., & Feldman. M. W. (2019). Understanding niche construction as an evolutionary process In T. Uller & K. N. Laland (Eds.), *Evolutionary causation: Biological and philosophical reflections* (pp. 127–156). MIT Press.

Lee, J. (2023). Enactivism meets mechanism: Tensions & congruities in cognitive science. *Minds and Machines*, *33*(1), 153–184.

Levins, R., & Lewontin, R. C. (1985). *The dialectical biologist*. Harvard University Press.

Lyon, P. (2006). The biogenic approach to cognition. *Cognitive Processing*, *7*(1), 11–29.

Machamer, P., Darden, L., & Craver, C. F. (2000). Thinking about mechanisms. *Philosophy of Science*, *67*(1), 1–25.

Maturana, H. R., & Varela, F. J. (1980). *Autopoiesis and cognition: The realization of the living*. D. Reidel.

Montévil, M., & Mossio, M. (2015). Biological organization as closure of constraints. *Journal of Theoretical Biology, 372*, 179–191.

Moreno, A. (2018). On minimal autonomous agency: Natural and artificial. *Complex Systems*, *27*(3), 289–313.

Moreno, A., & Mossio, M. (2015). *Biological autonomy: A philosophical and theoretical enquiry*. Springer.

Mossio, M. (2020). *Organisation biologique et finalité naturelle* [Unpublished habilitation]. Habilitation à Diriger des Recherches (HDR).

Mossio, M., & Bich, L. (2017). What makes biological organization teleological? *Synthese*, *194*(4), 1089–1114.

Nagel, T. (1986). *The view from nowhere*. Oxford University Press.

Nicholson, D. J. (2012). The concept of mechanism in biology. *Studies in History and Philosophy of Science Part C: Studies in History and Philosophy of Biological and Biomedical Sciences*, *43*(1), 152–163.

Rosenblueth, A., Wiener, N., & Bigelow, J. (1943). Behavior, purpose and teleology. *Philosophy of Science*, *10*(1), 18–24.

Saborido, C., Mossio, M., & Moreno, A. (2011). Biological organization and cross-generation functions. *The British Journal for the Philosophy of Science*, *62*(3), 583–606.

Schlosser, M. (2019). *Agency*. In E. N. Zalta (Ed.), *The Stanford encyclopedia of philosophy* (Winter 2019 ed.). Stanford University. https://plato.stanford.edu/archives/win2019/entries/agency/

Sultan, S. E. (2015). *Organism and environment: Ecological development, niche construction, and adaption*. Oxford University Press.

Sultan, S. E., Moczek, A. P., & Walsh, D. (2022). Bridging the explanatory gaps: What can we learn from a biological agency perspective? *BioEssays*, *44*(1), 2100185. https://doi.org/10.1002/bies.202100185

Taylor, R. (1966). *Action and purpose*. Prentice-Hall.

Varela, F. J. (1979). *Principles of biological autonomy*. North Holland.

Velleman, J. D. (1992). What happens when someone acts? *Mind*, *101*(403), 461–481.

Virenque, L., & Mossio, M. (2024). What is agency? A view from autonomy theory. *Biological Theory*, *19*, 11–15. https://doi.org/10.1007/s13752-023-00441-5

Walsh, D. M. (2015). *Organisms, agency, and evolution*. Cambridge University Press.

Walsh, D. M. (2017). Chance caught on the wing. Metaphysical commitment or methodological artifact? In P. Huneman & D. M. Walsh (Eds.), *Challenging the modern synthesis: Adaptation, development, and inheritance* (pp. 239–260). Oxford University Press.

Weber, A., & Varela, F. J. (2002). Life after Kant: Natural purposes and the autopoietic foundations of biological individuality. *Phenomenology and the Cognitive Sciences, 1*(2), 97–125.

West-Eberhard, M. J. (2003). *Developmental plasticity and evolution.* Oxford University Press.

Winning, J., & Bechtel, W. (2018). Rethinking causality in biological and neural mechanisms: Constraints and control. *Minds and Machines, 28*(2), 287–310.

14 Organismal Agency as a (Partly Psychological) Capacity

Bendik Hellem Aaby

14.1 Introduction

Organisms do a lot of different things. Some of the things they do seem merely to be happening to them. Processes like aging, falling asleep, digestion, and the like seem to fit this characterization. Other things that organisms *do* are not merely processes that happen to them; rather they are processes that result from organisms acting. Bonobos engage in polyadic grooming, plants secrete secondary metabolites to ward off herbivores, reptiles bask in the sun to thermoregulate, and so on. These activities seem to be instances of *action*. But are all things that organisms do—and not merely happening to them—instances of *agency*? This chapter aims to provide two arguments that build on each other.

First, I shall argue that organismal agency is the capacity of a biological entity, acting as a whole, to perform goal-directed behavior that aims at satisfying goals or ends that are ultimately derivable from said biological entity or group of entities (i.e., goals that are intrinsic). It also requires the ability of the biological entity to exert a degree of control over the relevant goal-directed behavior. In making this argument, I provide a traditional conceptual analysis of organismal agency by explicating what I take to be necessary conditions for it. I shall not, however, pretend to claim that these amount to jointly sufficient conditions for organismal agency. However, I think that there are some conditions (or variants thereof) that need to be in place in order for the concept of agency to be resilient to simple counterexamples. I will arrive at these conditions through a discussion of several examples of distinctions made in action theory focusing on human, rational agency.

Starting the analysis of the concept of agency with the kind of agency that we have the most intimate relationship with—*human (rational) agency*—and then working toward organismal agency is a methodological decision that might strike some readers as odd, anthropocentric, or otherwise misguided. A more popular methodological approach amongst those interested in agency beyond human animals (and a few species of mammals and birds) is to start with a minimal concept of agency and offer that as a concept imputable to a vast array of different organisms (e.g., Barandiaran et al., 2009; Burge, 2009; Fulda, 2017, van Hateren, 2022; Virenque, this volume). One strong reason to avoid the distinctions, concepts, and the general jargon of action theory is that it tends to "over-intellectualize" agency, even in the

DOI: 10.4324/9781003413318-18

case of human agency (Noë, 2005; Burge, 2022). In order to avoid unwarranted intellectualism, it might therefore seem much more sensible to "start from scratch," providing a basic notion of agency that is free of any psychological components that place unreasonably strong demands on the necessary conditions of the putative agents in question. However, when I take human agency as a paradigm in this chapter, I will also provide suggestions for "de-intellectualized" alternatives to the concepts, distinctions, and conditions that in the case of human agency are obviously chauvinistic requirements when applied to organisms more generally.

This brings me to the second argument of the chapter. I will argue that by approaching the question of organismal agency from an anthropocentric paradigm and then proceeding to "de-intellectualize" it, we end up with fairly similar conditions as many of the minimal theories of agency. In and of itself this does not amount to a very interesting argument. However, I will further argue that what makes agency a concept of special interest and importance is the fact that some of the conditions that make up the concept are partly psychological in nature. Finally, I argue that it is the fact that organismal agency is a partly psychological capacity that allows us to properly differentiate refrigerators, traffic lights, and robotic lawnmowers from organisms and groups of organisms.

In the next section, I outline portions of the conceptual terrain in which we find a plethora of different notions of agency. I will start with a broad, inclusive notion of agency and work my way to what I take—more or less—to be the defining features of human agency. The aim of this section is to establish and motivate why agency should be seen as the capacity to perform acts, in which acts are a special class of activities—which in human beings are quintessentially *intentional actions*. This then lays the groundwork for a subsequent "de-intellectualization" when applying the concept of agency to organisms more generally. While providing a "de-intellectualized" alternative to the conditions for human agency, I will point out the similarity between these alternatives and the conditions offered by minimalist theories of agency.

14.2 The Conceptual Landscape

Agency is a promiscuous concept in both vernacular and more technical parlance (Steward, 2011). In fact, as Kenny (1992, p. 33) points out, when used in an unrestricted sense, agency can be considered a universal phenomenon. The broadest construal of agency is reflected by the use of action verbs. Anything that we can talk about in an active voice can potentially be considered capable of agency. Thus, not only living things in general, and especially not just human beings in particular, can be considered agents. The chair *seats* the person, the river *carries* the kayak downstream, and aqua regia *dissolves* gold. The case of the river and the aqua regia, however, are different from the chair. Aqua regia is the reactive *agent* in a chemical process. It is a case in which gold is considered the patient and the solution of nitric and hydrochloric acid that constitutes aqua regia is considered an agent. The aqua regia is *doing something* to the gold—it is dissolving it. The process of being dissolved, on the other hand, is not something that the gold does but rather something that happens to it. The same is true of the river. It does something to the kayak, namely, carries it downstream. The chair, on the other hand, can only be said to be able to seat a person by virtue of being *acted on*. No chairs seat people without a person taking a seat. There is thus a difference between the chair, on the one hand, and the aqua regia and river on the other, which is in many cases disguised by the language we use when we describe what things are capable of doing. When we say that the chair *seats* people, we are merely describing the chair *as if* it was doing something, namely, seating people. But a more appropriate way of describing an instance of chair-seating would be to say that a person took a seat or was seated, thus relinquishing the agency of the chair and transferring it back to its appropriate referent—the person who took a seat in the chair. In the case of the aqua regia, it is not simply the case that we are using action verbs when it is strictly speaking not appropriate to do so. In that case, the aqua regia is actually *doing something*. It is actively exercising its power to dissolve gold. The chair is not actively exercising its power to seat people as the misleading use of active voice suggests. Rather its power to seat people is passively exercised by virtue of the person actively taking a seat. Thus, in the case of the chair, the attribution of agency consists of no more than a *way of talking* about chairs and their passive powers. Another way of saying this is that it is not in the *nature* of the chair to seat people, rather it is in the nature of people to use chairs and chair-like objects in order to take a seat. We can thus make our first initial distinction between possible forms of agency: agency as having powers whose exercise can be described with action verbs and agency as possessing and exercising *active* powers. We might call agency in the prior sense *descriptive* agency, as it amounts to no more than an ascription of agency in virtue of treating the subject in a sentence as an agent.[1] There is nothing in the nature of the chair that suggests that it is an agent. The latter sense, in which agency is understood as the capacity to possess and exercise active powers, might be called *natural agency* as it picks out a natural property of an object (namely, its active powers) as the justification for the attribution of agency (Kenny, 1992).[2]

   In the case of passive powers, such as the chair's power to seat people, the exercise of the power is wholly exogenous to the putative agent. It is in virtue of the person taking a seat that the chair can exercise its power to seat people. It is not something that the chair itself does; it is simply *being acted on* by the person. In the case of *active* powers, the exercise of the power is characterized both by endogenous and exogenous factors. In the case of the aqua regia, the salient endogenous factor is the appropriate mixture of nitric and hydrochloric acid (the optimal molar ratio is 1:3). The salient exogenous factor is the fact that gold must actually be submerged in the solution for the aqua regia to dissolve it. When these conditions are obtained, the aqua regia will *act on* the gold. Active powers can be contrasted with passive powers by the verbs "doing" and "happening to." The distinction between "happening to" and "doing" is also commonly used in contrasting behavior from non-behavior (e.g., Taylor, 1966; Dretske, 1988).[3] We do not say that seating people is typical chair-behavior because chairs are usually not seen as capable of behaving. Dissolving gold might be said to be typical aqua-regia-behavior, as chemical solutions are often seen as capable of reactive behavior given the appropriate circumstances. One might also use the verbs "acting on" and "being acted on," as well as the countable nouns "agent" and "patient," and uncountable nouns "agency" and "patiency" to pick out what or who is exercising their active or passive powers.

Let us quickly take stock of some crucial vocabulary before moving on. The concept of power refers to the dispositional properties of objects that allow them to undergo, engender, or otherwise be involved in change that can be exercised given the appropriate circumstances or opportunities (Hacker, 2007)—that is, powers are what cause things to happen. When it comes to spelling out what things bear causal powers—be it events, processes, substances, or systems—I will assume a form of agent-causation. This position allows for systems to be genuine causes of events or changes in the world. In the interest of brevity, I will not offer arguments for the viability of this position here (but see Hacker, 2007; Mayr, 2011). However, one remark should be noted. If one takes a neo-Aristotelian approach to agent-causation, one is immediately met with a counterexample. A neo-Aristotelian approach to agent-causation argues that since the causal capacities of (human or organismal) agents cannot be fully explained by the underlying physio-chemical mechanisms of the agent, we need to grant special intrinsic causal powers to the agent which are wholly distinct from the underlying physio-chemical mechanisms. This is problematic since it seems that agent-causation (at least from the neo-Aristotelian view) would then need some kind of mysterious non-naturalistic source for its causal powers (van Hateren, 2022). However, there is an option of offering criteria for when a system can appropriately be considered a source of causation that is compatible with a naturalistic view. On such an account, the "special" powers granted by agent-causation are *emergent* and *non-reductive* (see Potter & Mitchell, 2022). Such a view will be assumed, but not argued for, in this chapter.

14.3 What Kinds of Activities Constitute Agency?

When we are inquiring into organismal agency, we are clearly looking for something more than simply the possession and exercise of active powers. It is not just what organisms do *simpliciter* that is of interest when attributing agency to living things, but rather a specific class of activities. After all, all organisms can metabolize, develop, and age. Some organisms digest, gestate, shiver, shed, and so on. Sub-organismal parts also do things. Blood vessels expand and retract, hearts pump blood, and toenails grow. These are all things that organisms (and their sub-organismal parts) are *doing*. But these are generally not the kinds of activities we are interested in when we are asking whether an organism should be considered capable of agency. Thus, it is useful to stop and ask *why* these kinds of activities are usually not seen as indicative of agency. By answering this we can get a better grip on what sets agential behavior apart from non-agential behavior—that is, what it is with the capacity for agency that makes it special from other "doings" that organisms perform.

In the following sections, I will make use of some distinctions from action theory primarily aimed at explaining human agency to elaborate on what makes a behavior agential or not. However, with each of my examples, I provide a "de-intellectualized" version of the distinction such that they can in principle be applied to a broader range of organisms.

14.4 Agency and Volition

There are several reasons for why digesting, shivering, flinching, and the like are generally not considered agential behaviors. The first of these lies in the fact that such behaviors are not generally considered *voluntary*. In analyzing the concept of volition, we can make a distinction between behaviors that are voluntary and those that are non-voluntary. Non-voluntary behaviors can further be divided into behaviors that are non-voluntary because they are involuntary, which includes behaviors that could have been performed voluntarily, yet are not (e.g., breathing, blinking, shivering, flinching), and behaviors that are non-voluntary because we have no say in whether or not they are performed (e.g., digestion, heartbeat, aging).[4] Voluntary behaviors, on the other hand, are those that we are able to *try* to perform, *decide* to perform, *plan* or *intend* to perform, can be performed on *request*, and so on. The common denominator is that they are acts that we can either perform or elect not to perform.[5] In other words, they are instances in which we are capable of *refraining* from performing them (Bennett & Hacker, 2022).

If we return for the moment to the concept of power presented above, we can introduce a further distinction—between *one-way* and *two-way* powers—to highlight the difference between voluntary (cf. agential) behaviors and non-voluntary (cf. non-agential) behaviors. A one-way power is a power that is *always* exercised given that the appropriate opportunities or conditions for its exercise obtain. The aqua regia again serves as a good example. If gold is submerged in aqua regia, then (under normal circumstances) the aqua regia will dissolve it. Stimulus–response behaviors also seem to fit the bill of one-way powers. Two-way powers, on the other hand, are powers of which its possessor can exercise or *refrain* from exercise given an opportunity for its exercise. That is, it does not follow necessarily that the power will be exercised (under normal circumstances) even if the appropriate conditions for its exercise obtain. All voluntary human behavior constitutes an exercise of two-way powers. But I think it extends to many instances of animal behavior as well. For example, think of a dog's ability to swim. My dog Milo may or may not exercise his ability to swim on our daily walks along the coast. Each time we walk along the coast, Milo has the opportunity to exercise his ability to swim. Sometimes he does exercise it, and other times he refrains from doing so. The choice of Milo's exercising or refraining is presumably further explained by certain environmental states (e.g., outside temperature), certain internal states (e.g., body temperature), and certain conative states (e.g., wanting to cool down body temperature)—more on this below.

While much more can be said concerning two-way powers in relation to agency, abilities, and capacities, what I want to highlight here is the way in which it excludes the aforementioned actions that we do not take as being indicative of agency. Consider again some of the examples mentioned above, digestion and aging. If we consume foodstuffs, it is not in our power to refrain from digesting them. Under normal circumstances, the foodstuff will be digested come what may. The same is true of aging. It is not in our power to refrain from aging (at least not yet, and certainly not directly through volition). Thus, aging and digestion are *in themselves* not indicative of agency, as they constitute the exercise of one-way powers. As we saw above, this does not mean that we are unable to influence such processes by exercising other two-way powers. Human beings may age more gracefully by applying sunscreen each time they go out in the sun. Applying sunscreen is obviously a behavior that we may refrain from doing. The same can arguably be said about certain animal behaviors as well. Take brown tree snakes as an example. After successfully catching and eating large prey (large is considered about one-third of its body weight), brown tree snakes clearly have a digestive challenge ahead of themselves. Usually, brown tree snakes will show a drastic depression of activity for five– to seven days after consuming large prey (Siers et al., 2018). However, if exposed to scorching sun, brown tree snakes will presumably move to avoid overheating. Clearly, at least before the sun poses a life-threatening risk, it is in the power of brown tree snakes to refrain from finding shade. Thus, while digestion is not in itself indicative of agency, behaviors that aid the animal in its fulfillment might very well be. Or think of marine mammals. They need to resurface at different time intervals in order to take a breath. However, if, say, a leopard seal is chasing a squid, it might refrain from returning to the surface for a certain portion of time in order to successfully capture the squid. The leopard seal has to return to the surface at some point, or it will perish. Yet there is a time interval in which the leopard seal may prioritize the need for sustenance over the need for oxygen—when the need for oxygen is not yet critical—and thus refrain from acting on the (felt) need for oxygen and resurface.

A central question, thus, is what it takes for someone or something to be able to refrain from exercising a power. If we characterize it too broadly, any malfunction might count as an instance of refraining from exercising a power. Cardiac arrest is an instance of the heart not exercising its power to circulate blood. If we characterize "refraining" as simply being able to not exercise a power, then cardiac arrest indicates that the heart possesses a two-way power. Clearly, when talking about agency we are interested in a more specific notion of refraining. But if we go too far in the other direction and characterize it too narrowly, we might end up with a notion of refrainment as resulting only from rational deliberation or choice. It is not clear that such narrow characterization would be appropriate even for human beings, as much of our agential behaviors do not necessarily result from prior rational deliberation. When talking about human volitional movement, Bennett and Hacker (2022) write:

> A necessary condition for a movement to be voluntary is that it involves the exercise of a two-way power to do or refrain from doing. Behaviour is voluntary if one can engage in it at will. In this sense, it is behaviour which one can *control*

directly—that is, not by doing something else that causes it or stops it [...]. A fully voluntary movement is one which the agent controls in its inception, continuation and termination. Hence blinking is only partly voluntary, since one can blink at will, but cannot control its "continuation" or termination, and sneezing is only partly voluntary, inasmuch as one can inhibit it but not initiate it directly.

(p. 241; emphasis added)

In the above quotation, the notion of control is central to making sense of what makes human behaviors voluntary. Direct control over the inception, continuation, and termination of the behavior is necessary for an act to be considered fully voluntary. Thus, many human acts that we consider indicative of agency will not be fully, but only partly, voluntary. Breathing can be partly voluntary, as when you focus on your breathing, or refrain for a period of time from taking a breath, but most of the time it is involuntary. It seems natural, then, to take volition to be a concept that has somewhat vague borders. While there are instances that are clearly voluntary, involuntary, or non-voluntary, there are also cases that are borderline. I think the same is true of agency. There are cases that are clearly instances of agency—such as going to the fridge to get a beer because you wanted a beer. And there are cases that are clearly not instances of agency—heartbeats, digestion, and the like. Then there are cases in which it is unclear whether or not they are indicative of agency, namely, those behaviors that the putative agent can *partly* control or influence.

The notion of control deserves more attention. Control is often thought to imply some sort of responsibility. In action theory, an attribution of agency is often accompanied by the view that the putative agent bears a degree of responsibility over the action that they perform. An influential approach to (moral) responsibility is cashed out in terms of whether or not the agent exerts a degree of control over her actions (Fisher & Ravizza, 1998). While it is common to talk about moral responsibility in action theory, the more interesting kind of responsibility in the context of organismal agency is whether or not prior experiences of the organisms play a role in explaining why it behaves as it does or whether the behaviors are attributable to inherited behavioral dispositions (e.g., instincts or stimulus–response patterns, reflexes). Think of the behaviors of pets. It is common to absolve the misbehavior of one's pet if the behavior is thought to be instinctual. In an important sense, this absolution is not (at least not simply) a moral absolution, but rather an absolution in terms of the pet's agential capacity being overridden by impulses that are beyond the control of the pet. The cases in which reproach is justified in the context of pets (and small children for that matter) are cases in which it is in the the power of the animal (or child) to do otherwise. So, to answer the question that begat the discussion of volition in relation to agency, flinching, shivering, digestion, and the like are not generally considered indicative of agency because they cannot be directly controlled (i.e., they are non-voluntary, as such as digestion) or can in principle be partly controlled but are not in the particular instance (i.e., they are involuntary, such as shivering or flinching). In other words, these are not instances of agency because the putative agent is not assumed to be responsible for their execution. From the preceding discussion, I hope to have established that a *desideratum* for a concept of agency is that it should include as a necessary condition a degree of control—or at least something similar—on behalf of the putative agent over the actions it performs. Such a condition is helpful in distinguishing genuinely agential behaviors from non-agential ones. There are several alternative conditions that capture the sentiment and rationale associated with control discussed above available in the literature. Frankfurt's (1988) condition that an act should to some degree be under the influence of or guided by the agent captures the same idea. From more minimalistic theories of agency, Burge's (2009) conditions that agential behavior should issue from an organism's central behavioral capacities, as well as conditions like authorship and behavioral freedom (van Hateren, 2015, 2022), autonomy (Barandiaran et al., 2009), all seem to serve a similar purpose—and be introduced by similar considerations—to the idea of only treating actions that are (at least partly) voluntary as indicative of a capacity for agency in action theory.

14.5 Agency and the Activities of Parts

A second reason for why digestion and the beating of the heart in particular are not generally considered indicative of agency is that it is not something that the putative agent is doing but rather something the putative agent's parts are doing. This ties up to the intuition that agency is a capacity of the putative agent (or system) *acting as a whole* (e.g., Burge, 2009, 2022). The condition of acting as a whole is also covered in minimal theories of agency by other conditions such as individuality understood as an agent being distinguishable from its environment (see also Michelini, this volume) combined with autonomy understood as an agent doing something of its own accord in that environment (Barandiaran et al., 2009; see also Virenque, this volume). It might not be immediately obvious, however, why we should hold on to such an intuition. Thus, I wish to spend a little time defending it before moving on. Although there are several ways of defending it, I will rely on a rather simple rationale for why acting as a whole should be considered a necessary condition for agency. To begin with, think of functional descriptions of the activities of the heart or of the digestive system. The main function of the heart is to circulate blood through the cardiovascular system and the main function of the digestive system is to extract nutrients and minerals from consumed foodstuffs. These are not functions that the human body as a whole is performing. Rather they are functions of subsystems of the human body. The cardiovascular and digestive systems are of course functionally integrated subsystems, so their (mal) function often has consequences for the functioning of the human body as a whole. However, the attribution of function is still made to the subsystem itself. Another example of attribution to wholes and not parts is perception. Perception is paradigmatically something an individual engages in (Burge, 2010). It is not the eyes that perceive but individuals that possess eyes. The eyes are sense organs, and they are parts of the perceptual system. They can function and malfunction, which again can have consequences for the perceptual states that the perceptual system generates. But here too we must be careful to not attribute perceptual abilities to the eye itself but rather to the individual *as a whole* (Burge, 2022).

I think the same line of reasoning applies to agency. Agency is something attributable to a system acting as a whole. Agential behaviors are paradigmatically attributable to the putative agent, not to the activities of the agent's parts. When we are looking at the activities of the subsystems, we are looking at the functionally integrated activities of the system itself, not agency. We could, in principle, view digestion and the beating of the heart as agential behaviors. However, these behaviors would only be agential insofar as the heart and the digestive system were treated as putative agents. While it might be strange to use agential language to describe the activities of the heart and digestive system, there are other parts that an organism is composed of in which agential language might be more sensible. The behaviors of cells in eukaryotes might be a case in which agential language turns out to be more appropriate (e.g., Levin, 2023). However, this would constitute cell agency, and in that context, it is the behavior of the cell acting as a whole, which is what we treat as agential, and not the behaviors of the organism of which the cell is a part. Most philosophers of action, I believe, would balk at the idea of treating the cell as an agent. A central reason for such skepticism lies in a further intuition concerning the conditions of agency, namely, that it has a psychological or cognitive dimension. While cell signaling has been characterized as a proto-cognitive or cognitive process or capacity (e.g., Koseska & Bastiaens, 2017; Mathews et al., 2023), the notion of cognition in such cases is not sufficient to capture what action theorists generally have in mind when they talk about the psychological dimensions of agency.

14.6 Agency and Psychology

In action theory, the mainstream view of intentional action is some variant of Davidson's (1963) belief–desire account. In this view, an intentional action is a goal-directed behavior that results from, or is produced by, an ordered pair of beliefs and desires. For example, Jane desires a beer, Jane believes that there is beer in the fridge, thus Jane gets up from the couch to get a beer. On this account, the desire explains why Jane went to get a beer and not something else, and the belief explains why she went to the fridge and not somewhere else to obtain it. Put together, they explain why Jane getting a beer from the fridge constitutes an intentional action. One way to expound the psychological aspects of intentional agency is to use the notion of *direction of fit*. When treating beliefs and desires as an ordered pair, we can use the direction of fit to highlight the different roles they play as elements of an intentional action (Dancy, 2002; Searle, 2001). Beliefs are said to have a mind-to-fit-world direction of fit. It is the world (or state of affairs) that determines whether or not a belief is true. So, in order for a belief to be true, it needs to *fit* the world. Desires are different, they are states that concern not yet realized states of affairs. Desires are about what one would like the world to be like and thus have a world-to-fit-mind direction of fit. If the world does not fit one's desire, then one is unable to satisfy one's desire. In normal circumstances, one cannot simply update one's desire to one that fits the world as it presently is, as one can with false beliefs. In the case of unsatisfied desires, one needs to navigate the world to look for opportunities for their satisfaction. A slightly haphazard way of summarizing the difference in direction of fit between beliefs and desires is to say that desires are normally about the wants and needs of an individual, while beliefs are normally about what the individual takes the world to be like.

So, an intentional action according to the mainstream approach involves goal-directedness—what the intention is aimed at achieving. It involves a belief about what the world is like and finally a desire whose satisfaction constitutes the fulfillment of the goal or intention. An intentional action is thus composed of both cognitive and conative elements. The belief constitutes the cognitive element, it tells the individual something about what the world is like, while the desire and intention constitute the conative elements, concerning what drives the individual to take action. The conative states are informed by the cognitive states. That is, the opportunities for action—the means by which one can satisfy one's desire or achieve one's intention—are disclosed by the cognitive and perceptual abilities of the individual.

In thinking about non-human organisms, the belief–desire model is obviously over-intellectualized. However, I think that there are good reasons for why we should retain a condition that captures the cognitive and conative elements when talking about organismal agency. I think the most important contribution it makes is that it captures the common view that agency is a capacity that displays an organism's ability to act in a goal-directed manner. At the same time, it captures the intuition that when goal-directed behavior is agential, the goal-state and the means to satisfy the goal-state are attributable to the agent itself. In other words, it allows for goal-directed behavior to be the result of a negotiation between the experiences of the organism on the one hand and its (felt) needs and wants on the other.

In more minimalistic theories of agency, similar conditions are offered. For example, Walsh (2015) argues that agency requires the "capacity of a system to pursue goals, respond to conditions of its environment and its internal constitution in ways that promote the attainment, and maintenance, of its goal state" (p. 210). Pursuing goals can be thought of as both a cognitive and intentional element. Intentional in that it captures the "directedness" of the behavior and cognitive in that pursuing a goal requires recognizing environmental factors as affordances—which presumably implies some sort of information exchange and processing between the putative agent and its environmental resources. The conative element is captured by the idea that organisms exhibit agency when they respond to internal and external conditions in order to reach a goal. Presumably, the response to such conditions is what drives the organism to take action in appropriate circumstances.

14.7 Conditions for Organismal Agency

From the admittedly brief venture into action theory, we can formulate a list of four conditions that I take to be necessary in order to attribute the capacity for agency to an organism. I take the following to be these conditions:

a The organism must be able to possess goals and engage in goal-directed behaviors.[6]

b The goals should ultimately be derived from the needs and/or wants of the putative agent itself.
c The putative agential behaviors must be attributable to the organism *acting as a whole* and not to the activities of its parts.
d The putative agent (as a whole) should be able to exert a degree of control over the relevant behavior.

If we look at the typical conditions for agency offered in minimal theories of agency, summarized in van Hateren (2022): "goals, norms, freedom, individuality, directedness, and causal efficacy" (p. 92), we can see considerable overlap with the conditions arrived at through "de-intellectualizing" some of the key distinctions made in action theory. I take this to be good news. It shows that the notion of agency invoked in minimal theories of agency is closely related to the notion of agency, which we are most familiar with—i.e., human, rational agency. This, con-sequently, suggests that there is significant room for interdisciplinary dialogue between philosophers of biology (and in general those interested in agency in non-human organisms) and philosophers of action.

However, the crux of the argument that justifies the title of this chapter remains. So, in the remainder, I will offer an argument asserting that agency (minimal, rational, or otherwise) is a partly psychological capacity. What I mean by this is that some of the conditions I have laid out above—and the corresponding alterna-tive conditions found in the minimal theories of agency—carry implications that can only be fully explicated if they are understood as being (partly) psychological in nature.

14.8 Agency as a Partly Psychological Capacity

If we, for the sake of argument, assume that the aforementioned conditions (or some variants thereof) are necessary conditions for organismal agency, we could ask what satisfying these conditions amounts to. Let us start with the notion of control and its related alternative conditions (e.g., influence, guidance, responding to internal and external conditions, authorship). It seems to me that "controlling," "guiding," or "influencing" behavior, and "responding to internal and external states in ways that promote a goal-state" will be exceedingly difficult to explicate without relying on concepts that (at least implicitly) are partly psychological in nature. A response to an internal or external state that is associated with a goal-state seems to require not only a stimulus–response pattern but also some information processing. How would a system be able to tell if an incoming cue is relevant for the goal-state if the cue is not processed by the system at all? And how could a system provide an appropriate response to the cue without the cue being processed? When it comes to guidance, influence, and control, things are less obvious. For example, we could say that the angle of a riverbank influences the velocity of the water of a river, the meandering path of a river guides the water across the landscape, and a dam controls the flow of the water. Yet these are obviously not psychological uses of such terms, so why could it not be a similar thing for organismal agency? The difference is that the guidance, influence, and control of the water in a river are passively exercised. In the case of organismal agency, the guidance, control, or influence is related to the concept of authorship or autonomy that is supposed to point out that in the case of organismal agency, the agent is actively influencing, guiding, or controlling behavior. So, what does actively guiding, influencing, or controlling behavior amount to? It seems to me that to make any sense of that being a condition that relates to authorship or autonomy, there has to be a subject of some sort that is monitoring the behavior and manipulates it based on relevant information. Trying to explicate this in terms wholly devoid of psychological content (explicitly or implicitly) seems to either render these concepts passive or remove the subject to which the agential capacity is supposed to be imputed.

However, there are several artifactual things that seem to fit a lot of the conditions necessary for organismal agency. For example, the goal of a refrigerator is to achieve and maintain a set temperature, and it engages in flexible goal-directed behaviors to do this. The goal of traffic lights is to regulate traffic flow. The goal of robotic lawnmowers is to cut the grass in a given perimeter at regular time intervals, and they can display a lot of flexibility in doing this, even prioritizing their immediate needs (e.g., power) over their overarching goal. Now, someone exceedingly liberal with their use of the notion of "psychological" could argue that refrigerators, traffic lights with weight sensors, and robotic lawnmowers all have artificial psychological capacities. They could all be said to monitor their own internal states and respond to environmental conditions in ways that aim to fulfill a certain goal state, even in a flexible manner. Furthermore, in the case of the robotic lawnmower returning to its docking station to charge when power is low, we could argue that it displays a flexible hierarchical ordering of its goals. Could we therefore say that it possesses a degree of freedom over its behavior? Or that it guides, influences, or controls its behavior? Or that it has authorship over its behavior?

Even with an exceedingly liberal use of the term "psychological," we could still make a distinction between organisms and artifactual things. The difference between the robotic lawnmower and an organism comes down to where their goals ultimately derive from. In the case of the lawnmower, the goals are ultimately derived from the human beings that need it to do a certain task—namely, cutting grass. It does not matter to the lawnmower itself whether or not it fulfills this task— the lawnmower does not get increasingly desperate to fulfill its task when power is unsuspectingly draining rapidly. Borrowing from the discussion of the psychological elements of intentional action above, we can say that the conative locus of the lawnmower is ultimately external to it, while in organisms it is internal. If the lawnmower fails to finish cutting the grass or does not manage to return to its docking station, it is ultimately the need of the human being who produces an action targeted to mend the situation. While the lawnmower might exhibit a lot of elements of what it takes to be capable of agency, the fact that the conative locus is ultimately external makes us able to point to a categorical difference between organisms and artifactual things. Organisms strive to reach their goals because their goals are ultimately their own. And the fact that organisms strive to fulfill their goals—at least when they manifest as needs and/or wants—has a neat possible explanation, namely, that they are good candidates for being partly psychological in nature.

14.9 Conclusion

In this chapter I have approached the concept of organismal agency from an anthropocentric starting point. By doing this I have attempted to show that similar alternatives to the commonly cited conditions for minimal agency can all be obtained by "de-intellectualizing" central concepts and distinctions found in action theory. This, I think, indicates a rich potential for interdisciplinary dialogue. However, I have also argued that even though the conditions are "de-intellectualized," some are still arguably only fully explicable as partly psychological conditions making agency a partly psychological capacity.

Notes

1   A descriptive or metaphorical approach to agency does not have to be restricted to simply treating the subject of a sentence as an agent using active voice. We might also use agency in a more qualified yet still wholly descriptive sense, in which we describe what an entity is doing *as if* it were a specific type of agent. An example of this rendition of agency can be found in Okasha (2018), where organisms are treated as agents in an adaptationist manner, in which agential powers are ascribed in order to explain the goal-directed behavior we can observe them performing without attributing any individual-level access to reasons for such actions. In such a case, we are using "agency" in a metaphorical sense, when it is strictly speaking only literally ascribable to human beings (perhaps also to some of the so-called "higher" animals).

2   I use the term "object" in a broad sense, as more or less anything that can be ascribed properties and individuated in language. Of course, not all objects in this sense are putative agents even though they "do" something, can be individuated, or possess properties. A football match might cause a subsequent riot, and it can be ascribed properties (being entertaining or boring, 90 minutes in duration, and so on), yet it is not a putative agent. I do not wish to commit to a view in which objects have to be thought of as ontologically prior to anything else. Rather, objects are just a useful way of talking in abstract terms about the putative things for which the exercise of their powers might or might not constitute agency. I will return to the specific properties that I think are constitutive for a richer notion of agency below.

3   Burge (2009) argues that *primitive agency* begins with functional activities attributable to whole organisms and contrasts these with things that merely happen to organisms. I do not disagree with this point of view. As a stepping-stone for explaining the evolutionary origins of agency, it seems plausible to start with the functional activities of a whole organism. But for the sake of defining agency, I think we should restrict it further to be able to rule out some counterexamples, like that of reflexive and instinctual behaviors. Thus, what Burge calls primitive agency would for me be a *precursor* to agency and not an account of agency in general.

4   Of course, one can influence both one's digestion and heartbeat *indirectly* by, for example, taking a stroll after eating a heavy meal or engaging in exercise which raises the heartbeat. Some are also able to indirectly regulate their heartbeat through rhythmic

breathing (e.g., Pokrovskii & Polischuk, 2012). While the behaviors performed to *indirectly* influence such non-voluntary behaviors may be voluntary, this does not change the fact that both digestion and the beating of the heart themselves are non-voluntary.

5   Cases of acting in duress, under threat or obligation, etc. require a more fine-grained analysis, but for the sake of space and the context of the chapter—organismal agency—I have elected to omit a discussion of these. Nonetheless, there might be good reasons to include them, as there may be analogous (though thoroughly de-intellectualized) phenomena in the sphere of the behavioral repertoires of non-human organisms. This rough characterization of the concept of volition borrows heavily from Bennett and Hacker (2022, Chapter 9).

6   At first glance, this distinction is obviously zoocentric, as plants and fungi seem to do stuff through growth (developmental plasticity) or chemical processes (e.g., defense reactions against herbivory). However, for the sake of this chapter, I take behavior to apply very broadly, borrowing the intuitive distinction from Taylor (1966) and Dretske (1988) discussed above—i.e., behavior is the system doing something, as opposed to something happening to the system. If the reader is in vehement disagreement with such usage, they can swap out "behavior" for "activities." Defining behavior is a tricky question (e.g., Levitis et al., 2009), so the choice of concept here should be seen as a stylistic choice more than a substantial claim.

References

Barandiaran, X. E., Di Paolo, E., & Rohde, M. (2009). Defining agency: Individuality, normativity, asymmetry, and spatio-temporality in action. *Adaptive Behavior, 17*(5), 367–386.

Bennett, M. R. & Hacker, P. M. S. (2022). *Philosophical foundation of neuroscience, 2nd edition.* Wiley Blackwell.

Burge, T. (2009). Primitive agency and natural norms. *Philosophy and Phenomenological Research, 79*(2), 251–278.

Burge, T. (2010). *Origins of objectivity.* Oxford University Press.

Burge, T. (2022). *Perception: First form of mind.* Oxford University Press.

Dancy, J. (2002). *Practical reality.* Oxford University Press.

Davidson, D. (1967). Actions, reasons, and causes. *The Journal of Philosophy, 60*(23), 685–700.

Dretske, F. (1988). *Explaining behavior: Reasons in a world of causes.* MIT Press.

Fischer, J. M. & Ravizza, M. (1998). *Responsibility and control: A theory of moral responsibility.* Cambridge University Press.

Frankfurt, H. (1988). *The importance of what we care about.* Cambridge University Press.

Fulda, F. (2017). Natural agency: The case of bacterial cognition. *Journal of the American Philosophical Association, 3*(1), 69–90.

Hacker, P. M. S. (2007). *Human nature: The categorical framework.* Blackwell.

Kenny, A. (1992). *The metaphysics of mind.* Oxford University Press.

Koseska, A. & Bastiaens, P. I. H. (2017). Cell signaling as a cognitive process. *The EMBO Journal, 36*(5), 568–582.

Levin, M. (2023) Darwin's agential materials: evolutionary implications of multiscalecompetency in developmental biology. *Cellular and Molecular Life Sciences, 80*, 142. https://doi.org/10.1007/s00018-023-04790-z

Levitis, D. A., Lidicker Jr., W. Z., & Freund, G. (2009). Behavioural biologists do not agree on what constitutes behaviour. *Animal behavior, 78*(1), 103–110.

Mathews, J., Chang, A. J., Devlin, L., & Levin, M. (2023). Cellular signaling pathways as plastic, proto-cognitive systems: Implication for biomedicine. *Patterns, 4*(5), 1–18.

Mayr, E. (2011). *Understanding human agency.* Oxford University Press.

Nöe, A. (2005). Against intellectualism. *Analysis, 65*(4), 278–290.

Okasha, S. (2018). *Agents and goals in evolution.* Oxford University Press.

Pokrovskii, V. M. & Polischuk, L. V. (2012). On the conscious control of the human heart. *Journal of Integrative Neuroscience, 11*(2), 213–223.

Potter, H. D. & Mitchell, K. J. (2022). Naturalising agent causation. *Entropy, 24*(4), 472. https://doi.org/10.3390/e24040472

Searle, J. R. (2001). *Rationality in action.* The MIT Press.

Siers, S. R., Yackel Adams, A. A., & Reed, R. N. (2018) Behavioral differences following ingestion of large meals and consequences for management of a harmful invasive snake: A field experiment. *Ecology and Evolution, 8*(29), 10075–10093.

Steward, H. (2011). *A metaphysics for freedom.* Oxford University Press.

Taylor, R. (1966). *Action and purpose.* Prentice Hall.

van Hateren, J. H. (2015). The origin of agency, consciousness, and free will. *Phenomenology and the Cognitive Science, 14*, 979–1000.

van Hateren, J. H. (2022). Minimal agency. In L. Ferrero (Ed.), *The Routledge companion to the philosophy of agency* (pp. 91–100). Routledge.

Walsh, D. (2015). *Organisms, agency, and evolution.* Cambridge University Press.

Index